# THE BIOCHEMISTRY OF INORGANIC
## COMPOUNDS OF SULPHUR

# THE BIOCHEMISTRY OF
# INORGANIC COMPOUNDS
# OF SULPHUR

BY

## A. B. ROY

*Senior Fellow, John Curtin School of Medical Research*
*Australian National University, Canberra*

AND

## P. A. TRUDINGER

*Principal Research Scientist, Baas Becking Geobiological Laboratory*
*Bureau of Mineral Resources and Division of Plant Industry*
*Commonwealth Scientific and Industrial Research Organization, Canberra*

## CAMBRIDGE
### AT THE UNIVERSITY PRESS
### 1970

Published by the Syndics of the Cambridge University Press
Bentley House, 200 Euston Road, London N.W. 1
American Branch: 32 East 57th Street, New York, N.Y.10022

© Cambridge University Press 1970

Library of Congress Catalogue Card Number: 78–79056

Standard Book Number: 521 07581 5

*5 74.192*
*R 81 L*

Printed in Great Britain
at the University Printing House, Cambridge
(Brooke Crutchley, University Printer)

# CONTENTS

# PREFACE

The biochemistry of inorganic compounds of sulphur is an important topic which has many ramifications, the extent of which is only now being appreciated. Aside from their intrinsic academic interest, studies on the metabolism of inorganic sulphur compounds impinge directly upon such fields as the metabolism of steroids, diseases of the central nervous system, soil fertility, the corrosion of concrete, the origins of sulphide ore bodies and the early evolution of life.

These ramifications have provided a stimulus to research in the biochemistry of inorganic compounds of sulphur, a topic which might otherwise have suffered considerably in competition with the more fashionable fields of modern biology. As a result, the past decade or so has witnessed a remarkable increase in the published information on the metabolism of inorganic sulphur compounds. Several excellent reviews have, over the last few years, summarized various aspects of this subject but these have appeared in diverse medical, chemical, biological and industrial journals and we know of no recent publication embracing the broad field of the biochemistry of inorganic sulphur compounds.

The present monograph is an attempt to remedy this deficiency and to provide those with direct or fringe interests in sulphur biochemistry with a detailed account of the subject. Chapters on sulphur chemistry and methodology are included in the hope that the book may also prove to be a useful laboratory manual for research workers engaged in, and more particularly about to embark upon, experimental work in this field.

The book can conveniently be divided into four sections dealing with rather different aspects of the subject matter. After a general introduction, chapters 2 to 4 deal with the more chemical aspects of the subject; chapters 5 to 8 with investigations using more or less well characterized enzymes, chapters 9 to 11 with the metabolism of inorganic compounds of sulphur and finally chapters 12 and 13 consider what might be called the economic aspects of sulphur biochemistry.

Inorganic sulphur metabolism is, of course, intimately related to the metabolism of organic compounds of sulphur. The latter aspect, however, has been frequently discussed in general text books of biochemistry as well as in periodicals such as *Advances in Enzymology* and *Annual Reviews*. In this book, therefore, organic compounds of sulphur are largely ignored except in so far as is necessary for a full discussion of the metabolism of inorganic forms of the element.

We wish to thank our many friends and colleagues who have helped us in so many ways during the preparation of this book. In particular, Dr C. A. Appleby, Dr D. P. Kelly, Professor J. M. Swan and Dr A. E. R. Thomson deserve special mention for their most valuable criticisms of early drafts of several parts of the text. Dr R. G. Nicholls kindly helped us to prepare the subject index. We are also extremely grateful to Mrs E. Allen and Mrs L. Calis for their help, over several years, in ways too numerous to detail here, to Mrs A. Howard, Mrs V. Taylor and Mrs C. Pedersen for their skill in preparing the typescript, and to the latter and Miss A. Jerfy for their assistance with the onerous tasks of proof-reading and indexing. Without this assistance our book would certainly contain many more imperfections than those which, we fear, may still remain.

A.B.R.

P.A.T.

# A NOTE ON THE NOMENCLATURE OF SULPHUR-CONTAINING COMPOUNDS

The naming of sulphur-containing organic compounds is complex and there was no internationally agreed system of nomenclature until the recent appearance of rulings on the subject (*Pure and Applied Chemistry*, **11**, 1965) from which this brief summary has been prepared. It provides only a guide to the nomenclature of the compounds whose biochemistry is to be discussed in this book and it is not a general summary of the subject.

The following list gives the chemical groupings with which we will be concerned together with the suffixes and prefixes to be used in forming the systematic names of compounds containing them.

| Group | Suffix | Prefix |
|---|---|---|
| —SH | -thiol (or hydrosulphide)* | mercapto- |
| —S—SH | hydrodisulphide† | — |
| —S—$S_n$—SH | hydropolysulphide | — |
| —S—S— | disulphide | — |
| —S—$S_n$—S— | polysulphide | — |
| —SOH | -sulphenic acid | sulpheno- |
| —$SO_2H$ | -sulphinic acid | sulphino- |
| —$SO_3H$ | -sulphonic acid | sulpho- |
| —$NH.SO_3^-$ | *N*- sulphamate | sulphoamino‡ |
| —$O.SO_3^-$ | sulphate | — |
| —$S.SO_3^-$ | *S*- thiosulphate | *S*-sulpho- |

In some cases another form of nomenclature is useful, particularly when it is wished to stress that a compound contains a chain of sulphur atoms as occurs, for instance, in some of the derivatives of the more complex sulphur acids. This nomenclature takes the sulphanes, $HS.S_n.SH$ (disulphane, $n = 0$; trisulphane, $n = 1$; hexasulphane, $n = 4$; etc.) as

* The term thiophenol is permissible with derivatives of simple phenols.
† The term persulphide is frequently encountered in biochemical literature and will be used in this book.
‡ This name should strictly be applied only to the free acid and not to its salts.

x

the parent compounds and the more complex substances are named as derivatives of these. This nomenclature has been used in chapter 2 to stress the analogous reactions undergone by inorganic and organic compounds of sulphur, for example by disulphane and by alkyl disulphanes. (The latter compounds are usually called alkyl persulphides elsewhere in the text.) Examples of the use of this type of nomenclature are given in the following table.

| | | |
|---|---|---|
| $C_2H_5.S.SH$ | Ethyl disulphane | Ethyl hydrosulphide |
| $C_6H_5.S.S_n.SH$ | Phenyl polysulphane | Phenyl hydropolysulphide |
| $C_2H_5.S.S.C_2H_5$ | Diethyl disulphane | Diethyl disulphide |
| $C_6H_5.S.S.S.C_6H_5$ | Diphenyl trisulphane | Diphenyl trisulphide |
| $C_2H_5.S.SO_3^-$ | Ethyl sulphosulphane | S-Ethyl thiosulphate |
| $C_6H_5.S.S.SO_3^-$ | Phenyl sulphodisulphane | — |

The last example, phenyl sulphodisulphane, may be given the alternative name of benzene sulphenyl thiosulphate. This is not a name recognized by the International Union of Pure and Applied Chemistry but nevertheless such nomenclature is sometimes useful. It can also be applied to thiosulphate esters to give the analogous term, sulphenyl sulphites.

This same 'sulphane' nomenclature can be used to name some inorganic compounds when it is again wished to stress that they contain a chain of sulphur atoms. For example:

| | | |
|---|---|---|
| $HO_3S.S.S.SO_3H$ | Disulphodisulphane | Tetrathionic acid |
| $HO_3S.S.S.S.SO_3H$ | Disulphotetrasulphane | Hexathionic acid |

Unfortunately correct nomenclature has not been widely used in biochemical literature and if only the systematic names were used in this book then many of the compounds would appear in a most unfamiliar guise. Not many biochemists would at first sight recognize, for example, S-(2-amino-2-carboxyethyl) thiosulphate as S-sulphocysteine. In the following chapters of this book, therefore, the well-known trivial names have been used for many sulphur compounds (in a few instances more than one trivial name may have been used in different contexts) but the list at the end of this section will serve to provide the systematic names for these compounds. Other commonly used names, not necessarily adopted in this book, are also listed: in some cases these names are quite incorrect and if so, this is clearly indicated. It is to be hoped that in future works of this type it will be possible to use correct trivial names, if not systematic names, for sulphur-containing compounds of biochemical interest.

Finally, the nomenclature of sulphate esters requires some comment. The term $O$-sulphate is frequently encountered in names such as choline $O$-sulphate (for choline sulphate ester) and serine $O$-sulphate (for serine 3-sulphate) to stress that these compounds contain the $-O.SO_3^-$ grouping and are not to be confused with the corresponding salts of choline or of serine, nor with the sulphoamino derivative of the latter which contains the $-NH.SO_3^-$ group. Undoubtedly the term $O$-sulphate is wrong and a more correct designation would seem to be $O$-sulphonate, as has been used to describe the appropriate derivatives of oximes which contain the grouping $=N.O.SO_3^-$, but this does not seem to be in accord with the recognized system of nomenclature. Likewise the term $N$-sulphate is incorrectly used to describe the sulphoamino group, $-NH.SO_3^-$. Unfortunately, however, these two terms appear to have become established in the biochemical literature and they have, where necessary, been used in this book. For example, they have been used to distinguish between the $-O.SO_3^-$ and $-NH.SO_3^-$ groups which occur in heparin.

The following list gives the systematic names of the sulphur-containing compounds of biochemical interest which are considered in subsequent chapters of this book. Those trivial names given in italics are those which have been used in subsequent discussions.

| SYSTEMATIC NAME | TRIVIAL NAMES |
|---|---|
| Adenosine 3'-phosphate 5'-sulphatophosphate | *3'-Phosphoadenylyl sulphate*<br>3'-Phosphoadenosine<br>  5'-phosphosulphate |
| Adenosine 5'-sulphatophosphate | *Adenylyl sulphate*<br>Adenosine 5'-phosphosulphate |
| 2-Amino-2-carboxyethanesulphenic acid | *Alanine 3-sulphenic acid*<br>3-Sulphenoalanine<br>3-Sulphenylalanine*<br>Cysteine sulphenic acid* |
| 2-Amino-2-carboxyethanesulphinic acid | *Alanine 3-sulphinic acid*<br>3-Sulphinoalanine<br>3-Sulphinylalanine*<br>Cysteine sulphinic acid* |
| 2-Amino-2-carboxyethanesulphonic acid | *Alanine 3-sulphonic acid*<br>3-Sulphoalanine<br>3-Sulphonylalanine*<br>*Cysteic acid*<br>Cysteine sulphonic acid* |
| 2-Amino-2-carboxyethanethiosulphonic acid | *Alanine 3-thiosulphonic acid* |

| SYSTEMATIC NAME | TRIVIAL NAMES |
|---|---|
| 2-Amino-2-carboxyethyl hydrodisulphide | *Cysteine persulphide*†‡ |
|  | Thiocysteine* |
| 2-Amino-2-carboxyethyl sulphodisulphane | *Alanine sulphodisulphane* |
|  | Cysteine sulphenyl thiosulphate* |
| *S*-(2-Amino-2-carboxyethyl) thiosulphate | *S-Sulphocysteine* |
|  | Cysteine *S*-sulphonate* |
| Bis(2-Amino-2-carboxyethyl) trisulphide | *Thiocystine*† |
| 2-Aminoethanesulphinic acid | *Hypotaurine* |
| 2-Aminoethanesulphonic acid | *Taurine* |
| 2-Aminoethanethiol | *Cysteamine* |
| 2-Aminoethanethiosulphonic acid | *Thiotaurine*† |
| Bis(2-Aminoethyl) disulphide | *Cystamine* |
| 2-Aminoethyl hydrodisulphide | *Cysteamine persulphide*†‡ |
|  | Thiocysteamine* |
| 2-Carboxy-2-oxoethanesulphinic acid | *3-Sulphinopyruvic acid* |
|  | 3-Sulphinylpyruvic acid* |
| 2-Hydroxyethanesulphonic acid | *Isethionic acid* |

\* These names are incorrect and should not be used.

† The use of these trivial names is not to be recommended. They have been adopted in the present instance because of the lack of any suitable alternatives.

‡ The corresponding derivatives of other thiols, such as glutathione and dihydrolipoic acid, are named analogously.

## *The anions of some oxy-acids of sulphur*

| | |
|---|---|
| $SO_3^{2-}$ | sulphite |
| $SO_4^{2-}$ | sulphate |
| $S_2O_2^{2-}$ | thiosulphite |
| $S_2O_3^{2-}$ | thiosulphate |
| $S_2O_4^{2-}$ | dithionite (hydrosulphite) |
| $S_2O_5^{2-}$ | metabisulphite |
| $S_2O_6^{2-}$ | dithionate |
| $S_2O_7^{2-}$ | pyrosulphate (disulphate) |
| $S_2O_8^{2-}$ | persulphate |
| $S_3O_6^{2-}$ | trithionate |
| $S_4O_6^{2-}$ | tetrathionate |
| $S_5O_6^{2-}$ | pentathionate |
| $S_6O_6^{2-}$ | hexathionate |

# ABBREVIATIONS

| | |
|---|---|
| ADP | Adenosine 5'-diphosphate |
| AMP | Adenosine 5'-phosphate, adenylic acid |
| APS | Adenylyl sulphate |
| APSe | Adenylyl selenate |
| APW | Adenylyl tungstate (hypothetical) |
| ATP | Adenosine 5'-triphosphate |
| BAL | 2,3-Dimercaptopropanol |
| BVH | Reduced benzyl viologen |
| CTP | Cytidine triphosphate |
| DCIP | 2',6'-dichlorophenolindophenol |
| DNP | 2,4-Dinitrophenol |
| EDTA | Ethylenediaminetetraacetate |
| FAD, $FADH_2$ | Flavin-adenine dinucleotide (oxidized and reduced) |
| FMN, $FMNH_2$ | Flavin mononucleotide (oxidized and reduced) |
| GSH, GSSG | Glutathione and oxidized glutathione |
| G-6-P | Glucose 6-phosphate |
| GTP | Guanosine triphosphate |
| MVH | Reduced methyl viologen |
| $NAD^+$, NADH | Nicotinamide-adenine dinucleotide (oxidized and reduced) |
| $NADP^+$, NADPH | Nicotinamide-adenine dinucleotide phosphate (oxidized and reduced) |
| NCS | Nitrocatechol sulphate |
| NEM | $N$-Ethylmaleimide |
| NP | $p$-Nitrophenol |
| NPS | $p$-Nitrophenyl sulphate |
| PAP | Adenosine 3',5'-diphosphate |
| PAPS | 3'-Phosphoadenylyl sulphate |
| PCMB | $p$-Chloromercuribenzoic acid |
| PMS | $N$-Methylphenazonium methosulphate |
| $P_i$ | Orthophosphate ion |
| $PP_i$ | Pyrophosphate ion |
| tris | 2-Amino-2-hydroxymethyl-1,3-propanediol |
| UTP | Uridine triphosphate |

# LIST OF ENZYMES

The following list will serve to definitively identify most of the enzymes which are mentioned in the text. It cannot, however, detail all of them because some quite important enzymes (e.g. APS-reductase) have not been named by the Enzyme Commission. If, therefore, an enzyme is mentioned in the text but does not appear in the following list then it may be taken that it does not yet have a systematic name.

In the list are given the numbers assigned to the enzymes, their systematic names and some commonly used trivial names, the first of which is in each case that recommended by the Enzyme Commission. We have not adhered to this recommendation in all cases and the names used in this book are therefore given in italics.

| Number | Systematic names | Trivial names |
|---|---|---|
| 1.1.1.49 | D-Glucose-6-phosphate: NADP oxidoreductase | *glucose-6-phosphate dehydrogenase* |
| 1.2.1.12 | D-Glyceraldehyde-3-phosphate: NAD oxidoreductase (phosphorylating) | *glyceraldehyde-3-phosphate dehydrogenase* triose phosphate dehydrogenase |
| 1.6.4.2 | Reduced-NAD(P): oxidized glutathione oxidoreductase | *glutathione reductase* |
| 1.6.6.4 | Reduced-NAD(P): nitrite oxidoreductase | *nitrite reductase*\* |
| 1.6.99.2 | Reduced-NAD(P): (acceptor) oxidoreductase | reduced NAD dehydrogenase *NADH oxidase* |
| 1.7.99.1 | Ammonia: (acceptor) oxidoreductase | *hydroxylamine reductase* |
| 1.8.1.2 | Hydrogen sulphide: NADP oxidoreductase | *sulphite reductase*\* |
| 1.8.3.1 | Sulphite: oxygen oxidoreductase | *sulphite oxidase*\* |
| 1.9.3.1 | Ferrocytochrome-*c*: oxygen oxidoreductase | *cytochrome oxidase* |
| 1.9.6.1 | Ferrocytochrome-*c*: nitrate oxidoreductase | nitrate reductase (cytochrome) *nitrate reductase* |
| 1.12.1.1 | Hydrogen: ferredoxin oxidoreductase | *hydrogenase* |
| 2.7.1.25 | ATP: adenylylsulphate 3'-phosphotransferase | adenylylsulphate kinase *APS-kinase* APS-phosphotransferase |
| 2.7.4.3 | ATP: AMP phosphotransferase | *adenylate kinase* myokinase |
| 2.7.4.6 | ATP: nucleosidediphosphate phosphotransferase | *nucleosidediphosphate kinase* |
| 2.7.7.4 | ATP: sulphate adenylyltransferase | sulphate adenylyltransferase *ATP-sulphurylase* |

\* These names are also used to describe enzymes which catalyse similar transformations of their substrates but which use electron acceptors or donors other than those listed here.

# LIST OF ENZYMES

| Number | Systematic names | Trivial names |
|---|---|---|
| 2.7.7.5 | ADP: sulphate adenylyltransferase | sulphate adenylyltransferase (ADP) *ADP-sulphurylase* |
| 2.8.1.1 | Thiosulphate: cyanide sulphurtransferase | thiosulphate sulphurtransferase *rhodanese* |
| 2.8.1.2 | 3-Mercaptopyruvate: cyanide sulphurtransferase | *3-mercaptopyruvate sulphurtransferase* |
| 2.8.2.1 | 3′-Phosphoadenylylsulphate: phenol sulphotransferase | aryl sulphotransferase *phenol sulphotransferase* phenol sulphokinase |
| 2.8.2.2 | 3′-Phosphoadenylylsulphate: 3β-hydroxysteroid sulphotransferase | 3β-hydroxysteroid sulphotransferase *steroid sulphotransferase* steroid sulphokinase |
| 2.8.2.3 | 3′-Phosphoadenylylsulphate: arylamine sulphotransferase | *arylamine sulphotransferase* arylamine sulphokinase |
| 2.8.2.4 | 3′-Phosphoadenylylsulphate: oestrone sulphotransferase | *oestrone sulphotransferase* oestrone sulphokinase |
| 2.8.2.5 | 3′-Phosphoadenylylsulphate: chondroitin sulphotransferase | chondroitin sulphotransferase *mucopolysaccharide sulphotransferase* |
| 3.1.3.5 | 5′-Ribonucleotide phosphohydrolase | *5′-nucleotidase* |
| 3.1.3.6 | 3′-Ribonucleotide phosphohydrolase | *3′-nucleotidase* |
| 3.1.6.1 | Aryl-sulphate sulphohydrolase | *arylsulphatase* phenol sulphatase |
| 3.1.6.2 | Sterol-sulphate sulphohydrolase | sterol sulphatase *androstenolone sulphatase* steroid sulphatase 3β-steroid sulphatase |
| 3.1.6.3 | Sugar-sulphate sulphohydrolase | *glycosulphatase* |
| 3.1.6.4 | Chondroitin-sulphate sulphohydrolase | *chondrosulphatase* |
| 3.1.6.6 | choline-sulphate sulphohydrolase | *choline sulphatase* |
| 3.1.6.7 | Cellulose-sulphate sulphohydrolase | *cellulose sulphatase* |
| 3.6.1.1 | Pyrophosphate phosphohydrolase | inorganic pyrophosphatase *pyrophosphatase* |
| 3.6.1.3 | ATP phosphohydrolase | *ATPase* |
| 4.1.1.12 | L-Aspartate 4-carboxy-lyase | aspartate 4-decarboxylase *aspartate β-decarboxylase* |
| 4.2.1.5 | L-Homoserine hydro-lyase (deaminating) | homoserine dehydratase *cystathionase* |
| 4.2.1.20 | L-Serine hydro-lyase (adding indole) | *tryptophan synthase* |
| 4.2.1.22 | L-Serine hydro-lyase (adding hydrogen sulphide) | cysteine synthase *serine sulphhydrase* |
| 4.2.99.6 | Chondroitin-sulphate-lyase | chondroitin sulphate lyase *chondroitinase* |
| 4.4.1.1 | L-Cysteine hydrogensulphide-lyase (deaminating) | *cysteine desulphydrase* cysteine lyase |

# 1

## INTRODUCTION

Figure 1.1 summarizes the general pattern of the changes undergone by sulphur-containing compounds in nature and serves to stress the rather important role which living organisms play in the geochemistry of sulphur. This role is not coincidental but reflects the fact that sulphur-containing compounds participate in all living processes.

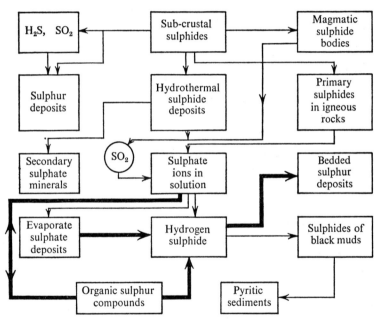

Fig. 1.1. The geochemical transformations of sulphur. Biological processes are outlined with thick lines. (Adapted from Day, 1963.)

Sulphur, as a constituent of amino acids such as cysteine and methionine and of cofactors such as biotin, thiamin, coenzyme A and lipoic acid, is essential for the structure and function of all living cells. The ultimate sources of this element are inorganic compounds of sulphur which, in the case of micro-organisms and plants, may be obtained directly from the environment and subsequently incorporated into organic forms. Animals, on the other hand, require preformed sulphur-containing amino acids and vitamins. In general, they obtain only small amounts of inorganic sulphur compounds from the environment and their requirements for such compounds are, for the most part, met by the oxidation of the sulphur-containing amino acids.

Aside from these general functions of sulphur, certain specialized groups of micro-organisms utilize either oxidative or reductive reactions on inorganic sulphur compounds in the energy-yielding processes of their metabolism.

The types of sulphur-containing compounds available to living organisms are many. In the hydrosphere they are mainly inorganic: in sea water, for instance, only sulphate is quantitatively important with a concentration of about $0.03$ M. Fresh water is more variable in composition and, depending upon many factors—geological, physical and biological—may contain either oxidized or reduced forms of sulphur: that is, principally sulphate or sulphide. In the lithosphere the chemical composition is much more complex, again depending upon the interplay of many factors, and both sulphate-containing and sulphide-containing rocks are widespread. In soil, organic derivatives of sulphur are important and it may come as a surprise that sulphate esters can account for a considerable proportion of the sulphur in soil (Freney, 1967 a). A further source of sulphur which must not be forgotten is the atmosphere in which the main sulphur-containing compound is sulphur dioxide, except in areas where volcanic gases are important when large amounts of hydrogen sulphide may also be present. This sulphur dioxide may be utilized by higher plants either by its direct absorption through the leaves or indirectly after its uptake by, and interconversion in, the soil. The importance of this process must not be underestimated because it generally appears to be the case that the sulphur present in harvested crops exceeds the available sulphur of the soil in which they were grown so that the difference must come from the atmosphere.

The biological changes in the various compounds of sulphur are best summarized in the form of a sulphur cycle, fig. 1.2, which will be considered in rather more detail below: at present it can be summarized by saying that, in general, micro-organisms and plants utilize the inorganic sulphur-containing compounds of the soil (or water) to form the sulphur-containing amino acids and other organic compounds required for their existence. These complex organic compounds are then utilized by the higher animals. Simple inorganic compounds are reformed by degradative processes which occur, to some extent at least, in all organisms.

The importance of these interconversions cannot be too highly stressed because it is upon the production of 'reduced' sulphur that the higher animals depend for their nutrition. The only exception to this generalization is that ruminants, and probably also birds, can utilize inorganic compounds of sulphur through their symbiotic association with micro-organisms in their digestive tracts. Plants generally utilize inorganic compounds of sulphur but the fertility with respect to this element of any particular soil can be greatly modified by the activities of the micro-flora present therein.

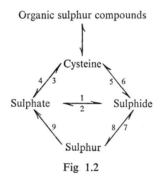

Fig. 1.2

It must be stressed that the general picture of sulphur metabolism given in fig. 1.2 does not indicate the *pathways* of the various transformations in living organisms but only their end-points. Nevertheless, a convenient introduction to the general problem of the biochemistry of inorganic compounds of sulphur can be made through a brief consideration of the various changes enumerated in this figure.

1. *Reduction of sulphate to sulphide.* This transformation is carried out on an extensive scale only by certain micro-organisms, the dissimilatory sulphate-reducers (e.g. *Desulfovibrio* and *Desulfotomaculum*). On a small scale it is, of course, carried out by many micro-organisms and plants although not to any significant extent by higher animals.

2. *Oxidation of sulphide to sulphate.* Again on a small scale this is widespread, occurring even in mammals: on a large scale it is carried out by organisms such as thiobacilli and the photosynthetic sulphur bacteria.

3. *Incorporation of sulphate into amino acids.* This process is again quite widespread although it is absent, at least from a practical point of view, from mammals apart from ruminants.

3

4. *Conversion of amino acids to sulphate.* This is the principal catabolic reaction undergone by the sulphur-containing amino acids and it probably occurs in all living organisms. In higher animals it is of unique importance because it is the main, if not the sole, source of inorganic compounds of sulphur.

5. *Conversion of amino acids to sulphide.* This reaction occurs in most micro-organisms. It can also occur in the higher animals but its extent and significance are not clear.

6. *Conversion of sulphide to amino acids.* This seems to be a quite ubiquitous process.

7. *Conversion of sulphur to sulphide.* This process can occur fairly readily by non-enzymatic reactions and so is probably of widespread, or potentially widespread, occurrence. Only in some bacteria and fungi does it normally occur to any extent and certainly the higher animals are unlikely ever to have to metabolize significant quantities of elemental sulphur.

8 and 9. *Oxidation of sulphide through sulphur to sulphate.* These transformations are characteristically bacterial in origin and they are of fundamental importance in organisms such as *Beggiatoa*, thiobacilli, and the photosynthetic sulphur bacteria.

In general terms, then, the subject matter of this book is concerned with the details of the above pathways and with their relation to other metabolic pathways in the organism. Some of these relationships are relatively clear, particularly in micro-organisms: for example, in certain photosynthetic bacteria sulphur compounds are accessory electron donors for the photosynthetic reaction while in certain chemoautotrophic organisms the oxidation of simple inorganic sulphur compounds is used to produce the energy required for synthetic processes. On the other hand, sulphate is used as an electron acceptor in the energy-yielding reactions carried out by the anaerobic dissimilatory sulphate-reducing bacteria. In micro-organisms generally and in plants the main role of sulphur metabolism seems to be the production of the sulphur-containing amino acids. Complications are caused, however, by the existence of a further metabolic cycle, the sulphate cycle represented in fig. 1.3, all the reactions of which are rather widely distributed in nature, occurring in micro-organisms, plants and animals. The enzymes involved in this cycle are rather specific and the details vary considerably from organism to organism but in general it can be said that it seems to reach its highest development (in the sense of the complexity of the enzyme systems) in the higher animals.

The relationship of this sulphate cycle to the general metabolic pathways of the organism, particularly if it be a higher animal, is obscure, the more so as the synthesis of sulphate esters involves the intermediate formation of an 'activated sulphate', a process which requires the utilization of two molecules of ATP per molecule of sulphate. The energy so used is not, in general, recoverable by the fission of the sulphate-ester linkage and certainly not if the process is hydrolytic.

The transformations outlined in figs. 1.2 and 1.3 must be regarded as closely integrated parts of one complex metabolic pathway because activated forms of sulphate participate not only in the formation of sulphate esters but also in the interconversions of sulphate and sulphite.

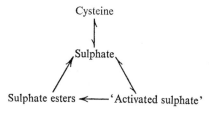

Fig. 1.3

Obviously much remains to be discovered about the function of sulphuric acid and its esters in biochemistry. It is difficult to avoid suggesting that our knowledge of this subject may only be at a stage not far removed from the understanding of the role of phosphoric acid and its esters prior to the work of Harden and Young in 1905. The true function of sulphate ions in biochemistry may have yet to be realized. One wonders what would be our present knowledge of the metabolic function of these ions if there had been available for their determination a method as simple and as sensitive as the colorimetric methods which have long been used for the determination of phosphate.

So far mention has been made only of the inorganic compounds of sulphur and of some simple organic derivatives thereof such as the sulphate esters. Living organisms, however, contain or can form many other organic compounds of sulphur, most of which contain C—S bonds: see, for example, the discussion by Freney (1967 a) of the types of sulphur compound found in soil. In general, such compounds will not be given detailed consideration in this book except for the special case of the thiosulphate esters, or Bunte salts, which contain the C—S bond but which are (or can be regarded as) sulphate esters of thiols. Other com-

5

pounds containing C—S bonds will be considered only in so far as their metabolism is connected with the metabolism of inorganic compounds of sulphur. Examples of such substances are the many types of sulphonic acid found in nature: for instance, taurine with its important role in the metabolism of liver and its many other more or less well authenticated functions (see review by Jacobsen & Smith, 1968); isethionic acid with its possible role in regulating the excitability of tissues (Jacobsen & Smith, 1968) and 6-sulpho-D-quinovose, a constituent of the widely distributed plant sulpholipid (Benson, 1963). Many other types of compound containing C—S bonds could be mentioned including, for instance, ergothionine, lipoic acid, coenzyme A, thiamin, biotin and penicillin (the latter three being sulphur heterocycles) to name only a few, but the biochemistry of these compounds will not be considered here and information on them must be sought elsewhere (see Young & Maw, 1958; Kun, 1961). Compounds containing N—S bonds are rare in nature and only one type, the sulphamates, which can be regarded as atypical sulphate esters, will be considered briefly. Naturally occurring heterocycles containing N—S bonds do not appear to be known.

A considerable part of the interest in the biochemistry of sulphur and its inorganic compounds stems from the view that energy metabolism based on this element may have preceded, in an evolutionary sense, those based on oxygen. It is commonly held that the primordial conditions on the earth were highly reducing (Abelson, 1966; Cloud, 1968) and that sulphur existed largely in the form of sulphides: from this has developed the idea that primitive life may have been based on the photosynthetic oxidation of sulphides to sulphur and sulphate. Such primitive photosynthetic sulphur-oxidizers thus served as precursors of the modern photosynthetic sulphur bacteria as well as of dissimilatory sulphur reducers and the chemosynthetic sulphur oxidizers. Certainly there is evidence, based on the distribution of stable sulphur isotopes in nature, that dissimilatory sulphate reduction was occurring some $3 \cdot 5 \times 10^9$ years ago which, in a geological sense, is soon after the emergence of life on the earth (about $4 \times 10^9$ years ago). Free oxygen, on the other hand, probably did not appear in the atmosphere until about $1–2 \times 10^9$ years ago and certainly only accumulated at a much later date, after the development of the photosynthetic activities of green plants. Presumably only after this time did the oxygen-requiring sulphur autotrophs develop. For a more detailed consideration of this topic the most interesting speculations of Peck (1966–7) should be consulted.

# 2

## THE CHEMISTRY OF SOME
## SULPHUR COMPOUNDS

In this chapter it is our intention to outline very briefly the structures and some properties of a number of sulphur compounds which are important in the metabolism of inorganic sulphur. The literature on sulphur chemistry is, of course, vast and for detailed treatments the reader is referred to standard texts such as those of Abegg, Auerbach & Koppel (1927), Mellor (1930), Gmelin (1960, 1963) and Durrant & Durrant (1962).

Our particular aims, in this chapter, are to highlight first those reactions which occur readily at physiological temperatures and pH values and thus may lead to secondary, non-specific events during metabolic experiments, and secondly some chemical reactions which may be models for enzymic reactions on sulphur compounds. Other reactions which are primarily of analytical interest are discussed in chapter 4.

Sulphur exhibits valencies of 2, 4 and 6 and oxidation states of $-2$, $0$, $+2$, $+4$ and $+6$, and its electronic structure in the ground state is depicted thus:

$$[\text{Ne}]^{10}\, 3s^2\, 3p_x{}^2\, 3p_y{}^1\, 3p_z{}^1\, 3d^0\, 3d^0\, 3d^0\, 3d^0\, 3d^0$$

A characteristic feature of sulphur is its readiness to form the S—S bond and consequently to form long chains of sulphur atoms. Branched sulphur chains are apparently not formed (cf. Foss, 1950, 1960).

Pauling (1949) considered that the $\sigma$ bond between two sulphur atoms is nearly pure $p$ in character. Every sulphur atom then still possesses two free electron pairs, one of which, the $s$-pair, is spherically distributed around the nucleus. The remaining pair exist as $3p\pi$ electrons on the 90° axes and cause a barrier to rotation around the sulphur–sulphur bond which arises from Coulombic repulsion between unshared $p$-electrons on adjacent sulphur atoms.

This concept has been challenged by Schmidt (1963) who considers it inadequate to explain many experimental results on the nucleophilic degradation of sulphur–sulphur bonds. He suggests that the 'free' $p\pi$ electrons of a particular sulphur atom are not, in fact, free but are involved

7

in sulphur–sulphur bond formation by overlap with an available *d*-orbital of the bonding partner. The sulphur–sulphur bond would then have some *pdπ*-bond character which may account for its stability (see also Foss, 1961 *a*) and possibly also its non-planarity (Krebs, 1957). Moffitt (1950) has shown that sulphur utilizes the 3*d* electron orbitals to attain the hexavalent state in compounds containing sulphur–oxygen bonds; these bonds therefore have some double bond character.

## 2.1 Structures of some inorganic compounds

### 2.1.1 Elemental sulphur

Elemental sulphur exists in a large number (about 30) of solid allotropes (Meyer, 1964) but only orthorhombic and colloidal sulphur will be considered here. The significance of the other allotropic forms with respect to metabolism is doubtful: indeed many of these forms are prepared and exist only under special conditions which are incompatible with biological activity.

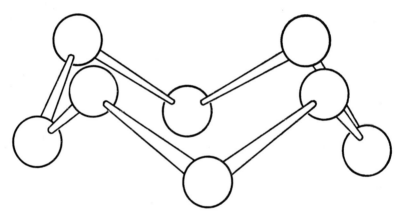

Fig. 2.1   The molecular structure of orthorhombic sulphur.

Orthorhombic sulphur (S$\alpha$) is the only thermodynamically stable form under normal temperatures and pressures: all other allotropes ultimately convert to it (Meyer, 1964). Its structure is that of a staggered eight membered ring (fig. 2.1; Warren & Burwell, 1935; Abrahams, 1955) with the following molecular constants (Donohue, 1961, 1965): average S—S bond length, 2·060 Å; average S—S—S bond angle, 108·0°; average S—S—S—S dihedral angle 99·3°. Orthorhombic sulphur is readily identified by means of its X-ray diffraction pattern (see index cards 8-247 and 8-248, powder diffraction file, American Society for testing and materials, Philadelphia, 1966).

Orthorhombic sulphur is essentially insoluble in water but dissolves in a number of organic solvents (table 2.1). Its solubility in some natural fats and oils may possibly be of biological significance (see, for example, Umbreit, Vogel & Vogler, 1942).

Colloidal sulphur exists in hydrophobic and hydrophilic forms according to the method of preparation (see Mellor, 1930). An early suggestion was that the hydrophilic form is stabilized by the presence of pentathionate in the micelles and Weitz, Gieles, Singer & Alt (1956) have shown that colloidal sulphur behaves, in many of its properties, as a polythionate containing 40–140 sulphur atoms.

TABLE 2.1    *Solubility of elemental sulphur (Linke, 1965)*

| | Solubility (g sulphur/100 g solvent) | | |
|---|---|---|---|
| | 20° | 100° | Other |
| $CS_2$ | 41·8 | 92 | . |
| $CCl_4$ | . | . | 0·86 g/100 g sat. soln. at 25° |
| benzene | 1·7 | 17·5 | . |
| linseed oil | 0·6(30°) | . | . |
| olive oil (S.G. 0·885) | 4·3(30°) | . | . |
| lanoline (anhydrous) | 0·38(45°) | . | . |

### 2.1.2  Oxyacids of sulphur

Sulphur forms a large number of oxyacids although most of these are known only in solution; their ions, however, are well characterized. Sulphate, sulphite, thiosulphate, polythionates and possibly dithionate have been shown to be involved in various phases of the reductive and oxidative metabolism of sulphur. The occasional suggestions that other sulphuroxy ions such as dithionite and persulphates may also be metabolic intermediates await confirmation and these compounds will not be discussed further in this chapter. The hypothetical acids, sulphoxylic (sulphur dihydroxide, $S(OH)_2$), thiosulphurous (disulphur dihydroxide, $S_2(OH)_2$) and sulphenic (sulphur hydrate, $SH_2O$) whose structures are somewhat uncertain, have often been proposed as transient intermediates in hydrolytic reactions on inorganic sulphur compounds and may possibly be formally analogous to transient intermediates (free or bound) arising during metabolism of sulphur compounds.

9

### Sulphate, thiosulphate, dithionate, sulphite

The structures of the sulphate, thiosulphate, sulphite and dithionate ions are shown in formulae *2.1* to *2.4*

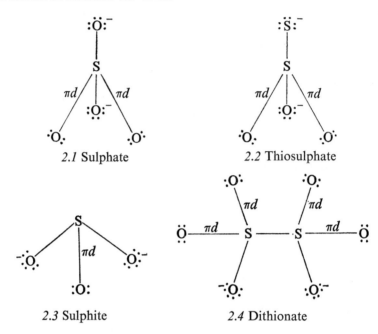

*2.1* Sulphate  *2.2* Thiosulphate

*2.3* Sulphite  *2.4* Dithionate

All are resonance structures in which the oxygen atoms are equivalent. The sulphate and thiosulphate groups are tetrahedral forms with the hexavalent sulphur at their centres as also are the two sulphonate groups of dithionate: sulphite is pyramidal.

Thiosulphate may be viewed as being derived from sulphate by substitution of an oxygen atom by sulphur. The non-equivalence of the two sulphur atoms in thiosulphate was elegantly demonstrated by Buch Andersen (1936) who synthesized thiosulphate from $^{35}$S-labelled elemental sulphur and unlabelled sulphite (equation 2.1) and showed that the labelled product decomposed in the presence of hydrogen or silver ions to give the distribution of $^{35}$S shown in equations 2.2 and 2.3.

$$\overset{*}{S}+SO_3^{2-} \rightarrow [S-SO_3^{2-}]^* \tag{2.1}$$

$$[S-SO_3^{2-}]^* \overset{H^+}{\rightarrow} \overset{*}{S}+SO_3^{2-} \tag{2.2}$$

$$[S-SO_3^{2-}]^* \overset{Ag^+}{\rightarrow} \overset{*}{S}{}^{2-}+SO_4^{2-} \tag{2.3}$$

## Polythionates

Polythionates have the general formula $S_nO_6^{2-}$ where $n$ is three or greater. While polythionates with up to 40 sulphur atoms have been reported (Weitz, Becker, Gieles & Alt, 1956) it is probable that only the low molecular weight polythionates ($n < 8$) are significant biologically.

A mass of evidence, including Raman spectra, X-ray fluorescence, diamagnetic susceptibility and X-ray cystallography has established that polythionates are linear chains of sulphur atoms (at oxidation $-2$) terminated by sulphonate groups (oxidation state of sulphur $+6$) (see reviews by Foss, 1950, 1960); the average bond lengths are; inner S—S, 2·04 Å; terminal S—SO$_3$, 2·11 Å. The polythionates are, in essence, disulphonic acid derivatives of sulphanes ($H_2S_n$; Schmidt, 1957 a) and the structure of the tetrathionate ion, the derivative of disulphane, for example, is shown in the following formula.

$$
\begin{array}{ccc}
\text{O} & & \text{O} \\
\vert \ \pi d & & \vert \ \pi d \\
{}^-\text{O—S—S—S—S—O}^- & & \\
\vert \ \pi d & & \vert \ \pi d \\
\text{O} & & \text{O}
\end{array}
$$

It may be noted in parenthesis that 'dithionic acid' is often erroneously described as the 'basic' member of the polythionate group. It is, however, not a derivative of sulphane and its physical and chemical properties clearly distinguish it from true polythionates (see for example Brodskii, 1954).

## 2.2   Reactions of sulphur compounds

Sulphur is, in general, a highly reactive element and the sulphur–sulphur bond can be cleaved by a variety of agents such as heat and radiation (equation 2.4), free radicals (equation 2.5), electrophilic agents (equation 2.6) and nucleophilic reagents (equation 2.7)

$$\text{RS—SR}' \xrightarrow[\text{radiation}]{\text{heat}} \text{RS}^{\cdot} + \text{R}'\text{S}^{\cdot} \tag{2.4}$$

(unimolecular homolytic cleavage)

$$\text{RS—SR}' + \text{R}'' \rightarrow \text{RS—R}'' + \text{R}'\text{S}^{\cdot} \tag{2.5}$$

(bimolecular homolytic cleavage)

$$\text{RS—SR} + \text{A}^+ \rightarrow \text{RSA} + \text{RS}^+ \tag{2.6}$$

(electrophilic cleavage)

$$\text{RS—SR} + \text{B}^- \rightarrow \text{RSB} + \text{RS}^- \tag{2.7}$$

(nucleophilic cleavage)

11

As pointed out by Kice (1968), the scission of the S—S bond is most effectively catalysed by the co-operative effect of an electrophilic and a nucleophilic reagent. The former converts one of the sulphur atoms into a better leaving group which is then displaced by the nucleophile attacking the other sulphur atom.

These reactions have been the subject of several excellent recent reviews (for example Parker & Kharasch, 1959; Pryor, 1962; Foss, 1961 b; Davis, 1964). Radiation-induced homolytic scission of disulphide groups, and the secondary reactions (e.g. bimolecular homolytic scission; equation 2.5) initiated by the free radicals generated by this process, are undoubtedly of biological significance, certainly in respect to the effects of radiation on organisms (Alexander, 1960) and possibly with regard to photosynthetic reactions. Electrophilic reactions at the disulphide bond may also be biologically important. For example, disulphide-exchange reactions have been attributed to electrophilic cleavage of sulphur–sulphur bonds by protons and sulphenium ions (Benesch & Benesch, 1958) and electrophilic cleavage by methyl carbonium ions may be involved in biological methylation reactions (Challenger, 1955). As far as the inorganic sulphur compounds are concerned, however, the most important reactions which occur at moderate temperatures and pH values and which involve the sulphur–sulphur bond appear to be bimolecular nucleophilic ($S_N2$) reactions. Reactions of this type have often been invoked to account for enzymic modifications to sulphur compounds and many such compounds react with nucleophilic reagents, which are commonly found in biological systems, at sufficiently rapid rates to cause non-specific chemical side reactions and thus complicate the results of biological experiments.

The reactivity of a number of nucleophilic ions towards the sulphur–sulphur bonds has been the subject of a number of studies (see for example Foss, 1947, 1961 b; Parker & Kharasch, 1959; Pryor, 1962) and the following order appears to be established for ions of biological significance.

$$HS^- > RS^- (C_2H_5S^-, C_6H_5S^-, RCH_2S^-) > CN^- > SO_3^{2-} > OH^-$$
$$> S_2O_3^{2-} > RSO_2S^- > SCN^- > Cl^-$$

### 2.2.1 Reactions of elemental sulphur

The $S_8$ ring of elemental sulphur is readily cleaved by a number of nucleophilic reagents. A reaction with hydroxyl ions, for example, occurs at temperatures close to 100° when sulphur disproportionates in sodium or ammonium hydroxide solutions to give sulphide and thiosulphate (equation 2.8, Pryor, 1962).

$$S_8 + 8OH^- \rightarrow 2S_2O_3{}^{2-} + 4HS^- + 2H_2O \tag{2.8}$$

Although the reaction presumably involves an initial hydrolysis of the $S_8$ ring to form a sulphane sulphenic acid ($^-S$—$(S_6)$—$SOH$) the exact mechanism by which the products are formed is uncertain. The possibility that some hydrolysis of sulphur may occur at physiological temperatures and pH, particularly with colloidal preparations or with 'nascent' sulphur formed by biological reactions, should be seriously considered.

### Reactions of sulphur with sulphite and cyanide

Rather better characterized are the reactions between sulphur and cyanide or sulphite which give rise to thiocyanate and thiosulphate respectively (equations 2.9 and 2.10).

$$S_8 + 8SO_3{}^{2-} \rightarrow 8S_2O_3{}^{2-} \tag{2.9}$$

$$S_8 + 8CN^- \rightarrow 8SCN^- \tag{2.10}$$

Foss (1950) and Bartlett & Davis (1958) proposed that cyanolysis of sulphur involves an initial, rate-limiting, nucleophilic attack by cyanide on the $S_8$ ring (equation 2.11) followed by a series of fast displacements of the weakly nucleophilic thiocyanate ion by cyanide (equation 2.12).

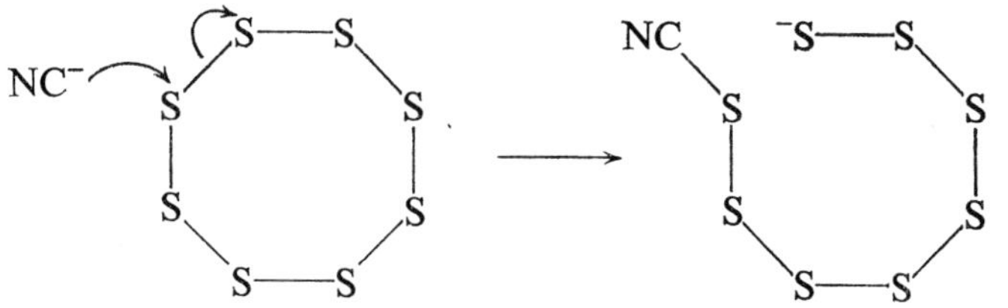

$$\text{(2.11)}$$

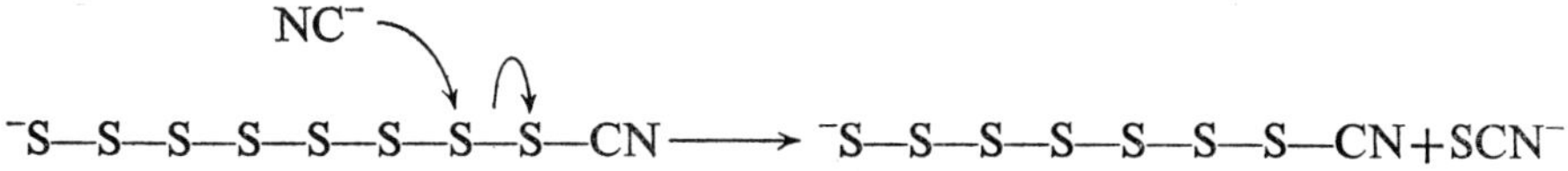

$$^-S\text{—}S\text{—}S\text{—}S\text{—}S\text{—}S\text{—}S\text{—}S\text{—}CN \longrightarrow {}^-S\text{—}S\text{—}S\text{—}S\text{—}S\text{—}S\text{—}S\text{—}CN + SCN^- \tag{2.12}$$

etc.

The reaction between sulphur and cyanide is rapid, at least in organic solvents, and the following second-order rate constants (in l mole$^{-1}$ sec$^{-1}$) have been recorded by Bartlett & Davis (1958): 33·6 in methanol at 25°;

7·97 in methanol at 4·5°; 4·28 in 8·3% (w/w) $H_2O$ in methanol at 4·5°; 3·40 in 24·3% (w/w) $H_2O$ in methanol at 4·5°.

According to Foss (1950) the reaction between sulphite and sulphur is analogous to the cyanolytic reaction. Initially the $S_8$ ring is cleaved and octasulphane monosulphonate is formed and subsequently a thio-sulphate ion is displaced by sulphite with the formation of hepta-sulphane monosulphonate (equation 2.13): further displacements of thio-sulphate ions result in the complete conversion of sulphur to thiosulphate.

$$^{-2}O_3S \diagdown$$
$$^-S\!-\!S\!-\!S\!-\!S\!-\!S\!-\!S\!-\!S\!-\!S\!-\!SO_3^- \longrightarrow {}^-S\!-\!S\!-\!S\!-\!S\!-\!S\!-\!S\!-\!S\!-\!SO_3^- + S_2O_3^{2-}$$

$$(2.13)$$

Schmidt (1965), however, has pointed out that the nucleophilic character of sulphane monosulphonates ($^-S\!-\!(S)_n\!-\!SO_3^-$) decreases with increasing values for $n$ and that reaction 2.13 should, therefore, be viewed as a displacement of the higher sulphane monosulphonate ions by sulphite (equation 2.14 see also Schmidt, 1963).

$$^{-2}O_3S \diagdown$$
$$^-S\!-\!S\!-\!S\!-\!S\!-\!S\!-\!S\!-\!S\!-\!S\!-\!SO_3^- \longrightarrow {}^-O_3S\!-\!S\!-\!S\!-\!S\!-\!S\!-\!S\!-\!S\!-\!S^- + S_2O_3^{2-}$$

$$(2.14)$$

It should be noted, however, that since the intermediates are identical in both instances the two mechanisms are for all practical purposes essentially equivalent. The reaction between sulphite and sulphur is also very rapid: Schmidt & Talsky (1959), for example, reported that 10–15 mg of sulphur was completely converted to thiosulphate within 30 sec at room temperature in a mixture composed of 1–3 ml of chloroform (in which the sulphur was dissolved), 30 ml of 4% aqueous sodium sulphite and 100 ml of acetone (or 50 ml of methanol).

### Reactions of sulphur with sulphide and thiols

Sulphide is a powerful $S$-nucleophile (see sequence above) and it has long been known that aqueous solutions of sodium sulphide dissolve sulphur with the formation of polysulphanes (equation 2.15)

$$S_8 \xrightarrow{\;HS^-\;} HS_n^-$$

$$(2.15)$$

(where $n = 2$, 3 etc. depending upon the sulphide concentration).

The reaction is extremely rapid and a second-order rate constant of greater than $1000 \, l \, mole^{-1} \, sec^{-1}$ at $25°$ in benzene has been recorded for the degradation of sulphur by triethylammonium hydrosulphide (Davis, 1958). Even at $-78°$ the rate constant of the reaction between sulphur and sulphide in alkaline methanol is $0·72 \, l \, mole^{-1} \, sec^{-1}$.

The linear polysulphanes undergo nucleophilic displacement reactions at much faster rates than does $S_8$ (Pryor, 1962) and as a consequence sulphide, by causing a rapid cleavage of the $S_8$ ring, catalyses the reactions of $S_8$ with other nucleophilic reagents in which the ring cleavage is the rate limiting step. Examples of such catalysis by sulphide are the sulphitolysis of sulphur (Levenson, 1954) and the reaction between sulphur and triphenylphosphine (Bartlett, Colter, Davis & Roderick, 1961; Bartlett, Cox & Davis, 1961; Davis, 1962).

The $S_8$ ring is also cleaved by thiols with the formation of organic disulphanes and polysulphanes (equation 2.16).

$$8RSH + S_8 \rightarrow 8RSSH \qquad (2.16)$$

Hylin & Wood (1959), for example, studied the reactions between sulphur and mercaptoethanol, 3-mercaptopyruvate or cysteine and showed that about 10% of the thiol compounds were converted to 'polysulphane' (as determined by cold cyanolysis) within 15 min at $37°$ and pH 9 in aqueous mixtures containing 6 mg of flowers of sulphur and $0·1$ M solutions of the thiol compounds; these rates were comparable with those of the reaction between sulphur and sulphide under the same conditions. Hylin & Wood also showed that the thiols (except 3-mercaptopyruvate) catalysed the cyanolysis of aqueous suspensions of sulphur, presumably by forming intermediate polysulphanes which are attacked by cyanide more readily than the $S_8$ ring itself.

De Marco, Coletta & Cavallini (1961) reported that thiols such as cysteine, cysteamine and 2-mercaptoethanol promote the formation of thiosulphonates from hypotaurine or alanine 3-sulphinate. They proposed that intermediate disulphanes are formed which transfer sulphur to the sulphino compounds (equation 2.17).

$$RSSH + R'SO_2H \rightarrow RSH + R'SO_2SH \qquad (2.17)$$

## 2.2.2 Reactions of sulphide

Sulphide undergoes reactions with sulphur and polythionates which are discussed elsewhere (see above and § 2.2.6). In alkaline solutions sulphide reacts with organic disulphides (e.g. cystine) apparently to form di-sulphanes (equation 2.18, Rao & Gorin, 1959).

$$RSSR + S^{2-} \rightarrow RSS^- + RS^- \qquad (2.18)$$

In the presence of excess sulphide further reactions between the di-sulphanes and sulphide take place (e.g. equation 2.19).

$$RSS^- + S^{2-} \rightarrow RS^- + S_2^{2-} \qquad (2.19)$$

Under acid conditions elemental sulphur is formed from cystine and sulphide, apparently by the breakdown of the disulphane intermediate. Thus the overall reaction in acid is a reduction of the disulphide to thiols by sulphide (equation 2.20).

$$RSSR + S^{2-} \rightarrow 2RS^- + S^0 \qquad (2.20)$$

Sulphide solutions oxidize in air, particularly in the presence of traces of heavy metals (Krebs, 1929), and on exposure to light; elemental sulphur is generally the main product but sulphate and other oxyacids may be formed in small amounts.

One of the most important properties of hydrogen sulphide from the biological point of view is its ionization: the first ionization constant $(H_2S \rightleftharpoons H^+ + HS^-)$ is $1 \cdot 15 \times 10^{-7}$ and the second ionization constant $(HS^- \rightleftharpoons H^+ + S^{2-})$ is $1 \cdot 0 \times 10^{-15}$. Thus at physiological pH values sulphide is largely in the form of $H_2S$ which evaporates rapidly from solution: this often causes experimental difficulties in the use of sulphide in bio-logical experiments (cf. Postgate, 1963 *d*).

### 2.2.3  Reactions of sulphite

Sulphite is a relatively strong nucleophile and, in particular, will displace thiosulphate from many compounds such as polythionates (see § 2.2.6).

### *Reaction of sulphite with disulphides*

Sulphite reacts with disulphide groups to give thiols and organic thio-sulphates (equation 2.21).

$$RSSR + SO_3^{2-} \rightleftharpoons RSSO_3^- + RS^- \qquad (2.21)$$

This reaction (with cystine) was first described by Clarke (1932) and by Lugg (1932) but other disulphides have since been shown to similarly react with sulphite (see reviews by Parker & Kharasch, 1959; Milligan & Swan, 1962).

As expected from the relative nucleophilicities of sulphite and $RS^-$ the equilibrium of reaction 2.21 at neutral or alkaline conditions is strongly in favour of disulphide and sulphite although the values for the equilibrium constants are strongly pH-dependent owing to the ionization of the reacting species (Stricks & Kolthoff, 1951; Stricks, Kolthoff & Kapoor, 1955). The rates of reaction 2.21 may be very fast; Cecil & McPhee (1955), for example, have reported an apparent second order rate constant at 25° and pH 6·9 for the reaction between cystine and sulphite of 400 l mole$^{-1}$ min$^{-1}$. The rates are, however, strongly influenced by the net charge in the vicinity of the sulphur–sulphur bond and are also slow at low pH values because of the formation of the weakly nucleophilic bisulphite ion. The second dissociation constant of sulphurous acid (HSO$_3^-$ $\rightleftharpoons$ H$^+$ + SO$_3^{2-}$) is, incidentally, 6·24 × 10$^{-8}$.

## Oxidation of sulphite

Sulphite solutions autoxidize readily to sulphate in air at physiological pH values and temperature. Postgate (1963 $d$) reported that a 0·1 M solution of sodium sulphite in physiological saline buffered with phosphate fell to 0·07 M after shaking in air for 1 hr at 37° and to 0·022 M after 2½ hr. The reaction is bimolecular but shows first-order kinetics with respect to sulphite when there is a constant supply of oxygen. Schroeter (1963, 1966) studied the kinetics of oxidation of sodium sulphite at 25° in a reactor containing 300 ml of solution which was stirred at 1000 rev/min. and aerated with CO$_2$-free air at 50 ml/sec: an average first-order constant of 2·88 × 10$^{-3}$ sec$^{-1}$ was obtained. Using pure oxygen, a first-order rate constant of 13 × 10$^{-3}$ sec$^{-1}$ was found by Fuller & Crist (1941).

The autoxidation of sulphite is catalysed by a number of metal ions: in our laboratory we have found the following ions to be active, Cu$^{2+}$, Zn$^{2+}$, Co$^{2+}$, Ni$^{2+}$, Mn$^{2+}$, Fe$^{2+}$ and Fe$^{3+}$. Fuller & Crist (1941) reported that the rate of sulphite oxidation is directly proportional to the cupric ion concentration when this exceeds 1 nM; the catalytic constant (in l mole$^{-1}$ sec$^{-1}$ at 25° with pure oxygen) was 2·5 × 10$^6$ which is about 5 × 10$^8$ times as great as the first-order rate constant for the uncatalysed reaction.

The effects of metal ions on the autoxidation of sulphite are abolished by complexing agents such as ethylenediamine tetraacetate (Schwab & Strohmeyer, 1956). Traces of metal ions are often present in reagents and biological materials so that the use of complexing agents is, wherever possible, recommended for biological experiments on sulphite metabolism. Since sulphite autoxidation is a free radical chain reaction it may also be inhibited by reagents which break the reaction chain (see Schroeter, 1966). Among such reagents are organic acids and alcohols, sugars, polysaccharides, amines and amides which inhibit sulphite oxidation at concentrations as low as 1 $\mu$M.

### 2.2.4 Reactions of thiosulphate

Thiosulphate is relatively stable around pH 7 and, in fact, can be autoclaved in aqueous solutions with little or no decomposition: the decomposition may, however, be accelerated by metal ions. In alkaline solutions thiosulphate decomposes to sulphide and sulphate or sulphite depending on the conditions (cf. Gmelin, 1960).

Below about pH 4–5 thiosulphate decomposes rapidly in a complex manner, the end product depending upon the acidity of the solutions (Pollard & Jones, 1958). In strongly acid media, sulphur dioxide, sulphur and polythionates are produced; at medium acidity high molecular weight sulphanes are found while in weakly acid media thiosulphate is converted almost entirely to sulphur and bisulphite (equation 2.22).

$$8HS_2O_3^- \rightarrow S_8 + 8HSO_3^- \qquad (2.22)$$

The mechanism of thiosulphate degradation in acid is not absolutely certain but perhaps the most consistent hypothesis is that of Davis (1958) which involves nucleophilic displacement of sulphite ions from hydrogen thiosulphate by thiosulphate ions (equations 2.23 to 2.27); although sulphite is more nucleophilic than thiosulphate it is removed from the equilibrium by conversion to bisulphite or sulphurous acid

$$S_2O_3^{2-} + H^+ \rightleftharpoons HS_2O_3^- \qquad (2.23)$$

$$HS_2O_3^- + S_2O_3^{2-} \rightarrow HS_3O_3^- + SO_3^{2-} \qquad (2.24)$$

$$HS_3O_3^- + S_2O_3^{2-} \rightarrow HS_4O_3^- + SO_3^{2-} \qquad (2.25)$$

and so on until $\quad HS_8O_3^- + S_2O_3^{2-} \rightarrow HS_9O_3^- + SO_3^{2-} \qquad (2.26)$

$$HS_9O_3^- \rightarrow S_8 + HSO_3^- \qquad (2.27)$$

The rate of the overall reaction, 2.22, is governed by the concentrations of hydrogen ions and thiosulphate according to the following approximate expression:
$$\text{rate} \propto [H^+]^{\frac{1}{2}}[\text{thiosulphate}]^{\frac{2}{3}}$$

The specific rate constants for reactions 2.23 and 2.27 have been reported to be 0·31 l mole$^{-1}$ min$^{-1}$ and 0·14 min$^{-1}$ respectively (Zaiser & La Mer, 1948; Dinegar, Smellie & La Mer, 1951; Davis, 1958). The formation of polythionates and hydrogen sulphide in acidified solutions of thiosulphate is thought to be due to the interaction of the sulphane monosulphonate intermediates (e.g. equation 2.28; Davis, 1958; see also Schmidt, 1965).

$$HO_3SSS^- + {}^-O_3SSH \rightarrow HO_3SSSSO_3^- + HS^- \tag{2.28}$$

It should be mentioned that the mechanism of acid degradation of thiosulphate proposed by Davis (1958) has been criticized by Agarwala, Rees & Thode (1965) who studied the kinetic isotope effects during the course of the decomposition reaction. The results were more in accord with a simple bimolecular reaction.

## Reactions with thiols and disulphides

Thiols such as cysteine and glutathione reduce thiosulphate to sulphide and sulphite (Steigmann, 1945; Szczepkowski, 1958; Kaji & McElroy, 1959). The reaction is rapid at high temperature at pH values near neutrality or at low temperatures at pH values less than 4. Significant reduction may occur, however, at room temperature and pH 7. According to Szczepkowski (1958) the reaction at pH 7 involves the intermediate formation of an organic thiosulphate (equations 2.29 and 2.30) which can be detected chromatographically.

$$RSH + SSO_3^{2-} \rightleftharpoons RSSO_3^- + HS^- \tag{2.29}$$

$$RSSO_3^- + RSH \rightleftharpoons RSSR + HSO_3^- \tag{2.30}$$

In acid media a preliminary decomposition of thiosulphate to sulphite and sulphur may be involved (equations 2.31 and 2.32; Kaji & McElroy, 1959).

$$S\text{—}SO_3^{2-} \rightarrow S^0 + SO_3^{2-} \tag{2.31}$$

$$S^0 + 2RSH \rightarrow RSSR + H_2S \tag{2.32}$$

Although thiosulphate is a weak nucleophile it can apparently displace thiols from disulphides at elevated temperatures (60°) to form substituted sulphodisulphanes: a reaction of this type between cystine and thiosulphate (equation 2.33) was reported by Szczepkowski (1958).

$$RSSR + SSO_3^{2-} \rightarrow RSSSO_3^- + RS^- \tag{2.33}$$

## Chemical reactions of thiosulphonates

Thiosulphonates could be regarded as organic derivatives of thiosulphate substituted at the hexavalent sulphur. The reactions of these compounds at physiological pH values and temperatures have not been studied extensively. It would appear, however, that substitution labilizes the sulphur–sulphur bond. The rates of reaction of organic thiosulphonates with cyanide, for example, are greatly in excess of that between cyanide and thiosulphate: Mintel & Westley (1966 a) report the following relative rates: thiosulphate, 1; ethanethiosulphonate, 21; methanethiosulphonate, 43; p-toluenethiosulphonate, 83; benzenethiosulphonate, 103; p-bromobenzenethiosulphonate, 140. Sulphite displaces sulfinate ions from organic thiosulphonates (equation 2.34; Foss, 1947).

$$RSO_2S^- + SO_3^{2-} \rightarrow RSO_2^- + S_2O_3^{2-} \qquad (2.34)$$

### 2.2.5 Reactions of dithionate

Dithionate is a stable ion and appears to undergo no spontaneous reactions which might interfere in biological experiments. Extreme conditions are required for its oxidation to sulphate.

### 2.2.6 Reactions of polythionates

Polythionates are highly reactive and their reactivity increases with increasing chain length. Since their structures contain the weakly-nucleophilic sulphane monosulphonate groups the polythionates are very prone to nucleophilic reactions with stronger nucleophiles.

As in the case of the reaction between sulphite and elemental sulphur alternative mechanisms have been proposed for those nucleophilic reactions which may be illustrated by the degradation of polythionates by sulphite. Tetrathionate and higher polythionates are rapidly and completely converted to trithionate and thiosulphate by excess sulphite (Colefax, 1908; Raschig, 1920; Foerster & Centner, 1926). Christiansen & Drost-Hansen (1949) showed that the reaction between tetrathionate and [35]S-labelled sulphite results in labelled trithionate and unlabelled thiosulphate indicating that the reaction involves a displacement of thiosulphate group by sulphite (equation 2.35)

$$^-O_3S\!-\!S\!-\!S\!-\!SO_3^- + \overset{*}{S}O_3^{2-} \rightarrow {}^-O_3S\!-\!S\!-\!\overset{*}{S}O_3^- + S_2O_3^{2-} \quad (2.35)$$

Foss (1947) assumed that the degradation of higher polythionates by sulphite involves the sequential displacement of thiosulphate groups with the intermediate formation lower polythionates. This is illustrated by the reaction of sulphite with hexathionate (equations 2.36 to 2.39).

$$SO_3^{2-} + {}^-O_3S\!-\!S\!-\!S\!-\!S\!-\!S\!-\!SO_3^- \rightarrow {}^-O_3S\!-\!S\!-\!S\!-\!S\!-\!SO_3^- + S_2O_3^{2-}$$
$$(2.36)$$

$$SO_3^{2-} + {}^-O_3S\!-\!S\!-\!S\!-\!S\!-\!SO_3^- \rightarrow {}^-O_3S\!-\!S\!-\!S\!-\!SO_3^- + S_2O_3^{2-} \quad (2.37)$$

$$SO_3^{2-} + {}^-O_3S\!-\!S\!-\!S\!-\!SO_3^- \rightarrow {}^-O_3S\!-\!S\!-\!SO_3^- + S_2O_3^{2-} \qquad (2.38)$$

$$\text{SUM } S_6O_6^{2-} + 3SO_3^{2-} \rightarrow S_3O_6^{2-} + 3S_2O_3^{2-} \qquad (2.39)$$

According to Schmidt (1965), however, the displacement of higher sulphane monosulphonates by sulphite should occur more readily than that of thiosulphate (monosulphane monosulphonate) and he proposed the following mechanism (equations 2.40 to 2.43) for the reaction of hexathionate with sulphite which involves the sequential displacement of sulphane monosulphonates of decreasing chain length.

$$^{-2}O_3S + {}^-O_3S-S-S-S-S-SO_3^- \longrightarrow {}^-O_3S-S-S-S^- + {}^-O_3S-S-S-SO_3^- \quad (2.40)$$

$$^{-2}O_3S + {}^-S-S-S-SO_3^- \longrightarrow {}^-O_3S-S-S^- + S_2O_3^{2-} \quad (2.41)$$

$$^{-2}O_3S + {}^-S-S-SO_3^- \longrightarrow {}^-O_3S-S^- + S_2O_3^{2-} \quad (2.42)$$

$$\text{SUM } S_6O_6^{2-} + 3SO_3^{2-} \rightarrow S_3O_6^{2-} + 3S_2O_3^{2-} \quad (2.43)$$

It may be noted that in the case of tetrathionate both mechanisms are formally equivalent and the same end result is predicted by both hypotheses where sulphite is limiting. With higher polythionates, however, the two hypotheses predict different products when sulphite is limiting. For example, with pentathionate according to the Foss hypothesis (equations 2.37 and 2.38) mixtures of pentathionate, tetrathionate, trithionate and thiosulphate should be obtained whereas by the Schmidt mechanism trithionate, thiosulphate and disulphane monosulphonate will be formed. In fact, traces of sulphite catalyse the disproportionation of pentathionate into complex mixtures of sulphur compounds (including higher polythionates) a result which, according to Schmidt (1965) would be produced by secondary reactions of the highly unstable disulphane monosulphonate. It should be noted, however, that complex mixtures could also arise as the result of the Foss mechanism due to disproportionation reactions catalysed by thiosulphate (see p. 22). The kinetics of sulphite degradation of polythionates were studied by Foerster & Centner (1926) and by Stamm, Seipold & Goehring (1941). Foerster and Centner reported the following bimolecular rate constants (in 1 mole$^{-1}$ sec$^{-1}$) at 0° and near neutral pH values: 0·17 for tetrathionate ($\mu = 0.26$–$0.52$) and 3·9 for pentathionate ($\mu = 0.006$–$0.16$).

## Reactions of polythionates with thiosulphate

Thiosulphate catalyses the disproportionation of tetrathionate and higher polythionates (Colefax, 1908; Foerster & Hornig, 1923; Kurtenacker & Kaufmann, 1925; Kurtenacker, Mutschin & Stastny, 1935). The first reaction is thought to be a displacement of sulphite (cf. Foss, 1961 c) to form the next higher polythionate (e.g. equation 2.44 for tetrathionate).

$$^-O_3S-S-S-SO_3^- + S_2O_3^{2-} \rightleftharpoons {}^-O_3S-S-S-S-SO_3^- + SO_3^{2-} \quad (2.44)$$

The equilibrium is then displaced by a further reaction of sulphite with another polythionate molecule (equation 2.45).

$$^-O_3S-S-S-SO_3^- + SO_3^{2-} \rightleftharpoons S_2O_3^{2-} + {}^-O_3S-S-SO_3^- \quad (2.45)$$

That this mechanism is correct is indicated by the fact that in the presence of formaldehyde to remove sulphite, pentathionate and hexathionate, but not trithionate, are formed in mixtures of tetrathionate and thiosulphate (Skarżyński & Szczepkowski, 1959). The rates of sulphite displacement from polythionates by thiosulphate are relatively slow and these reactions should not be very significant in short-term biological experiments (cf. London, 1964) except perhaps with hexathionate and higher polythionates. In growth experiments lasting several days, however, thiosulphate-catalysed polythionate disproportionations could be appreciable in some circumstances. Foss & Kringlebotn (1961) reported the following bimolecular rate constants for sulphite displacement from polythionates by thiosulphate at 25°, pH 6·8 and $\mu = 1·15$: tetrathionate $1·3 \times 10^{-3}$ 1 mole$^{-1}$ sec$^{-1}$; trithionate $1·1 \times 10^{-4}$ 1 mole$^{-1}$ sec$^{-1}$. Fava & Bresadola (1955) reported a rate constant of $3·9 \times 10^{-3}$ 1 mole$^{-1}$ sec$^{-1}$ for the tetrathionate reaction at 50°, pH 7·15 and $\mu = 0·94$ while Fava & Pajaro (1954) found the rate constant for the trithionate reaction to be $0·66 \times 10^{-4}$ 1 mole$^{-1}$ sec$^{-1}$ at 20°, pH 7 to 10 and $\mu = 0·816$. The latter value was based on isotope exchange between thiosulphate and trithionate. At 51° the rate constant for the reaction with trithionate was $6·2 \times 10^{-4}$ 1 mole$^{-1}$ sec$^{-1}$.

## Reactions with hydroxyl ions

In neutral and moderately acid solutions and at moderate temperatures polythionates up to hexathionate appear to be stable (cf. Pollard, Nickless & Glover, 1964 a; Schmidt & Sand, 1964 c). The often claimed instability of tetrathionate, pentathionate and hexathionate can be attributed to sulphite and thiosulphate impurities which often arise during the preparation of these compounds and which catalyse the decomposition of polythionates as discussed in the preceding sections (Schmidt, 1965).

In alkaline solution polythionates decompose with the formation of various products depending upon the conditions. In strongly alkaline solutions thiosulphate is a major product but lower polythionates, sulphite, sulphate and elemental sulphur are also formed. The numerous hydrolytic reactions ascribed to polythionates have been tabulated in Gmelin (1960). Pentathionate and hexathionate (and presumably higher polythionates) are particularly sensitive to hydroxyl ions and traces of the latter catalyse their decomposition to tetrathionate and elemental sulphur (equations 2.46 and 2.47).

$$S_6O_6^{2-} \xrightarrow{OH^-} S_5O_6^{2-} + S^0 \qquad (2.46)$$

$$S_5O_6^{2-} \xrightarrow{OH^-} S_4O_6^{2-} + S^0 \qquad (2.47)$$

These reactions apparently involve nucleophilic displacements by hydroxyl ions (Foss, 1945) and in terms of the concepts of Schmidt (1957 $b$, 1963, 1965) the mechanism for pentathionate decomposition may be described by equations 2.48 to 2.50.

$$HO^- + {}^-O_3S—S—S—S—SO_3^- \rightarrow HOS—SO_3^- + S—S—SO_3^{2-} \qquad (2.48)$$

$$S—S—SO_3^{2-} \rightarrow S^0 + S—SO_3^{2-} \qquad (2.49)$$

$$S—SO_3^{2-} + HOS—SO_3^- \rightarrow {}^-O_3S—S—S—SO_3^- + OH^- \qquad (2.50)$$

According to Goehring, Helbing & Appel (1947) 16 mmoles of $K_2S_5O_6$ in 100 ml of 0·02 M-NaOH are almost completely decomposed within 1 hr at 25°. Under the same conditions the decomposition of $K_2S_4O_6$ is less than 12%. Christiansen, Drost-Hansen & Nielsen (1952) found that the hydrolysis of pentathionate between pH 7 and pH 12 is first order with respect to pentathionate and hydroxyl ions and shows a positive salt effect. At pH 10·5 the rate constant with respect to pentathionate is $2·72 \times 10^{-2}$ min$^{-1}$.

### Reactions of polythionates with sulphide and thiols

In alkaline solution polythionates react rapidly and quantitatively with excess sulphide to form thiosulphate and in the case of tetrathionate and higher polythionates, elemental sulphur (equation 2.51) (see Goehring, 1952)

$$S_nO_6^{2-} + S^{2-} \rightarrow 2S_2O_3^{2-} + (n-3)S^0 \qquad (2.51)$$

The mechanism of this reaction is probably analogous to that of the hydrolytic reaction (equations 2.48 and 2.49; Schmidt, 1965). The reaction also occurs rapidly in solutions buffered with acetate (Hansen, 1933) but in acidified solutions the degradation of polythionates by $H_2S$ is slower and more complicated; relatively small amounts of thiosulphate and large amounts of sulphur are formed (Kurtenacker & Kaufmann, 1925).

Tetrathionate is reduced spontaneously by a number of thiols (equation 2.52) including cysteine and glutathione (see for example, Footner & Smiles, 1925; Gilman, Philips, Koelle, Allen & St John, 1946; London 1964).

$$2RS^- + S_4O_6^{2-} \rightarrow RSSR + 2S_2O_3^{2-} \tag{2.52}$$

Trithionate is reduced to sulphite and thiosulphate (equation 2.53).

$$2RS^- + S_3O_6^{2-} \rightarrow RSSR + S_2O_3^{2-} + SO_3^{2-} \tag{2.53}$$

The reactions presumably involve nucleophilic displacements of thiosulphate by the stronger nucleophile, $RS^-$ (equations 2.54 and 2.55).

$$RS^- + {}^-O_3S—S—S—SO_3^- \rightarrow RS—S—SO_3^- + S_2O_3^{2-} \tag{2.54}$$

$$RS^- + RS—S—SO_3^- \rightarrow RS—SR + S_2O_3^{2-} \tag{2.55}$$

Szczepkowski (1958) obtained chromatographic evidence for the formation of a substituted sulphodisulphane ($R—S—S—SO_3^-$) during the reduction of tetrathionate by cysteine: he also detected a substituted sulphotrisulphane ($R—S—S—S—SO_3^-$) which indicated that displacement of sulphite from tetrathionate also occurs (equations 2.56).

$$RS^- + {}^-O_3S—S—S—SO_3^- \rightarrow R—S—S—S—SO_3^- + SO_3^{2-} \tag{2.56}$$

$S$-Sulphocysteine ($RSSO_3^-$) was also formed, possibly by the reaction between thiosulphate and cysteine described earlier (p. 19). $S$-Sulphocysteine and the analogous sulphenyl thiosulphate, alanine sulphodisulphane ($RSSSO_3^-$), are formed in mixtures of cysteine and trithionate (Stelmaszyńska & Szczepkowski, 1961); in addition to these compounds alanine sulphotrisulphane ($RSSSSO_3^-$) and elemental sulphur arise from the reaction between cysteine and pentathionate or hexathionate.

## 2.2.7 Reactions of sulphur compounds with metabolic inhibitors

In view of the reactivity of many inorganic sulphur compounds, the possibility of chemical reactions occurring between sulphur-containing substrates (or products) and other additives to biological systems must be considered. By way of illustration, the reactivity of sulphur compounds towards a few commonly used metabolic inhibitors is considered in this section.

The well-known tendency of heavy metals to react with sulphides, thiosulphate and polythionates, and to catalyse sulphite oxidation, often precludes the use of these metals in biological experiments on sulphur metabolism. Cyanide reacts instantaneously at room temperature, and at pH values near neutrality, with tetrathionate and higher polythionates (see chapter 4). It also reacts rapidly with colloidal sulphur and poly-sulphides although thiosulphate and trithionate are not attacked signi-ficantly by cyanide under physiological conditions.

Azide and tetrathionate ions interact rapidly at 25° according to equation 2.57 (Hofman-Bang, 1949, 1951)

$$S_4O_6{}^{2-} + 2N_3{}^- \rightarrow 2S_2O_3{}^{2-} + 3N_2 \tag{2.57}$$

while sulphide, thiosulphate, trithionate, tetrathionate, pentathionate and hexathionate catalyse the oxidation of azide by oxidizing agents such as iodine and bromine (Raschig, 1908, 1915; Metz, 1929; Dodd & Griffith, 1949; Griffith & Irving, 1949; Hofman-Bang, 1949, 1950).

The reactions between sulphur compounds and thiol-binding reagents are of particular interest in view of the 'thiol' nature of some of these compounds and the speculations that cellular thiols and disulphides may be concerned in sulphur metabolism (see chapters 8, 9, 10 and 11). $N$-Ethylmaleimide forms adducts with sulphide, thiosulphate and sulphite (Gregory, 1955; Margolis & Mandl, 1958; Ellis, 1964 $a$). Using 5 ml of a mixture containing 10 $\mu$moles of $N$-ethylmaleimide and 100 $\mu$moles of $Na_2S_2O_3$ at pH 7, Trudinger (1965) reported that about 7% of the former reacted (as determined spectrophotometrically) within 20 min at 1° and that about 25% reacted during the same time at 30°. Ellis (1966) studied the reactions between $N$-ethylmaleimide and inorganic sulphur com-pounds in 0·1 M-phosphate, pH 7·4, at 20° by measuring the decrease in the absorbance (due to $N$-ethylmaleimide) at 300 m$\mu$. Mixtures of mM $N$-ethylmaleimide with mM sulphide or sulphite reacted almost com-pletely within 5 min but no reaction between mM-$N$-ethylmaleimide and 1–5 mM-thiosulphate was detected. Trudinger (1965) also studied the reactions of thiosulphate with iodoacetamide and with $p$-chloromercuri-benzoate. With iodoacetamide, evidence was obtained for the formation of acetamide 2-thiosulphate by the reaction shown in equation 2.58. Such elimination of halogen ions from organic halides by thiosulphate is of course well-known (cf. the general synthesis of Bunte salts from organic halides (Milligan & Swan, 1962)).

$$S_2O_3{}^{2-} + ICH_2CONH_2 \rightarrow {}^-O_3SS.CH_2CONH_2 + I^- \tag{2.58}$$

In 5 ml of a mixture containing 10 $\mu$moles of iodoacetamide and 100 $\mu$moles of $Na_2S_2O_3$ at pH 7, nearly 10% of the former had reacted in

20 min at 1° and nearly 50% at 30° by a process exhibiting second-order kinetics. Tetrathionate did not react with iodoacetamide. Chromatographic evidence indicated that $p$-chloromercuribenzoate reversibly formed a complex with thiosulphate (cf. Schellenburg & Schwarzenbach, 1962).

### 2.2.8 Exchange reactions of sulphur compounds

Increasing use is being made of [35]S-labelled compounds in the study of inorganic sulphur metabolism. It is important therefore that possible chemical exchange of sulphur between labelled and unlabelled compounds be taken into account. Table 2.2 lists the exchange reactions of some of the biologically important inorganic sulphur compounds. The most significant reactions are those between polythionates and thiosulphate which, in the case of polythionates containing four or more sulphur atoms, are essentially instantaneous (Fava, 1953). Exchange between sulphite and thiosulphate is probably not significant at physiological temperatures: the second-order rate constant at 25° is 0·5 l mole$^{-1}$ sec$^{-1}$ (Fava & Pajaro, 1956). The rate of exchange between sulphite and sulphonyl groups is increased dramatically (as much as 1000 times) by aryl substitution at the sulphane sulphur of thiosulphate (Fava & Pajaro, 1956; Fava & Iliceto, 1958). Sulphite may indirectly induce exchange reactions in the presence of higher polythionates due to the displacement of thiosulphate from the polythionates as described in § 2.2.6.

### 2.3 Sulphate esters

In the present connection the term 'sulphate ester' is taken to mean a salt of a half-ester of sulphuric acid with a hydroxyl-containing compound. Such substances have the general formula $R.OSO_3^-M^+$, where R may be any one of a very wide range of residues; aliphatic, aromatic, alicyclic or heterocyclic. Although the potassium or sodium salts are the most commonly prepared, for many purposes the much less soluble salts with organic bases such as $p$-toluidine (Barton & Young, 1943; Laughland & Young, 1944; Paterson & Klyne, 1948) or the aminoacridines (Dodgson, Rose & Spencer, 1955) are very useful. Such organic salts are particularly valuable for the isolation of sulphate esters from biological materials and at least some of them have the valuable property of being soluble in non-polar solvents: for example, steroid pyridinium sulphates are soluble in chloroform (McKenna & Norymberski, 1960) (see also p. 66). It should be noted that the compound called, for example, phenyl sulphate in this book (R = $C_6H_5$ and $M^+ = K^+$ in the above formula) is more

TABLE 2.2 *Exchange reactions of inorganic sulphur compounds*
*(After Stranks & Wilkins, 1957)*

| Species | Conditions | Exchange rate |
|---|---|---|
| *Exchanging species* | | |
| $S^{2-}$, $\overset{*}{S_8}$ | | >95% 1 hr (90–100°) |
| $S_2O_3^{2-}$, $\overset{*}{HS^-}$ | 1 N-NaOH | >95% 24 hr at 100° <br> None at room temperature |
| $S\!-\!\overset{*}{S}O_3^{2-}$, $SO_3^{2-}$ <br> $S\!-\!SO_3^{2-}$, $\overset{*}{S}O_3^{2-}$ | pH 5–14 | $k = 2 \cdot 3 \times 10^6 \, e^{-14500/RT}$ <br> 1 mole$^{-1}$sec$^{-1}$ |
| $\overset{*}{S}\!-\!SO_3^{2-}$, $S_3O_6^{2-}$ <br> $S\!-\!\overset{*}{S}O_3^{2-}$, $S_3O_6^{2-}$ | pH 7–10 | $k = 6 \cdot 2 \times 10^{-4}$ 1 mole$^{-1}$ sec$^{-1}$ <br> at 51° ($\mu = 0 \cdot 816$) |
| $S_4O_6^{2-}$, $\overset{*}{S}\!-\!SO_3^{2-}$ <br> $S_5O_6^{2-}$, <br> $S_6O_6^{2-}$, $S\!-\!\overset{*}{S}O_3^{2-}$ | $H_2O$ | Essentially instantaneous |
| $\overset{*}{S}O_3^{2-}$, $S_3O_6^{2-}$ | | Second order |
| *Non-exchanging species* | | |
| $H_2\overset{*}{S}$, $SO_2$ <br> $H_2S$, $\overset{*}{S}O_2$ | $CHCl_3$ and other conditions | |
| $SCN^-$, $\overset{*}{S_8}$ | Alcohol, acetone, 5–21 days | |
| $SCN^-$, $\overset{*}{S^{2-}}$ | Alkaline, 2 days 100°, 23 days RT | |
| $SCN^-$, $\overset{*}{S}\!-\!SO_3^{2-}$ | | |
| $\overset{*}{S_2}O_3^{2-}$, $S_2O_4^{2-}$ | pH 6, 75 min; pH 13, 44 hr | |
| $\overset{*}{S}\!-\!SO_3^{2-}$, <br> $HOCH_2SO_2Na$ | 5 days | |
| $H\overset{*}{S}O_3^{2-}$, $S_3O_6^{2-}$ | pH 7, 96 hr | |
| $SO_3^{2-}$, $\overset{*}{S}O_4^{2-}$ | 0·1 N-HCl, or alkali, 35 hr 100° | |
| $\overset{*}{S}O_4^{2-}$, $S_3O_6^{2-}$ | 26 hr 25° | |
| $\overset{*}{S}O_4^{2-}$, $S^{2-}$ | Alkaline solution, 36 hr 100° | |
| $\overset{*}{S}O_4^{2-}$, $S_2O_8^{2-}$ | Variety of conditions | |
| $\overset{*}{S}O_4^{2-}$, $S_4O_6^{2-}$ | Neutral or weakly acid, 44 hr | |

correctly known as potassium phenyl sulphate but for simplicity the name of the cation is generally omitted. Further, in virtually all biochemical experiments the salts are fully ionized and only the anion itself takes a direct part in metabolic changes. The corresponding free acids (e.g. phenyl hydrogen sulphate) are probably stronger acids than sulphuric acid itself (Burkhardt, Ford & Singleton, 1936) and are unstable so that they cannot normally be isolated. Some sulphate esters can be obtained as metal-free internal salts, for example choline sulphate ester (choline $O$-sulphate; $R = C_2H_4.N^+.(CH_3)_3$ in the above formula).

As already stated, sulphate esters are commonly prepared in the form of sodium or potassium salts. These in many cases are more water-soluble than are the parent compounds but the absolute solubilities of the esters range widely from those of the highly water-soluble methyl sulphate or glucose sulphate to that of the almost water-insoluble cholesteryl sulphate. The latter is, however, appreciably soluble in dimethylform-amide. The alkali metal salts are, unlike those of certain organic bases, poorly soluble in organic solvents although those esters containing non-polar residues (e.g. polycyclic compounds) are appreciably soluble in aqueous ethanol. A useful solvent for many sulphate esters is $n$-butanol, the more so as it can be used to extract them from aqueous solution.

It has already been pointed out that sulphate esters can be derived from many different types of hydroxyl-containing compound. One type of ester of particular interest would be the enol-sulphate of a ketone: no good evidence for the existence of such compounds has been provided but Oertel, Treiber & Rindt (1967) claim to have prepared such a deriva-tive of androst-4-ene-3,17-dione (formula 2.5). Such enol-sulphates would be expected to be extremely unstable and further work is required to substantiate their occurrence.

(2.5)

Obviously little can be said of the detailed chemistry of the sulphate esters because this will depend upon the chemistry of the parent com-pounds and their only properties in common will be those of the ester linkage. Suter (1944) has given a very useful summary of much of the early work on these substances. Most of the treatment in this chapter will be devoted to reactions causing fission of the ester linkage and the only other general reaction to be mentioned is the formation of the corresponding double esters of sulphuric acid, $R.O.SO_2O.R'$. For example, treatment of a sulphate ester $R.OSO_3^-$, with diazomethane will give the corresponding methyl ester, $R.O.SO_2.O.CH_3$. Such compounds are obviously much less polar than the parent sulphate esters and they have proved useful derivatives of, for example, steroid sulphates (McKenna & Norymberski, 1957 b; Pasqualini, Zelnik & Jayle, 1962).

### 2.3.1 Acid hydrolysis of sulphate esters

All sulphate esters are quite readily hydrolysed by acid, the relative ease of hydrolysis being governed by the nature of the organic residues: in general terms alkyl sulphates are less readily hydrolysed than are aryl sulphates, the velocity constants differing by a factor of several hundred (Kice & Anderson, 1966). There are, of course, exceptions to this generalization as is shown by the very rapid hydrolysis of the five-membered cyclic ester, ethylene sulphate (2.6) (Kaiser, Panar & Westheimer, 1963).

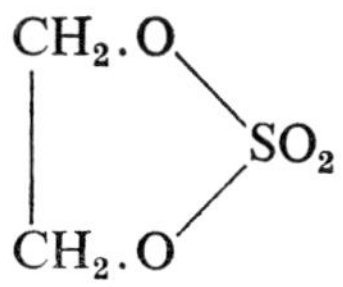

$$CH_2.O\diagdown$$
$$\qquad\qquad SO_2 \qquad\qquad (2.6)$$
$$CH_2.O\diagup$$

With both alkyl and aryl esters it is the O—S bond which is split during acid hydrolysis (Burwell, 1952; Burstein & Lieberman, 1958 b; Spencer, 1958; Batts, 1966 b).

The relation between the structure of an aryl sulphate and its rate of acid hydrolysis was investigated in considerable detail over thirty years ago by Burkhardt who studied this reaction with a long series of substituted phenyl sulphates (Burkhardt, Ford & Singleton, 1936; Burkhardt, Evans & Warhurst, 1936; Burkhardt, Horrex & Jenkins, 1936). In general terms, the more acid the hydroxyl group (or the more electron-withdrawing a substituent in the benzene ring), the more readily is hydrolysed the corresponding sulphate ester. p-Nitrophenyl sulphate is therefore a rather acid-labile compound which can, for instance, decompose autocatalytically during its recrystallization if the solution is not kept neutral. Havinga, de Jongh & Dorst (1956) have further shown that the hydrolysis of the nitrophenyl sulphates can be initiated by a photochemical reaction, the rate of which is essentially independent of pH: this is, of course, followed by the usual autocatalytic acid hydrolysis. The nitrophenyl sulphates are obviously rather more reactive than the simple aryl sulphates and their properties are in some ways more analogous to those of a mixed anhydride of sulphuric acid than to those of an ester (see also Fendler & Fendler, 1968).

Kice & Anderson (1966) have shown that the acid hydrolysis of aryl sulphates takes place by an A-1 type of mechanism, probably as shown in reaction 2.59.

$$R.OSO_3^- + H^+ \underset{}{\overset{fast}{\rightleftharpoons}} \underset{\underset{H}{|}}{R.O^+} \text{---} SO_3^- \overset{slow}{\longrightarrow} R.OH + SO_3 \qquad (2.59)$$
$$\downarrow$$
$$H_2SO_4$$

This mechanism is quite consistent with the earlier findings of Burkhardt and it also serves to explain the much greater lability to acid of aryl sulphates as compared with alkyl sulphates. This can arise through the stabilization of the transition state in the former case by resonance inter-action of the phenolic oxygen with the aromatic ring, a phenomenon which is obviously impossible in the alkyl sulphates. Batts (1966 $b$) has likewise shown that the acid hydrolysis of alkyl sulphates is an A-1 type reaction and has proposed a mechanism similar to that above. It is therefore of some interest that the hydrolysis of alkyl selenates is not an A-1 but an A-2 reaction: this was clearly shown by Bunton & Hendy (1963) who proposed the mechanism shown in reaction 2.60. This essen-tially represents an attack of water on the selenium atom and here again it is the O—Se bond which is split.

$$R.OSeO_3^- + H^+ \overset{fast}{\rightleftharpoons} R.OSeO_3H$$

$$R.OSeO_3H + H_2O \overset{slow}{\longrightarrow} R.O^- + H_2O^+.SeO_3H \overset{fast}{\longrightarrow} R.OH + H_2SeO_4 \qquad (2.60)$$

Rees (1963) has shown that it is possible to use the rates of acid hydrolysis of carbohydrate sulphates as a guide to the nature of the hydroxyl group which is esterified. In 0·25 N-HCl at 100° the half-lives of esters derived from the three possible types of hydroxyl group were as follows: 0·1 to 0·5 hr, equatorial secondary hydroxyl (e.g. glucose 3-sulphate; mucopoly-saccharide from shark cartilage); 1·0 to 1·5 hr, axial secondary hydroxyl (e.g. galactose 4-sulphate, κ-carrageenan); 1·5 to 2·5 hr, primary hydroxyl (e.g. glucose 6-sulphate, laminarin sulphate). It should be noted that here also the bond which is split during acid hydrolysis is the O—S bond (Guiseley & Ruoff, 1961). It might be mentioned that this use of rates of hydrolysis to characterize carbohydrate sulphates complements the use of infra-red spectroscopy for the same purpose (see § 4.4.2).

It must be noted that the acid hydrolysis of a sulphate ester may not simply give the parent alcohol or phenol: the reaction may take a more complex course and artefacts may be produced. This has been of particular interest in the case of steroid sulphates: for example, the hydrolysis of androstenolone sulphate (2.7) in hydrochloric acid can give not only androstenolone itself (2.8) but also the corresponding 3,5-diene (2.9) and 3β-chloro- (2.10) derivatives. These and other reactions involving fission of the C—O bond have been described by Ramseyer, Williams & Hirschman (1967).

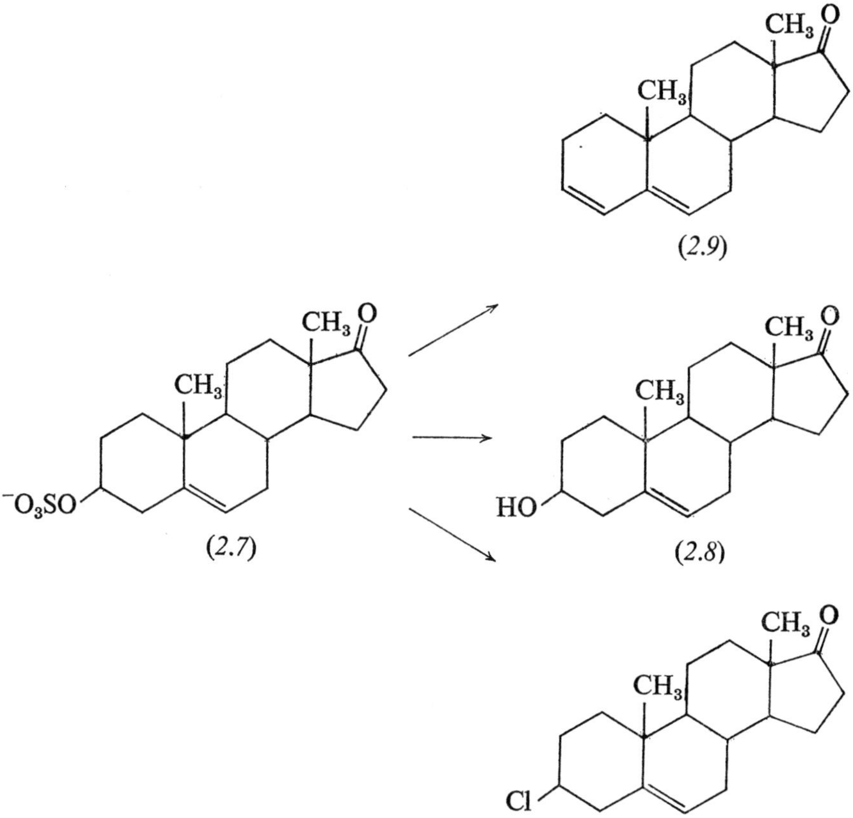

(2.9)

(2.7)

(2.8)

(2.10)

## 2.3.2 The alkaline decomposition of sulphate esters

Most alkyl and aryl sulphates are fairly stable to alkali and require rather vigorous conditions to bring about their hydrolysis. For instance, the half-time for the hydrolysis of sec-butyl sulphate in 5 M-NaOH at 100° is 20 hr compared with the half-time of 1 hr for the hydrolysis in 0·5 N-H$_2$SO$_4$ (Burwell, 1952). Further, this alkaline hydrolysis is accompanied by a considerable amount of inversion of the configuration of the alcohol which indicates that fission of the C—O bond has occurred. Burwell showed that the alkaline hydrolysis was independent of the concentration of OH$^-$ ions and suggested that the reaction was a displacement of an SO$_4^{2-}$ ion by water.

*p*-Nitrophenyl sulphate slowly hydrolyses in alkaline solution (Fendler & Fendler, 1968) under which conditions it shows nitrophenylating properties analogous to the methylating properties of methyl sulphate (Burkhardt, Ford & Singleton, 1936): again this reaction involves fission of the O—S bond (Spencer, 1958). Dodgson & Spencer (1957 *b*) reported that the hydrolysis of *p*-nitrophenyl sulphate in alkaline solution was catalysed by proline or hydroxyproline but this interpretation of their experimental findings is not strictly correct because it is now clear that the reaction product was not *p*-nitrophenol but *N*-(*p*-nitrophenyl)-proline. A number of cyclic secondary amines will behave similarly to proline and indeed crystalline *N*-(*p*-nitrophenyl) derivatives of pyrrolidine and piperidine have been prepared by allowing the bases to stand in 1 N-NaOH containing 0·03 M-*p*-nitrophenyl sulphate at room temperature for a few hours. Benkovic & Benkovic (1966) have described these and similar reactions in considerable detail.

Some carbohydrate sulphates are, unlike most alkyl and aryl sulphates, readily decomposed in alkaline solution (Percival, 1949) with the accompaniment of more or less drastic changes in the sugar. This is not a hydrolytic reaction but a base-catalysed elimination of sulphate which leads to the formation of an anhydro ring as is shown by the fact that, for instance, treatment of methyl galactoside 6-sulphate gives the corresponding 3,6-anhydrohexoside while di-*O*-isopropylidine galactose 6-sulphate, which does not have a free hydroxyl group in the 3 position, is stable to 2 N-NaOH at 100° for many hours. Much of this work has recently been reviewed (Turvey, 1965; Rees, 1966) and need not be considered further here. As essentially similar elimination of sulphate

32

occurs with the formation of a four-membered oxide ring when a bile alcohol sulphate, such as scymnol sulphate, is treated with alkali (see Haslewood, 1967).

### 2.3.3 Hydrolysis of steroid sulphates in neutral solution

The sulphate esters of $3\beta$-hydroxy-$\Delta^5$ steroids are quite readily hydrolysed in hot neutral solution (Roy, 1956 $a$). This reaction is also accompanied by an alteration in the structure of the steroid and gives the corresponding $i$-steroid (reaction 2.61) which under acid conditions reverts to the normal steroid.

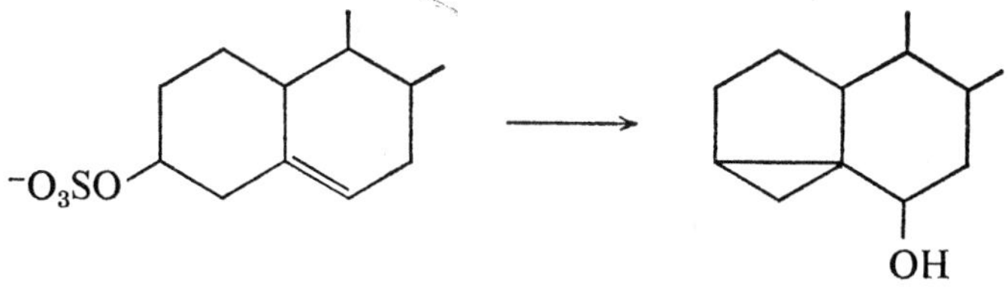

$$-\text{O}_3\text{SO} \qquad\qquad\qquad \longrightarrow \qquad\qquad\qquad \text{OH} \tag{2.61}$$

### 2.3.4 The solvolysis of sulphate esters

Many sulphate esters can readily undergo solvolysis and this procedure has been much used in the analysis of steroid sulphates (see Bradlow, 1969). The reaction is very rapid in many non-polar solvents, especially ethers, and its velocity is greatly retarded by an increase in the polarity of the solvent which is contrary to the findings with many other solvolytic reactions. McKenna & Norymberski (1957 $a$) studied the fission of cholesteryl sulphate in a number of solvents, including dioxan and tetrahydrofuran, and suggested the following mechanism (2.62)

$$\text{R.OSO}_3^- + \text{X.O.Y} \rightarrow \text{R.O}^- + \begin{matrix} \text{X} \\ \diagdown \\ \phantom{x} \end{matrix}\text{O}^+ - \text{SO}_3^- \\ \begin{matrix} \diagup \\ \text{Y} \end{matrix}$$

$$\tag{2.62}$$

$$\begin{matrix} \text{X} \\ \diagdown \\ \phantom{x} \end{matrix}\text{O}^+ - \text{SO}_3^- + \text{H}_2\text{O} \rightarrow \text{X.O.Y} + \text{H}^+ + \text{HSO}_4^- \\ \begin{matrix} \diagup \\ \text{Y} \end{matrix}$$

Burstein & Lieberman (1958 $a$) examined the reaction under rather different conditions and concluded that the unionized steroid hydrogen sulphate, or the isomeric dipolar ion, was the reactive species under their conditions as is shown in reaction 2.63.

$$R.OSO_3^- + H^+ \rightleftharpoons R.OSO_3H \rightleftharpoons R.\underset{\underset{H}{|}}{O^+}\!\!-\!\!SO_3^-$$

(2.63)

$$R.\underset{\underset{H}{|}}{O^+}\!\!-\!\!SO_3^- + X.O.Y \rightarrow R.OH + \overset{X}{\underset{Y}{\diagup}}\!\!\diagdown\!O^+\!\!-\!\!SO_3^-$$

More recently Batts (1966 *a*) has studied in detail the solvolysis of methyl and ethyl sulphates and has shown that the rates of hydrolysis of these are increased by a factor of $10^7$ when the solvent is changed from pure water to moist dioxan. The mechanism suggested by Batts was essentially similar to that put forward by Burtstein & Lieberman.

For analytical purposes the technique used by Burstein & Lieberman (1958 *b*) for urine is useful: the solution of the steroid sulphate is made 2 N with $H_2SO_4$ (or 1 N-$H_2SO_4$ and 20% NaCl) and extracted with an equal volume of ethyl acetate to quantitatively extract the steroid hydrogen sulphate. The organic phase is separated, filtered through glass wool to remove any water droplets and kept at 37° for 24 hr, by which time quantitative solvolysis of compounds such as androsterone sulphate has occurred. A similar solvolysis occurs if the steroid sulphate be dissolved in dioxan containing 10% trichloroacetic acid. The first of these procedures fails when applied to corticosteroid 21-sulphates which are sufficiently polar for the free acid not to be extractable by ethyl acetate. In such cases the aqueous solution is made 0·1 N in HCl and 20% in NaCl before extracting with tetrahydrofuran: the organic phase is then separated and treated as before. Drayer & Giroud (1965) slightly modified this technique by making the tetrahydrofuran extract 0·01 N in perchloric acid before allowing the solvolysis to proceed. Such techniques have proved very useful in the steroid field where conventional methods of acid hydrolysis often give artefacts (see § 2.3.1). They do not appear to have been widely used elsewhere.

## 2.4 The mustard oil glycosides

The mustard oil glycosides are sulphate esters of a rather unusual type which contain the $=\!N.O.SO_3^-$ grouping and so have obvious analogies to the oxime *O*-sulphonates (Smith, 1948) which are in fact sulphate esters (see formula *7.9*, § 7.4). The formula of a typical representative of this class of compounds, sinigrin, is given below (formula *2.11*). It is not possible here to enter into a general discussion of the chemistry of these compounds, a topic which has been admirably discussed in recent years by Challenger (1959) and by Kjaer (1960). As is discussed in § 7.4, the above formula has only relatively recently been shown to be correct (Ettlinger & Lundeen, 1956, 1957) despite the fact that these compounds have been studied for over a hundred years.

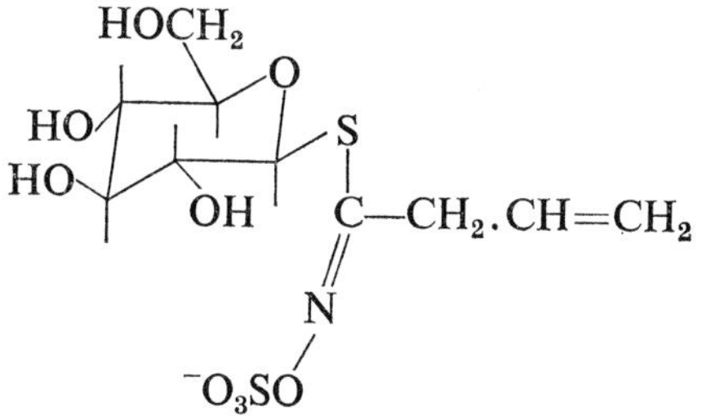

$$
\begin{array}{c}
\text{HOCH}_2 \\[2pt]
\text{(pyranose ring with } \text{HO}, \text{HO}, \text{OH substituents)} \\[2pt]
\underset{\substack{\| \\ N \\ / \\ {}^{-}O_3SO}}{C}\!-\!CH_2.CH{=}CH_2
\end{array}
\qquad (2.11)
$$

Being sulphate esters, albeit of a rather special kind, the mustard oil glycosides are hydrolysed by acid and by alkali but the reaction products are complex and vary tremendously with changes in the experimental conditions. This aspect will not be considered further here.

## 2.5 The sulphamates

The sulphamates form a group of compounds having certain analogies with the sulphate esters, their general formulae being $R.NH.SO_3^{-}M^{+}$ where once again the group R may take many different forms. These compounds are, however, not of widespread distribution in nature and only two small groups of them are known.

The first group comprises heparin and related compounds which contain as one of the units of their structure the sulphamate of D-glucosamine, 2-deoxy-2-sulphoamino-D-glucose. The sulphoamino linkage in these carbohydrates is much more acid-labile than the O-sulphonate (usually incorrectly called O-sulphate) linkages which are also present so that the two types of ester can in general be readily differentiated by the much easier hydrolysis of the former (Foster & Huggard, 1955).

The second group is made up of the aryl sulphamates which are metabolic products of certain aromatic amines such as 2-naphthylamine. Once again these compounds are quite acid-labile and the aryl sulphamates are, for instance, hydrolysed during the usual diazotization procedure adopted for the determination of aromatic amines. Scott & Spillane (1967) have investigated the mechanism of acid hydrolysis and have suggested that it is an A-2 type reaction. This is in rather striking contrast to the hydrolysis of aryl sulphates (§ 2.3.1) or aryl thiosulphates (§ 2.6) which are both A-1 type hydrolyses.

A rather useful reaction undergone by the aryl sulphamates is their conversion to aminosulphonic acids. For example, phenyl sulphamate readily yields sulphanilic acid when salts of the former are heated (reaction 2.64). Although this type of reaction does not occur under strictly

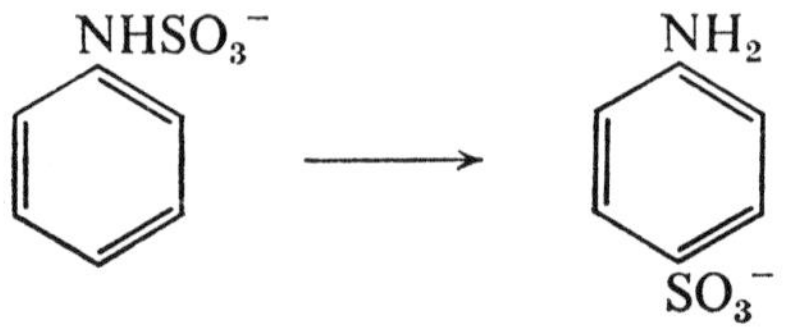

$$\underset{\text{NHSO}_3^-}{\bigcirc} \longrightarrow \underset{\text{SO}_3^-}{\overset{\text{NH}_2}{\bigcirc}} \qquad (2.64)$$

anhydrous conditions it is usually considered to be an intramolecular change. Spillane & Scott (1967) have, however, produced evidence to show that it is in fact an intermolecular reaction which involves the hydrolysis of the sulphamate to give sulphuric acid which is in turn responsible for the subsequent sulphonation of the other reaction product, the arylamine.

## 2.6 Thiosulphate esters

These compounds have the general formula $R.SSO_3^-M^+$ and can be regarded either as esters of thiosulphuric acid and a hydroxyl-containing compound or as esters of sulphuric acid and a thiol. They are commonly known as Bunte salts or, quite incorrectly, as S-sulphonates. They should be called S-alkyl (or S-aryl) thiosulphates or, alternatively, alkyl (or aryl) sulphenyl sulphites. The chemical properties of the Bunte salts have been reviewed in considerable detail by Milligan & Swan (1962).

Probably the commonest example of this type of compound in biochemistry is that incorrectly known as cysteine S-sulphonate: a better trivial name for this compound (2.12) is S-sulphocysteine while its systematic name is S-(2-amino-2-carboxyethyl) thiosulphate.

$$\begin{array}{c} CH_2.SSO_3^- \\ | \\ H.C.NH_2 \\ | \\ COOH \end{array} \qquad (2.12)$$

S-Sulphocysteine must be clearly distinguished from alanine 3-thiosulphonate, a derivative of cysteic acid and not an ester, which is

36

considered below in § 2.8. The analogous thiosulphate esters derived from cysteamine and from glutathione are also known to occur naturally.

Aryl thiosulphates are well known chemically (Baumgarten, 1930; Dornow, 1939) but they do not appear to be found naturally.

The mechanism of the acid hydrolysis of Bunte salts (to give a thiol and sulphuric acid) has recently been investigated by Kice, Anderson & Pawlowski (1966) who suggest that here, as in the hydrolysis of sulphate esters, the reaction is an A-1 type. Once again the initial reaction is a protonation so that the mechanism can be represented as in (2.65).

$$R.SSO_3^- + H^+ \rightleftharpoons R.\underset{\underset{H}{|}}{S^+}{-}SO_3^- \rightarrow R.SH + \underset{\underset{H_2SO_4}{\downarrow}}{SO_3} \tag{2.65}$$

Despite the general similarity between the hydrolysis of sulphate esters and of thiosulphate esters there are some important differences. For example, alkyl and aryl thiosulphates are hydrolysed at comparable rates in acid whereas under similar conditions aryl sulphates are much more rapidly hydrolysed than alkyl sulphates. Again, the introduction of an electron-withdrawing substituent into phenyl sulphate greatly increases its lability to acid: the introduction of a similar substituent into phenyl thiosulphate stabilizes it against acid hydrolysis. Attempts to explain these facts have been made by Kice et al. (1966).

Finally it should be pointed out that aryl thiosulphates are much more alkali-labile than are the aryl sulphates: treatment in 0·3 N-KOH at 100° for 5 min converts an aryl thiosulphate into the corresponding sulphenic acid and thence to the disulphide (Dornow, 1939; Lecher & Hardy, 1955).

## 2.7 The sulphatophosphates

These compounds are mixed anhydrides of sulphuric acid and phosphoric acid which contain the very reactive grouping *2.13*.

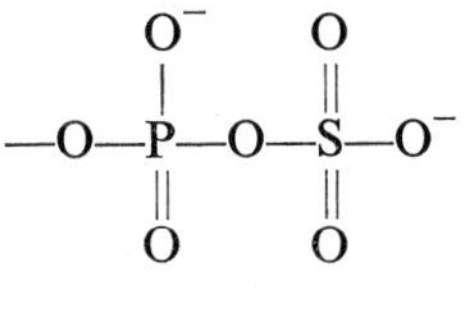

$$\underset{\underset{O}{\|}}{\overset{\overset{O^-}{|}}{-O{-}P}}{-}O{-}\underset{\underset{O}{\|}}{\overset{\overset{O}{\|}}{S}}{-}O^- \tag{2.13}$$

This is highly acid-labile, its half-life in 0·1 N-HCl at 37° being only about 6 min (Robbins & Lipmann, 1957). It is, on the other hand, relatively stable to alkali, being unchanged on standing for 1 hr in 0·1 N-NaOH (Robbins & Lipmann, 1957) and only partially hydrolysed by treatment with 0·1 N-NaOH at 100° for 2 hr (Baddiley, Buchanan, Letters & Sanderson, 1959). The two most important representatives of this class of compound are adenylyl sulphate (adenosine 5'-sulphato-phosphate, adenosine 5'-phosphosulphate, APS, formula *2.14*) and 3'-phosphoadenylyl sulphate (adenosine 3'-phosphate 5'-sulphatophosphate, 3'-phosphoadenosine 5'-phosphosulphate, PAPS, formula *2.15*). The related nucleotides uridylyl sulphate, cytidylyl sulphate and guanidylyl sulphate have been prepared (Ishimoto & Fujimoto, 1961; Michelson & Wold, 1962).

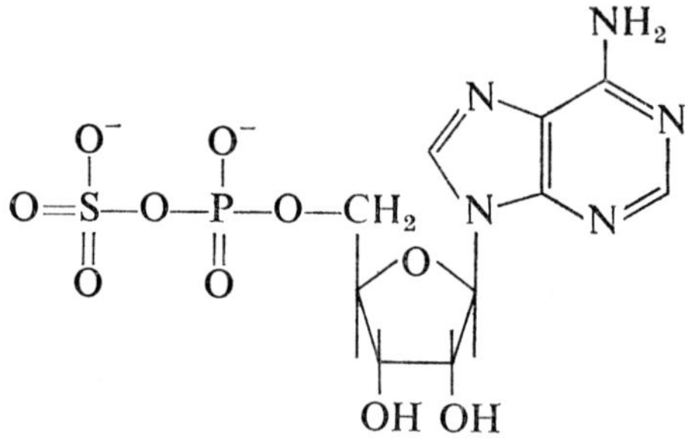

$$(2.14)$$

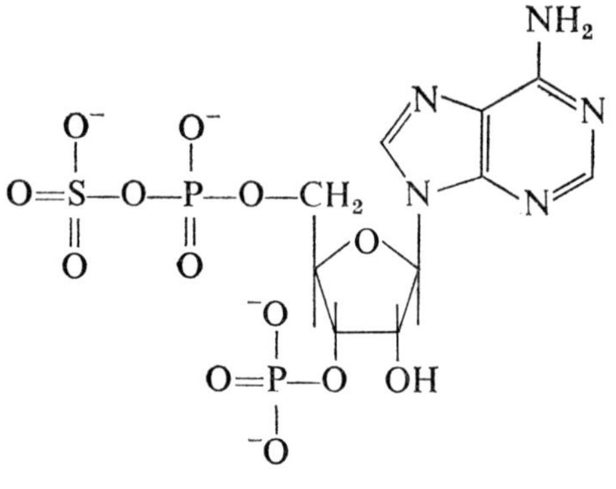

$$(2.15)$$

An interesting, and potentially important, reaction of PAPS has been described by Adams (1962) who showed that the transfer of the sulphate group from this compound to amino sugars, or to other hydroxyl-containing compounds such as ethanol or tris, was catalysed by charcoal. The extent of the reaction was small and the transfer was only detected because of the very high specific activity of the [$^{35}$S]PAPS which was used as the donor. Nevertheless the possible occurrence of this transfer must be kept in mind when sulphate esters or PAPS are isolated from media containing ethanol or tris by procedures involving the use of activated charcoal.

Two further compounds related to adenylyl sulphate are the isomeric adenosine 5'-sulphatopyrophosphates which have been described by Yount, Simchuk, Yu & Kottke (1966) as potential analogues of ATP. The isomers differ in whether the sulphate group is esterified with the $\alpha$ or the $\beta$ phosphate of the ADP. The structure of the $\beta$ isomer is shown in formula 2.16.

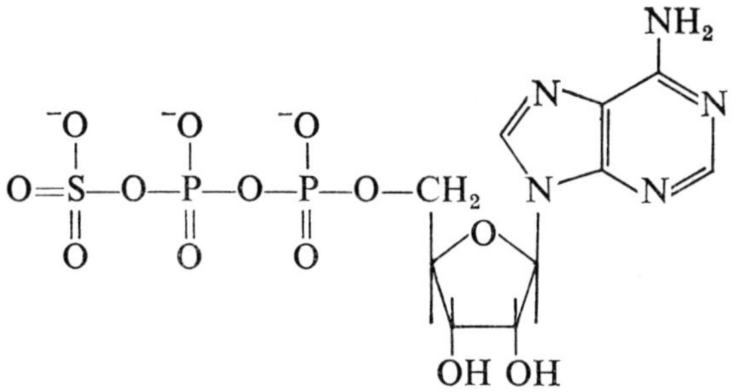

(2.16)

In view of the participation of $Mg^{2+}$ ions in many of the enzymatic reactions undergone by PAPS it is interesting that the sulphatophosphates apparently do not strongly chelate these ions. Yount et al. (1966) give the following values for $K_f'$ at pH 7.4 (where $K_f' = [LMg^{2+}]/[L].[Mg^{2+}]$):

adenylyl sulphate 24

adenosine sulphatopyrophosphate 42

These figures should be compared with that of 4,500 for $ATP^{4-}$ or, perhaps more realistically, with those of 28 and 31 for $ADP^{2-}$ and $ATP^{3-}$ respectively. The corresponding value of $K_f'$ for PAPS does not appear to have been determined but it seems likely to be of the same order as that for APS.

## 2.8 Sulphonic acids and related compounds

These compounds differ from those previously considered essentially in that they contain a direct carbon to sulphur linkage and they therefore hardly enter the scope of the present discussion except in so far as some of them may be intermediates in the oxidation of cysteine to $SO_4^{2-}$ ions. The following substances are among the more biologically important of those which will be considered later.

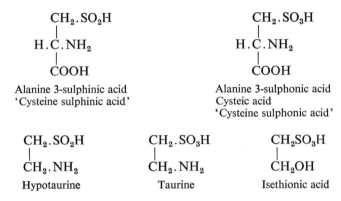

CH₂.SO₂H
|
H.C.NH₂
|
COOH

Alanine 3-sulphinic acid
'Cysteine sulphinic acid'

CH₂.SO₃H
|
H.C.NH₂
|
COOH

Alanine 3-sulphonic acid
Cysteic acid
'Cysteine sulphonic acid'

CH₂.SO₂H
|
CH₂.NH₂

Hypotaurine

CH₂.SO₃H
|
CH₂.NH₂

Taurine

CH₂SO₃H
|
CH₂OH

Isethionic acid

Perhaps the only common property of these compounds is that they cannot yield sulphate ions on treatment with either acid or alkali under conditions where sulphate esters are rapidly hydrolysed.

Related to alanine 3-sulphonic acid is the compound alanine 3-thiosulphonic acid which had already been mentioned: this has the structure shown below.

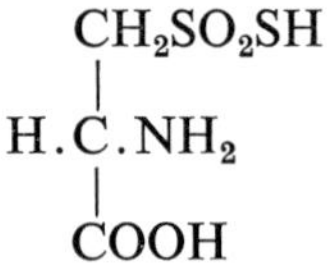

CH₂SO₂SH
|
H.C.NH₂
|
COOH

Alanine 3-thiosulphonic acid

Other examples of compounds containing C—S bonds will be encountered in subsequent chapters but the chemistry of these substances cannot be discussed here. Details of their behaviour must be sought in more general texts (e.g. Connor, 1943; Suter, 1944; Challenger, 1959; Kharasch, 1961; Kharasch & Meyers, 1966).

Only one other sulphonic acid must be mentioned here: this is 6-sulpho-α-D-quinovose (6-deoxy-6-sulpho-α-D-glucose) which is a component of the very widely distributed sulpholipid of green plants.

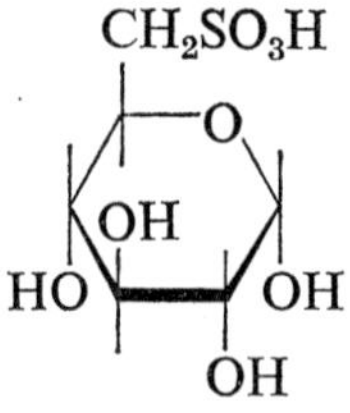

CH₂SO₃H

6-Sulpho-α-D-quinovose

40

In the sulpholipid the sugar is present in a glycosidic linkage with a diacyl glycerol (Benson, 1963). Although such a compound would perhaps not normally be regarded as a derivative of sulphate there is evidence that it is in fact both formed from (Abraham & Bachhawat, 1963) and can be degraded to (Martelli & Benson, 1964) sulphate in biological systems.

## 2.9 Steroid sulphatides

These substances are of considerable potential interest but so far little is known of them and they have been studied only by Oertel who first claimed their isolation in 1961 (Oertel, 1961 a). They are derived from diglycerides and have been assigned the structure shown below.

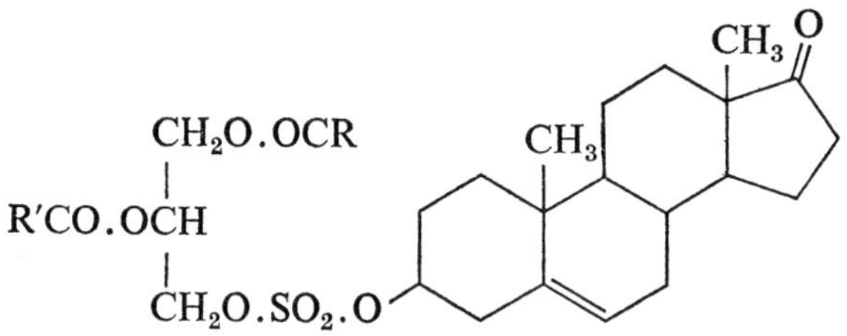

$$CH_2O.OCR$$
$$R'CO.OCH$$
$$CH_2O.SO_2.O$$

Androstenolone sulphatide

The analysis of such materials prepared biosynthetically (in preparations from guinea pig liver) have shown that palmitic, stearic, nonadecanoic, oleic and linoleic acids are the fatty acids present in greatest amounts (Oertel, 1963 a; Oertel & Groot, 1965).

Oertel (1961 b) claimed to have synthesized androstenolone sulphatide by the reaction of silver androstenolone sulphate and dipalmitoyl-L-α-iodopropyleneglycol; no details were given and the product was characterized only by analysis and by chromatography. It was reported that the sulphatide was rather unstable, forming androstenolone sulphate during chromatography on silicic acid or on paper in acid solvents. This instability is perhaps not unexpected in view of the observation by McKenna & Norymberski (1957 b) that both cholesteryl methyl sulphate and cholestanyl methyl sulphate decomposed on neutral alumina and readily underwent reaction with nucleophilic reagents.

It must be admitted that the structure of these steroid sulphatides has not been established beyond doubt and the possibility of their having a non-covalent nature should not be completely disregarded. Complications might be caused by the tendency of steroid sulphates to form lipid-soluble complexes as shown, for instance, by their interaction with phospholipids (Burstein, 1962). Perhaps it is not without interest that the original preparations of steroid sulphatides did contain phosphorus (Oertel, 1961 *a*) although it is now stated that the latter is present in the analogous steroid phosphatides (Oertel, 1963 *b*).

Despite these reservations, there is no doubt that the steroid sulphatides, whatever be their structure, have biological properties quite different from the parent steroid sulphates. It is to be hoped that further studies on these most interesting derivatives will soon be forthcoming.

# 3

## THE PREPARATION OF
## SOME BIOLOGICALLY IMPORTANT
## COMPOUNDS OF SULPHUR

This chapter is not meant to be an exhaustive and critical discussion of the methods available for the preparation of inorganic compounds of sulphur and the simple organic derivatives thereof (e.g. compounds which can be regarded as esters of sulphuric acid): its purpose is rather to provide a guide to the scattered literature on such compounds and to give examples of their preparation. Wherever possible, general methods have been given. A wealth of information on the preparation of organic derivatives of sulphur (some of interest here) is given in volumes 6/2 (1963) and 9 (1955) of Houben–Weyl's *Methoden der Organischen Chemie*.

The choice of compounds included in this chapter will doubtless be subject to criticism both on the grounds of the types of compounds considered and the particular preparative methods chosen but our guiding principle has been the consideration of those substances which will be discussed in later chapters of this book.

### 3.1   Elemental sulphur

Flowers of sulphur or precipitated sulphur may be used for studies of the growth of sulphur-oxidizing organisms but for metabolic experiments uniform hydrophilic suspensions of sulphur are necessary. Suzuki (1965 *a*) prepared such suspensions by exposing precipitated sulphur in $0.05\%$ Tween 80 to sonic vibrations for 30 min: the suspension was then dialysed to remove soluble contaminants.

A hydrophilic preparation of sulphur suitable for metabolic experiments is readily prepared by acidifying sodium thiosulphate solutions. To a solution (30 ml) of 3 M-$Na_2S_2O_3$ immersed in an ice bath is added dropwise, with constant stirring, 10 ml of concentrated $H_2SO_4$. The milky suspension of sulphur is precipitated by the addition of an equal volume of saturated NaCl and the product collected by centrifuging. The precipitate is washed twice by suspending it in 50 ml of water and reprecipitating with 50 ml of saturated NaCl. The sulphur is finally re-suspended in water and, if necessary, dialysed to remove NaCl. This suspension consists of very finely divided or colloidal sulphur particles which flocculate on the addition of salts but nevertheless remain highly reactive in biological systems.

## 3.2 Sodium dithionate ($Na_2S_2O_6 . 2H_2O$)

Dithionates are usually prepared by the oxidation of sulphur dioxide with $MnO_2$ (equation 3.1)

$$MnO_2 + 2SO_2 \rightarrow MnS_2O_6 \tag{3.1}$$

as in the following method (Pfansteil, 1946) for the preparation of sodium dithionate.

Pure $SO_2$ is passed into 500 ml of water in a 1 l round-bottom flask which is immersed in an ice bath. The solution is stirred vigorously and 80 g of finely powdered $MnO_2$ is added in 1–2 g portions over 2–3 hours. The flow of $SO_2$ is maintained during the addition of the $MnO_2$ and the temperature of the mixture is kept below 10°. After the addition of the $MnO_2$ is complete the stirring is continued until there is no further colour change. Excess $SO_2$ is then removed *in vacuo* at 40° and the gelatinous residue is filtered off and washed with warm water. The filtrate and wash-ings are combined and stirred at 35 to 40° with solid $BaCO_3$ until the evolution of $CO_2$ stops. After stirring for a further 10 min the mixture is neutralized (litmus) with solid $Ba(OH)_2$. A sample is filtered and the filtrate acidified with dilute HCl before adding $BaCl_2$ solution: if a precipitate is formed, indicating the presence of sulphite and sulphate, hot saturated $Ba(OH)_2$ is added to the main part of the mixture and the acid $BaCl_2$ test repeated. When the test for sulphate is negative the reaction mixture is filtered and the precipitate washed with 50 ml of water.

Sodium carbonate (about 65 g) is added to the combined filtrate and wash water until the mixture is permanently alkaline to litmus. The mixture is stirred vigorously during the addition and the temperature raised to 45°.

The precipitate of $BaCO_3$ is filtered off and washed with 150 ml of water at 50°. If the filtrate is acid to litmus it is treated with more $Na_2CO_3$ and again filtered. The filtrate is concentrated on a water bath to about one quarter of its original volume and then allowed to stand at 10° or less. (Note that any precipitate formed at the beginning of the concentration step is discarded.) The crystalline $Na_2S_2O_6.2H_2O$ is filtered off and recrystallized from water. The yield is about 88% based on $MnO_2$ used.

## 3.3 Polythionates

A number of methods have been described for the preparation of polythionates. Perhaps the most elegant of these involve the displacement of chloride ions from sulphur dichloride or disulphur dichloride by sulphite or thiosulphate: sulphur dihydroxide [$S(OH)_2$] and disulphur dihydroxide [$S_2(OH)_2$], formed by reactions between the sulphur chlorides and water, are thought to be intermediates in the synthesis of polythionates. Good yields of polythionates with a high degree of purity are obtained.

The polythionates are best prepared as the potassium salts which crystallize easily and are readily purified. Regardless of the method used for the preparation, prolonged exposures to high temperatures should be avoided to minimize decomposition and dismutation of the polythionates. Trithionate will decompose if exposed to acid but tetrathionate, and higher polythionates, are stabilized by acid conditions.

### 3.3.1 Potassium trithionate ($K_2S_3O_6$)

The preparation is essentially that of Stamm & Goehring (1942) and can be represented by equation 3.2

$$2KHSO_3 + SCl_2 \rightarrow K_2S_3O_6 + 2HCl \qquad (3.2)$$

Sulphur dioxide is bubbled through 800 ml of 5 M-KOH at $-5°$ in a 3–4 l stoppered flask until the pH reaches approximately 7. A solution of 100 g of $SCl_2$ in $1·5$ l of light petroleum at $-20°$ is added in 200 ml portions to the bisulphite solution. The liquid turns yellow and it must be decolorized by thorough shaking between each addition. The temperature should not rise above $10°$ during the reaction. After all the $SCl_2$ has been added the mixture is kept at $0°$ until separation of the trithionate is complete, usually in about an hour. The crystalline slurry is filtered and the residue washed with acetone and dried. The yield is 120 g of material containing 86% of $K_2S_3O_6$ and contaminated with KCl and sulphur. To purify this material the salt is dissolved in about 350 ml of water at $35°$, hot-filtered and cooled rapidly to $0°$. Pure $K_2S_3O_6$ separates. After removal of this material a further crop of crystals can be obtained by adding an equal volume of acetone to the mother liquors and cooling to $0°$. The crystals of purified potassium trithionate are washed with acetone and dried to give a yield of about 85 g.

Trithionate can also be prepared by reacting tetrathionate with sulphite (Fava & Divo, 1952) (equation 3.3).

$$S_4O_6^{2-} + SO_3^{2-} \rightarrow S_3O_6^{2-} + S_2O_3^{2-} \tag{3.3}$$

### 3.3.2 Potassium tetrathionate ($K_2S_4O_6$)

A number of different methods are available for the preparation of this compound.

*Method* 1. This is based on that of Stamm & Goehring (1942) and is represented in equation 3.4.

$$2H_2SO_3 + S_2Cl_2 \rightarrow H_2S_4O_6 + 2HCl \tag{3.4}$$

Water (750 ml) is saturated with $SO_2$ at $0°$ in a 2–3 l glass-stoppered flask. To this is added, in 100 ml portions, 75 g of $S_2Cl_2$ in 500 ml of light petroleum precooled to $-15°$. The liquid turns yellow with each addition and it must be decolorized by shaking and cooled to $0°$ before adding the next portion of $S_2Cl_2$. The mixture should still have the odour of $SO_2$ when all the $S_2Cl_2$ has been added.

The aqueous layer is separated from the light petroleum in a separating funnel and is flushed with air for several hours until no $SO_2$ can be detected. The solution is then cooled to $0°$ and taken to pH 6–7 with an ice-cold solution of 15% KOH in ethanol: approximately 150 g of KOH is required. About 165 g of crude tetrathionate containing 10% of KCl is precipitated. This is filtered off, dissolved in 120 ml of water at $70°$ and the solution hot-filtered into a beaker immersed in ice-water. It should be noted that Foss (1945) has recommended recrystallization from 0·5 N-HCl rather than from water. The crystalline product (about 120 g of pure $K_2S_4O_6$) is filtered off, washed with ethanol and dried *in vacuo* over $P_2O_5$. A further crop of impure potassium tetrathionate can be obtained by adding an equal volume of ethanol to the mother liquors from the above crystallization.

*Method* 2. (Trudinger, 1961*b*.) This is represented in equation 3.5

$$2S_2O_3^{2-} + I_2 \rightarrow S_4O_6^{2-} + 2I^- \tag{3.5}$$

Powdered iodine (about 52 g) is added slowly with vigorous stirring to an ice-cold solution of 50 g of $Na_2S_2O_3 . 5H_2O$ in 100 ml of water until the solution is faintly yellow: 100 ml of a saturated solution of potassium acetate is then added, followed by 800 ml of absolute ethanol. The crystals of potassium tetrathionate are filtered off and the material recrystallized from 0·5 N-HCl as in method 1. The yield is about 50 g of 99–100% $K_2S_4O_6$.

*Method* 3. This is a modification of method 2 which is suitable for milli-gram quantities of thiosulphate and which is particularly useful for the preparation of [$^{35}$S]tetrathionate (Trudinger, 1964 *a*). Sodium thiosulphate is dissolved in a small volume of ice-cold water and titrated with a strong solution of iodine until the mixture is faintly coloured. A slight excess of barium acetate is added, followed by ethanol to give a final concentration of 50% (v/v). The precipitated $BaS_4O_6$ is centrifuged off and washed with ethanol. It is then dissolved in a small volume of water and reprecipitated with ethanol, this step being repeated three or four times. The yield is about 85% of the theoretical. Solutions of potassium tetrathionate are obtained by the double decomposition of the barium salt with $K_2SO_4$ or by treating the barium salt with Dowex 50 ($K^+$).

### 3.3.3   Potassium pentathionate ($K_2S_5O_6.1.5H_2O$)

This is best prepared by the method of Goehring & Feldmann (1948) as represented in equation 3.6.

$$2H_2S_2O_3 + SCl_2 \rightarrow H_2S_5O_6 + 2HCl \qquad (3.6)$$

The following solutions are prepared: (1) 51 g of $SCl_2$ in 200 ml of $CCl_4$ at $-15°$ in a 2 l stoppered wide-necked flask; (2) 250 g of $Na_2S_2O_3.5H_2O$ in 400 ml of water at $0°$; (3) 200 ml of concentrated HCl plus 200 ml of water at $0°$. Solutions (2) and (3) are rapidly and simultaneously poured into solution (1) after which the flask is stoppered and shaken vigorously. The mixture, the temperature of which should not exceed $0°$, decolorizes within 20 seconds and the aqueous layer will show a slight turbidity due to sulphur. As rapidly as possible about 200 ml of 0·3 M-$FeCl_3$ (at $0°$) is added until the aqueous layer is a pale yellow colour (the dark colour of the intermediate ferric thiosulphate appears briefly and then fades): this layer is then separated and immediately concentrated to about 170 ml at 12 mm pressure and a temperature of 35 to 40°. Sodium chloride separates and is filtered off after which the filtrate is cooled to 0° and titrated dropwise with ice-cold 20% (w/v) KOH in methanol until the greenish black ferric hydroxide begins to separate at a pH of about 3. The mixture must be stirred continuously during the titration and the temperature kept below 10°. It should be noted that brown hydrated iron oxide forms on each addition of KOH but immediately redissolves.

When the titration is complete the mixture is cooled to $0°$ and the crystalline slurry filtered: the residue is washed with acetone until the yellow colour disappears. The product (about 102 g of 85% $K_2S_5O_6 \cdot 1.5$ $H_2O$ contaminated with KCl) is dried at room temperature. Recrystallization is carried out by dissolving 50 g of the crude pentathionate in 100 ml of $0.5$ N-HCl at $60°$: the mixture is rapidly reheated to $50°$ and filtered through a heated funnel into a beaker immersed in ice-water. Star-shaped crystals of pure $K_2S_5O_6 \cdot 1.5H_2O$ separate and are filtered off, washed with ethanol and dried *in vacuo* over $P_2O_5$. The total yield is about 46 g.

Potassium pentathionate may also be prepared by the reaction of thiosulphate with hydrochloric acid in the presence of arsenious acid (Stamm, Seipold & Goehring, 1941). The mechanism of this reaction is not fully understood.

### 3.3.4 Potassium hexathionate ($K_2S_6O_6$)

Goehring & Feldmann (1948) have also described the preparation of potassium hexathionate (equation 3.7).

$$2H_2S_2O_3 + S_2Cl_2 \rightarrow H_2S_6O_6 + 2HCl \tag{3.7}$$

Solutions of 100 g of $Na_2S_2O_3 \cdot 5H_2O$ in 150 ml of water and of 160 ml of 19% HCl (both at $0°$) are added simultaneously to a solution of 27 g of $S_2Cl_2$ in 100 ml of $CCl_4$ at $-15°$. The mixture is decolorized by shaking and then about 15 ml of ice-cold $0.6$ M-$FeCl_3$ is added until the aqueous layer becomes faintly yellow. This is then separated and immediately concentrated to about 50 ml at 12 mm pressure at $35°$. A precipitate of NaCl forms and is filtered off: the filtrate is cooled to $0°$ and titrated to pH 1–2 (by indicator paper) with cold methanolic KOH (see procedure for pentathionate, § 3.3.3). About 42 g of material containing 81% of $K_2S_6O_6$ is obtained by filtering the above solution and washing the residue twice with 40 ml of acetone. The material is purified by dissolving 20 g of it in 30 ml of 2N-HCl, heating the solution rapidly to $60°$, filtering and immediately cooling the filtrate in ice water. The crystals of potassium hexathionate (96–97% pure) are filtered off, washed with ethanol and dried *in vacuo* over $P_2O_5$. The yield is about 22 g in all.

Potassium hexathionate may also be synthesized by the reaction of thiosulphate with nitrite in acid solution (Stamm, Seipold & Goehring, 1941). Once again, the mechanism of this reaction is not understood.

## 3.4 Inorganic polysulphides

Inorganic polysulphides have not been widely used in biological experiments. In solution these compounds exist as equilibrium mixtures of which the average composition depends upon the relative concentrations of 'sulphide' and 'sulphur'. Methods for the preparation of salts of the polysulphides (from $S_2{}^{2-}$ to $S_6{}^{2-}$) are described by Fehér (1963).

## 3.5 The preparation of sulphate esters

Many methods for the preparation of sulphate esters were described by Burkhardt & Lapworth (1926), Burkhardt & Wood (1929), Burkhardt, Horrex & Jenkins (1936) and by Suter (1944). At present most methods of biochemical interest involve either a direct sulphurylation with chlorosulphonic acid in the presence of a nitrogenous base, such as dimethylaniline, or sulphurylation by the adduct of $SO_3$ with triethylamine, dimethylaniline, pyridine (pyridine-$SO_3$) or some similar nitrogenous base. Other procedures, such as the use of potassium pyrosulphate, are now usually of only historical interest although some are of value in the preparation of steroid sulphates. For instance Levitz (1963) prepared oestrone sulphate (labelled with [3]H and [35]S) by the reaction of oestrone with pyridine sulphate (Sobel, Drekter & Natelson, 1936) in the presence of acetic anhydride as a dehydrating agent while Joseph, Dusza & Bernstein (1966) used sulphamic acid to prepare a number of steroid sulphates. Several methods for the preparation of steroid sulphates have been described by Bernstein, Dusza & Joseph (1969) and the physical properties of the esters have been tabulated by Bernstein, Dusza & Joseph (1968).

Sulphation with $H_2SO_4$ is useful for the preparation of the sulphate esters of hydroxy amino acids (Dogson, Rose & Tudball, 1959; Dodgson Lloyd & Tudball, 1961) because the attempted preparation of such compounds with chlorosulphonic acid (see § 3.5.1) will form not only sulphate esters but also sulphamates. Mumma (1966) prepared several sulphate esters by the reaction of the alcohols with $H_2SO_4$ in the presence of dicyclohexylcarbodiimide: this method is likely to be a useful one, especially as labelled esters can thus be prepared directly from $H_2{}^{35}SO_4$. More recently Mumma (1968) has shown that the mild (e.g. by $Br_2$) oxidation of ascorbic acid 3-sulphate in the presence of a steroid yields the appropriate steroid sulphate. This may prove to be a very useful method of preparing the sulphate esters of labile steroids and presumably of other compounds (see also Chu & Slaunwhite, 1968, and §§ 6.1 and 6.9).

49

### 3.5.1  Methods using chlorosulphonic acid

For the preparation of the sulphate esters of phenols and of thiophenols, or of the sulphamates of certain aromatic amines, the following method (Burkhardt & Lapworth, 1926) is very useful. To 125 ml of dry $N:N$-dimethylaniline (1 mole) in 150 ml of dry, ethanol-free chloroform at $-10°$ is added slowly with constant stirring 26 ml of chlorosulphonic acid (0·4 mole). The temperature must be kept below about 5° throughout the addition. To the resulting yellow pasty mass of the sulphur trioxide adduct of dimethylaniline is then added 0·3 mole of the phenol dissolved (or suspended) in the minimum of chloroform and the mixture is stirred for 1 hr at 5° before being left standing overnight at room temperature. The chloroform is then removed by distillation under reduced pressure and the residual solution poured into a solution of 68 g of KOH in about 500 ml of water. The dimethylaniline is separated and the aqueous phase concentrated to crystallization point *in vacuo*: the potassium salt of the sulphate ester which is so obtained can be purified by recrystallization from water or from aqueous ethanol. It should be noted that the potassium salts of sulphate esters are, in general, more stable than the sodium salts. If the parent phenol is very soluble the methods required to obtain the sulphate ester free from inorganic material may be more troublesome.

### 3.5.2  Methods using pyridine-sulphur trioxide

Methods using the isolated adduct of sulphur trioxide with a nitrogenous base have the advantage that the reaction mixtures are free from $Cl^-$ ions which are sometimes difficult to remove completely from the final products of the reaction with chlorosulphonic acid. The preparation of pyridine-$SO_3$ from sulphur trioxide has been described by Baumgarten (1926), and more recently by Dodgson & Spencer (1957 *a*), and the reactions of this compound have been described in detail by Gilbert (1962). A useful preparation from chlorosulphonic acid rather than sulphur trioxide is the following. To a mixture of 20 ml of dry pyridine and 60 ml of ethanol-free chloroform at 0° is added dropwise a mixture of 8 ml of chlorosulphonic acid and 40 ml of chloroform. The temperature must be kept below 5° throughout the addition. The white precipitate of pyridine-$SO_3$ is filtered off and washed thrice with 15 ml portions of cold chloroform to remove any pyridine hydrochloride. The resulting solid should be dried and kept over $P_2O_5$ *in vacuo* under which conditions it is stable for many months. This reagent (m.p. 175°) is quite readily decomposed by hot water but it is relatively unaffected by cold water and can be washed with this to remove salts. The above method can also be used to prepare the sulphur trioxide adduct of other bases such as trimethylamine.

Many methods have been described for the use of pyridine-$SO_3$ as a sulphating agent, particularly for the preparation of steroid sulphates (Sobel & Spoerri, 1941; Paterson & Klyne, 1948; Roy, 1956 $a$; McKenna & Norymberski, 1957 $a$) but perhaps the most satisfactory is that of Emiliozzi (1960) which was devised primarily for the preparation of the sulphate esters of androsterone and related 17-oxosteroids. Androsterone (1 g) and pyridine-$SO_3$ (1 g) are suspended in 5 ml of pyridine and warmed to 60°–70° until solution occurs. The mixture is allowed to cool (when crystallization takes place) and 0·5 ml of triethylamine is added. After standing for 48 hr at room temperature the solvents are distilled off *in vacuo* below 60°. The residue, dried to constant weight, is then dissolved in 40 ml of saturated $Na_2SO_4$ and on adjusting the pH of this solution to 7 androsterone sulphate separates. The ester may be re-crystallized from water. Essentially similar methods have been described for the preparation of gestogen sulphates (Zelnik, Desfosses & Emiliozzi, 1960) and corticosteroid 21-sulphates (Pasqualini, Zelnik & Jayle, 1962).

### 3.5.3 Preparation of carbohydrate sulphates

Carbohydrate sulphates can be prepared by methods comparable to the above but greater difficulties are encountered in the purification of the products (Lloyd, 1960, 1961, 1962 $a, b$) because of the likelihood of sulphation occurring at several of the available hydroxyl groups. To avoid this problem suitably blocked derivatives of the carbohydrates may be used in definitive synthesis. General discussions of the problems of the synthesis of monosaccharide sulphates have recently appeared (Turvey, 1965; Rees, 1966) and detailed procedures for the preparation of several examples have been given by Peat, Turvey, Clancy & Williams (1960), by Guisely & Ruoff (1961) and by Whistler, Spence & BeMiller (1963) so that these methods need not be considered here. The adduct of sulphur trioxide with trimethylamine has also been used as a sulphuryl-ating agent in the carbohydrate field, particularly for the synthesis of the sulphates of $N$-acetylhexosamines (Meezan, Olavesen & Davidson, 1964), and recently Whistler, King, Ruffini & Lucas (1967) have described the use of the sulphur trioxide adduct of dimethylsulphoxide in the pre-paration of cellulose sulphate. This latter reagent may prove to be a very valuable one because of the solubility of several polysaccharides in dimethylsulphoxide and the fact that many other polysaccharides 'swell' in it to a considerable extent even if they do not dissolve.

In the above methods for the preparation of carbohydrate sulphates strictly anhydrous conditions must be maintained otherwise yields are negligible and for this reason Lloyd (1961) included 'Drierite' in such reaction mixtures. Lloyd, Wusteman, Tudball & Dodgson (1964) made use of this fact in their preparation of the sulphamate of D-glucosamine (potassium 2-deoxy-2-sulphoamino-D-glucose) directly from the parent sugar by sulphation with pyridine-SO$_3$ in aqueous solution: under these conditions sulphation of hydroxyl groups was negligible. This method is more convenient than those definitive methods which require the use of suitably blocked derivatives of glucosamine as the starting material.

It should finally be pointed out that most of the methods described above can readily be adapted for the preparation of [35]S-labelled sulphate esters. For this purpose Tudball (1962 a) has described a simple method of converting waste Ba[35]SO$_4$ (see § 4.2.1) into [[35]S]chlorosulphonic acid. The product is of rather low specific activity but a considerable saving of isotope can be achieved.

### 3.5.4 Sulphurylation with potassium persulphate

Of the other methods for the preparation of sulphate esters only one need be discussed: this is the use of potassium persulphate to produce the sulphate esters of nitrocatechols and nitroquinols from nitrophenols (Smith, J. N., 1951) and of o-aminophenols from arylamines (Boyland, Manson & Sims, 1953; Boyland & Sims, 1958). Indoxyl sulphate and related compounds have also been synthesized by this method (Boyland, Sims & Williams, 1956). This type of reaction is of particular value for the preparation of 2-hydroxy-5-nitrophenyl sulphate (nitrocatechol sulphate) which is much used as a substrate for the arylsulphatases (see § 7.1). Earlier preparations of this compound were contaminated with nitropyrogallol disulphate (Roy & Kerr, 1956) but the following method gives a pure product (Roy, 1958). A solution of 350 g of KOH, 350 g of potassium persulphate and 150 g of p-nitrophenol in 5 l of water is kept at 37° for 48 hr. After adjusting the pH to 4 with H$_2$SO$_4$ and filtering off the precipitate which forms on cooling, the solution is extracted with ether to remove free phenols before being readjusted to pH 10 with KOH. The precipitate which forms on cooling is again filtered off, the filtrate concentrated to about 750 ml *in vacuo* and diluted with 2 volumes of acetone to precipitate inorganic material which is removed, before again concentrating *in vacuo* to crystallization. The resulting crude dipotassium nitrocatechol sulphate is dissolved in the minimum of water and the pH adjusted to about 4 with acetic acid to precipitate the monopotassium salt of nitrocatechol sulphate which is recrystallized twice from water. To reform the dipotassium salt, the monopotassium salt is dissolved in the minimum of water containing a 20% excess of KOH and the product again recrystallized several times from water. The final product (yield about 15 g) contains a variable amount of water of crystallization and the anhydrous form is given by prolonged drying *in vacuo* over $P_2O_5$. It should be noted that the dipotassium salt is a phenoxide and therefore explosive on strong heating.

### 3.6    The preparation of thiosulphate esters

The classical method for the preparation of alkyl thiosulphates, or Bunte salts, is through the reaction of sodium thiosulphate with an alkyl bromide (Westlake & Dougherty, 1941) but this procedure is not without its difficulties and a valuable modification uses thallous thiosulphate in place of sodium thiosulphate (Lecher & Hardy, 1955). This method cannot be used to prepare aryl thiosulphates and a more general preparation is that of Baumgarten (1930) which utilizes the reaction between a thiophenol (or an alkanethiol) and pyridine-$SO_3$. In a typical preparation equimolecular proportions of a thiophenol and pyridine-$SO_3$ were refluxed in carbon tetrachloride for 1–2 hr and the mixture then chilled when the thiosulphate ester separated. This was purified by recrystallization from ethanol (Kice, Anderson & Pawlowski, 1966).

Lecher & Hardy (1955) described several methods for the preparation of Bunte salts but only one further one need be considered here: this is based on the well known fission of disulphides by sulphite ions. For example, *m*-nitrophenyl thiosulphate was prepared by mixing 0·06 mole of $SO_2$ in methanol with 0·06 mole of KOH in methanol and refluxing the mixture with 0·015 mole of *m*-nitrophenyl disulphide for $1\frac{1}{2}$ hr: after the further addition of 0·015 mole of KOH in methanol the mixture was hot-filtered and the filtrate chilled: the thiosulphate separated in about 80% yield.

As already noted, a thiosulphate ester which is of particular interest is *S*-sulphocysteine [*S*-(2-amino-2-carboxyethyl) thiosulphate] (see § 2.6) and a number of preparations of this compound have been described. All are based on the fission of the disulphide bond of cystine with $SO_3{}^{2-}$ ions and a useful method is that described by Nakamura & Sato (1963*a*).

L-Cystine (12 g), $Na_2SO_3$ (63 g) and $CuSO_4.5H_2O$ (0·5 g) were dissolved in 800 ml of 0·2 N-$NH_4OH$ and the mixture aerated at room temperature for 9 hr: at this time a further 63 g of $Na_2SO_3$ were added and the aeration continued for a further 9 hr. After adjusting the pH to 6·8 with 2 N-acetic acid the reaction mixture was concentrated *in vacuo* to about 400 ml., the temperature being kept below 40°. After filtration, $SO_4^{2-}$ and $SO_3^{2-}$ ions were precipitated by the addition of $BaCl_2$ and $Ba(OH)_2$, the latter being used to keep the pH below 7, and removed by centrifuging. The supernatant was taken to dryness *in vacuo* and the residue extracted with 100 ml of cold ethanol before drying *in vacuo*. The dry residue (about 20 g) was dissolved in water and passed through a column (3·2 × 32·5 cm) of Amberlite CG-120 ($H^+$) to remove cations: the eluate was adjusted to pH 5·4 with 2 N-NaOH and applied to a column (3·5 × 57·5 cm) of Dowex-1 × 8 (formate, 100–200 mesh) and eluted with a pyridine-formate buffer, pH 5·4 (44·6 ml of 85% (w/v) formic acid and 162 ml of pyridine diluted to 1 l with water). The fractions containing S-sulpho-cysteine, detected by the ninhydrin reaction following paper chromato-graphy, were combined, evaporated *in vacuo* to remove organic solvents and neutralized with 2 N-NaOH. After further concentration the addition of an equal volume of ethanol caused the crystallization of S-sulpho-L-cysteine. The material was recrystallized from 80% ethanol to give 7 g of pure S-sulphocysteine.

Sörbo (1958 *a*) pointed out that it was difficult to remove traces of copper from S-sulphocysteine obtained by earlier variants of the above method and devised a rather simple preparation which did not involve the use of $Cu^{2+}$ ions as an oxidizing agent.

Waley (1959) has described the preparation of S-sulphoglutathione by a method essentially similar to that for S-sulphocysteine.

### 3.7 The preparation of sulphatophosphates

The enzymatic reactions involved in the synthesis of the sulphatophos-phates will be discussed later (see chapter 5) and here we will be con-cerned only with the methods available for the preparation of such compounds, in particular of APS and of PAPS. Especially in the case of the latter compound enzymatic methods of synthesis are often simpler and preferable to chemical ones.

The chemical synthesis of APS, adenylyl sulphate, is the simpler problem because the starting material is the readily accessible adenosine 5'-phosphate. Baddiley, Buchanan & Letters (1957) described the preparation of APS (in about 5% yield) by the sulphurylation of adenylic acid with pyridine-$SO_3$ in sodium bicarbonate solution. Reichard & Ringertz (1959) described a method using sulphuric acid and carbodiimide in pyridine as the sulphurylating agent but this procedure suffered from the disadvantage of the formation of a number of derivatives of adenosine sulphated on the ribose: as already pointed out above, in aqueous solution such O-sulphurylation is negligible with pyridine-$SO_3$. Cherniak & Davidson (1964) used trimethylamine-$SO_3$ in a mixture of dioxane, dimethylformamide and pyridine as the sulphurylating agent and obtained a 75% yield of adenylyl sulphate from adenylic acid: this method would therefore appear to be the one of choice for the preparation of APS. Another useful method is the anion-exchange reaction used by Michelson & Wold (1962) to give adenylyl sulphate in 40% yield. Uridylyl sulphate was also synthesized by this method.

The synthesis of PAPS is more troublesome because the starting material is the relatively inaccessible adenosine 3′,5′-diphosphate. Baddiley, Buchanan & Letters (1958) described the synthesis of the latter by the phosphorylation of adenosine with dibenzyl phosphorochloridate and the subsequent sulphurylation with pyridine-$SO_3$ in sodium bicarbonate solution to give PAPS (Baddiley, Buchanan, Letters & Sanderson, 1959; Baddiley & Sanderson, 1963). The yield was low (less than 10%), probably because of the formation of some 3′-sulphatophosphate which decomposed to give adenosine 2′(3′),5′-diphosphate. The latter was in fact isolated from the reaction mixture. Fogarty & Rees (1962) have also described a preparation of adenosine 3′,5′-diphosphate using 2-cyanoethylphosphate. Another method for the synthesis of PAPS has been developed by Cherniak & Davidson (1964): this makes use of adenosine 2′(3′),5′-diphosphate, also prepared from adenosine by its reaction with dibenzyl phosphorochloridate (Cramer, Kenner, Hughes & Todd, 1957; Moffatt & Khorana, 1961), as the starting material and so eliminates some of the side reactions. This nucleotide was sulphurylated with triethylamine-$SO_3$ and the cyclic phosphate ester hydrolysed with $T_2$-ribonuclease from Takadiastase. The yield of PAPS from the diphosphate was between 40 and 65%: again this appears to be the method of choice for the preparation of PAPS.

Enzymatic methods for the preparation of PAPS have been described by Robbins (1963) and by Banerjee & Roy (1966): these methods, using the sulphate-activating system of rat liver, can readily give PAPS in 0·1 m-mole quantities in a yield of about 5% from the ATP which is the starting material. More recently Balasubramanian, Spolter, Rice, Sharon & Marx (1967) have described a similar method using the sulphate activating system of a mouse mastocytoma: a very thorough investigation was made to define the optimum conditions for the reaction and the only disadvantage of the method is the relative unavailability of the tumour. Such methods can also readily be adapted to prepare [$^{35}$S]PAPS (Robbins, 1963).

### 3.8 Sulphenyl thiosulphate esters

The preparation of sulphenyl thiosulphate esters (or sulphodisulphanes, often quite wrongly known as $S$-thiosulphonates) was first described by Foss (1947) who obtained solutions of these compounds by the following general reactions (equations 3.8 and 3.9).

$$R.S.SCN + S_2O_3^{2-} \rightarrow R.S.S.SO_3^- + SCN^- \qquad (3.8)$$

$$R.S.NR_2 + S_2O_3^{2-} + 2H^+ \rightarrow R.S.S.SO_3^- + R_2NH_2^+ \qquad (3.9)$$

Alanine 3-sulphenyl thiosulphate (alanine sulphodisulphane) was prepared by Szczepkowski (1961 $a,b$) by the oxidative condensation of cysteine with thiosulphate (equation 3.10).

$$\begin{array}{ccc} CH_2.SH & CH_2.S.S.SO_3^- & \\ | & | & \\ H.C.NH_2 + S_2O_3^{2-} + I_2 \rightarrow H.C.NH_2 & +2I^- & (3.10) \\ | & | & \\ COOH & COOH & \end{array}$$

Cysteine (15 m-moles) in 10 ml of water was mixed with 100 m-moles of $Na_2S_2O_3.5H_2O$ in 30 ml of water and shaken with 100 m-moles of iodine and 20 m-moles of $BaCO_3$ at 0°. The barium salt of alanine sulphodisulphane was precipitated with 900 ml of acetone and converted to the sodium salt by double decomposition with $Na_2SO_4$. Excess $Na_2SO_4$ was precipitated with methanol and the sodium salt of alanine sulphodisulphane precipitated by the addition of ether. It can be recrystallized from methanol.

The analogous derivative of glutathione may be prepared in a similar manner (Szczepkowski, 1961 $b$).

### 3.9 Thiosulphonic acids

Thiosulphonic acids are readily prepared by reacting sulphonyl chlorides with sulphide (equation 3.11).

$$R.SO_2Cl + S^{2-} \rightarrow R.SO_2S^- + Cl^- \qquad (3.11)$$

The following general method is that of Traeger & Linde (1901) modified by Mintel & Westley (1966 a). Sodium sulphide (13·6 g of $Na_2S.9H_2O$) is dissolved in 50 ml of water and the solution heated to 95 to 100°. The sulphonyl chloride (10 g) is then added dropwise and the mixture refluxed overnight with continuous stirring. The colorless solution is evaporated to dryness *in vacuo* and the dried material extracted with hot ethanol before recrystallizing twice from hot ethanol. It is dried at 25°.

*Alanine 3-thiosulphonic acid* (Barium salt)

$$CH_2.SO_2SH$$
$$|$$
$$H.\overset{|}{C}.NH_2$$
$$|$$
$$COOH$$

Alanine 3-thiosulphonic acid was prepared by De Marco, Coletta, Mondovi & Cavallini (1960) by reacting elemental sulphur with alanine 3-sulphinate (cysteine sulphinic acid). A solution of 612 mg (4 m-moles) of alanine 3-sulphinic acid is mixed with 200 mg of elemental sulphur (about 6 m-moles) and 20 ml of pyridine and refluxed for 5–10 min, until all the sulphur is dissolved. The solution is evaporated to dryness on a boiling water bath under reduced pressure and 10 ml of 8% $Ba(OH)_2$. $8H_2O$ added to the dry residue. Excess sulphur and small amounts of $BaCO_3$ are filtered off and 10 vols of 95% ethanol added to the filtrate. After 12–14 hr at 0° the precipitate is filtered, washed with 100% ethanol and dried with ether. About 800 mg of slightly hygroscopic, pasty white powder is obtained: this is about 80–85% pure by nitrogen analysis. The nature of the impurities was not reported by De Marco *et al*.

The formation of alanine 3-thiosulphonate by dismutation of cystine disulphoxide with $H_2S$ was reported by Sörbo (1956).

## 3.10 Persulphides

The formation of persulphides by reactions between sulphur and thiols was reported by Hylin & Wood (1959). Cavallini, De Marco, Mondovi & Mori (1960) attempted to prepare thiocysteine (2-amino-2-carboxyethyl hydrodisulphide) by this procedure but found that it rapidly changed to other compounds of greater stability. Rao & Gorin (1959) reported that thiocysteine is formed from cystine and sulphide under strongly alkaline conditions. In general the organic persulphides are highly unstable and no representative of this class having biological interest has been prepared in pure form.

### 3.11 Thiocystine

This compound, bis(2-amino-2-carboxyethyl)trisulphide (*3.1*), is formed by reacting sulphur dichloride with *N*-acetylcysteine which itself may be prepared by the method of Pirie & Hele (1933). The following procedure

$$\begin{array}{ccc} CH_2.S.S.S.CH_2 \\ | \qquad | \\ H.C.NH_2 \quad H.C.NH_2 \\ | \qquad | \\ COOH \qquad COOH \end{array} \qquad (3.1)$$

is that of Fletcher & Robson (1963). *N*-Acetylcysteine (2·9 g) is suspended in 50 ml of $CHCl_3$ and a solution of 0·9 ml of $SCl_2$ in 10 ml of $CHCl_3$ is added: the mixture is stirred for 40 hr at 45° and the $CHCl_3$ is then distilled off to give a dry residue which is hydrolysed for 4 hr with 6 N-HCl. After filtering, the acid solution is evaporated *in vacuo* to give a light brown tarry residue containing thiocystine, cystine, cysteine and an unidentified product. This is redissolved in 5 ml of 2·5 N-HCl and chromatographed on a $4 \times 19$ cm column of Zeo-Karb 225($\times 4$) and eluted with 2·5 N-HCl. Cysteine, cystine and thiocystine are eluted by approximately 440–640 ml, 700–920 ml and 1740–2280 ml of 2·5 N-HCl respectively, the compounds being detected by the phosphotungstate reduction method of Fletcher & Robson (1962). The fractions containing thiocystine are evaporated to dryness and the residue rechromatographed on a $2·5 \times 6$ cm column of Zeo-Karb 225($\times 4$): in this case the thiocystine is eluted with about 100–200 ml of 2·5 N-HCl. The residue obtained by taking the appropriate fractions to dryness is dissolved in 10 ml of water and the pH brought to 3·5 by the cautious addition of pyridine: water (1:3, v/v): an excess of pyridine leads to the formation of some cystine. After standing for two days at 4° the thiocystine is filtered off, washed with ice water and with ethanol, and then dried over phosphorus pentoxide. The yield is about 285 mg.

Thiocystine may also be prepared by the addition of sulphur to cysteine (Cavallini, DeMarco, Mondovi & Mori, 1960; Fletcher & Robson, 1963).

# 4

## THE ANALYSIS OF SOME
## SULPHUR COMPOUNDS

The methods described in this chapter appear to be particularly suited to the measurement of some sulphur compounds which are important in inorganic sulphur metabolism. No attempt has been made to review or compare the numerous analytical methods which have been reported in the literature, most of which are designed for the analysis of specific inorganic or organic samples and which often have limited applicability. The absence of highly specific tests for many sulphur compounds often necessitates a preliminary separation of complex mixtures to enable accurate results to be obtained. In some cases the use of [35]S-labelled substrates combined with electrophoretic or chromatographic isolation of the products is the only satisfactory method of analysis.

### 4.1 Chemical methods of analysis

### 4.1.1 Total sulphur

Total sulphur is generally estimated as sulphate after oxidation of the sample with perchloric and nitric acids. The sulphate may be determined gravimetrically as barium sulphate or colorimetrically by the benzidine or reduction methods described in § 4.1.2.

A number of procedures for the wet combustion of sulphur-containing samples have been reported in the literature; the following technique is that of Evans & St John (1944). The sample is heated at about 100° in a Kjeldahl flask with 20 ml of concentrated $HNO_3$ until the material is dissolved; then 5 ml of 72% $HClO_4$ per gram of sample is added and heating at 100° continued for 24 hr. (The predigestion with $HNO_3$ avoids the possibility of explosions occurring when the $HClO_4$ is added.) After 24 hr the temperature is raised to remove any residual $HNO_3$ and the digestion with the remaining $HClO_4$ continued for up to 15 hr. Further additions of $HClO_4$ are made if necessary. When digestion is complete the mixture is cooled, diluted with water and filtered through a fine filter which is washed thoroughly with water.

The rate of oxidation by $HClO_4$ depends upon the nature of the material to be analysed and preliminary trials should be carried out to determine the time required for complete oxidation.

A rapid technique for wet combustion (30 min or less) has been described by Bethge (1956); the risk of explosion with some biological materials is, however, fairly high.

### 4.1.2 The determination of sulphate ions

The determination of small amounts of sulphate ions is difficult: larger amounts, of the order of tens of milligrams of sulphate, present no major problems and standard methods are available. There are four main groups of micro methods: those based on the precipitation of $SO_4^{2-}$ ions as benzidine sulphate followed by the determination of the benzidine; those based on the colorimetric determination of chloranilic acid formed by the reaction of barium chloranilate with $SO_4^{2-}$ ions; those based on the turbidimetric, or nephelometric, determination of barium sulphate; and those based on reduction of sulphate to sulphide. Several other types of method have been described but none of them has been widely used except for those which involve the precipitation of sulphate ions by 4-chloro-4′-aminodiphenyl. The availability of so many different procedures for the determination of sulphate ions undoubtedly reflects the fact that none of them is entirely satisfactory under all circumstances.

### *Precipitation of benzidine sulphate*

Of the multitude of methods using the precipitation of benzidine sulphate, a very reliable one is that of Dodgson & Spencer (1953) and the following modification of this has proved useful. To 1 ml of reaction mixture containing between 0·1 and 0·7 $\mu$mole of $SO_4^{2-}$ ions is added 5 ml of ethanol: any precipitated proteins or mucopolysaccharides are then removed by centrifuging. A 5 ml sample of the supernatant is transferred to a tapered centrifuge tube, the inner tip of which has been ground to provide a slightly roughened surface, and to this solution is added 1 ml of 25% (w/v) trichloroacetic acid and 1 ml of 5% (w/v) benzidine in ethanol. After mixing, the stoppered tube is left standing overnight at 5° and the benzidine sulphate is then centrifuged down and washed with ethanol using the carefully standardized technique of Dodgson & Spencer (1953). The precipitate of benzidine sulphate is dissolved in 3 ml of 1 N-HCl and the solution chilled to 5°. To this is added 1 ml of 0·1% (w/v) $NaNO_2$ followed, after 3 min, by 5 ml of 0·5% (w/v) thymol in

2 N-NaOH and the absorbance of the resulting solution at 510 m$\mu$ is determined in cells with a 1 mm or 2 mm light path. Although this method gives reliable results it must be noted that if less than about 0·1 $\mu$mole of $SO_4^{2-}$ ions is present in the sample taken for analysis then quantitative precipitation of benzidine sulphate will not occur under the above conditions: to overcome this difficulty Dodgson & Spencer (1953) added $SO_4^{2-}$ ions in the solution of trichloroacetic acid. Before this method is applied to any new problem it is essential that careful control determinations be carried out because not only does the rate of crystallization of the benzidine sulphate vary somewhat with the composition of the reaction mixture but certain ions (such as $Ca^{2+}$, $Fe^{3+}$ and $PO_4^{3-}$ ions) interfere.

Kent & Whitehouse (1955) have described a method essentially similar to the above except that in it the precipitate of benzidine sulphate is collected and washed by filtration on a sintered-glass disk: such a procedure is perhaps easier if only a few determinations are to be performed but more difficult when large numbers must be carried out. Spencer (1960 a) has developed an ultra-micro modification of the method which is capable of determining about 0·2 $\mu$mole of $SO_4^{2-}$ ions in a volume of 20 $\mu$l: so far this procedure does not appear to have been widely used.

### Use of barium chloranilate

Probably the most useful of the methods using barium chloranilate is that described by Lloyd (1959). Again great care is required to ensure reproducible conditions when the procedure is applied to biological materials. It appears to show little advantage over the benzidine method except that it is much less time consuming. The sensitivity is such that 0·05 to 0·8 $\mu$mole of $SO_4^{2-}$ ions in a volume of 0·2 ml can be determined. Again Spencer (1960 a) has described an ultra-micro modification of this procedure.

### Turbidimetric methods

Of the turbidimetric methods involving the production of an unstable suspension of $BaSO_4$ under carefully standardized conditions, a useful method is that of Bray, Humphris, Thorpe, White & Wood (1952). Of the many other methods which depend upon the production of a more or less stable suspension of $BaSO_4$, the most useful is that of Dodgson (1961 a) which uses gelatine as a stabilizing agent. Certainly the latter method (or series of methods) is the more sensitive of the two and although its accuracy is not comparable to that of the benzidine method it is both rapid and useful provided that its limitations be kept in mind. To 0·2 ml of solution, containing between 0 and 0·2 $\mu$mole of $SO_4^{2-}$ ions, is added 3·8 ml of 4% (w/v) trichloroacetic acid and 1 ml of $BaCl_2$ reagent which is made up by dissolving 5 g of $BaCl_2$ in 100 ml of a 5% solution of gelatine prepared under carefully standardized conditions. Not all samples of gelatine are satisfactory for use in this procedure, the governing factor being their content of $SO_4^{2-}$ ions which must be significant without being too great. After mixing, the solution is allowed to stand for 10 min and its absorbance at 360 m$\mu$ measured against a reagent blank in cells of 2 cm light path. If the solutions contain material absorbing in the ultraviolet then the absorbance at 500 m$\mu$, or indeed at any suitable wavelength, may be measured although the sensitivity of the method is thereby decreased. It is reported that the suspension of $BaSO_4$ produced under the above conditions is stable for 1 hr.

## Precipitation by chloroaminodiphenyl

The use of 4-chloro-4'-aminodiphenyl as a precipitant for $SO_4^{2-}$ ions followed by the spectrophotometric determination of the excess reagent was described by Jones & Letham (1954). To 0·5 ml of solution containing from 0·3 to 1·0 $\mu$mole of $SO_4^{2-}$ ions is added 0·5 ml of water containing a trace of 'Cetavlon' followed by 0·5 ml of a 0·19% (w/v) solution of chloroaminodiphenyl in 0·1 N-HCl. After mixing, the solution is allowed to stand for at least 2 hr and then centrifuged to sediment completely the chloroaminodiphenyl sulphate, a process which is aided by the 'Cetavlon' in the reaction mixture. A sample (0·3 ml) of the supernatant is removed, diluted to 25 ml with 0·1 N-HCl and its absorbance at 245 m$\mu$ determined. The decrease in absorbance of the sample as compared with the control allows calculation of the concentration of the $SO_4^{2-}$ ions. In this method, as in the benzidine methods, $PO_4^{3-}$ ions interfere as also do $Cl^-$ ions if present in high concentrations.

## Methods based on the determination of barium

Several such methods have been developed. Picou & Waterlow (1963) precipitated sulphate with $^{133}BaCl_2$ and measured the radioactivity in a scintillation counter. To 0·5 ml of a solution containing 0·5–20 $\mu$moles of sulphate is added 1–2 ml of $BaCl_2$ reagent (0·03M-$BaCl_2$ in 1 N-HCl containing 60% ethanol and a suitable concentration of $^{133}BaCl_2$). After 1 hr the precipitate of $BaSO_4$ is centrifuged off, the centrifuge tube drained, the surface of the precipitate and the walls of the tube rinsed with 1 ml of non-radioactive $BaCl_2$ reagent and the radioactivity of the unwashed precipitate of $BaSO_4$ determined by counting in the centrifuge tube itself. If the $BaSO_4$ precipitate were washed before counting the results were less reproducible. No detailed study of this method, for which an accuracy of $\pm 2\%$ is claimed, has been made but it may be a valuable one when used under carefully controlled conditions.

Roe, Miller & Lutwak (1966) used a similar method but determined the barium by atomic absorption spectrophotometry. Lanthanum chloride was added to prevent interference by phosphate ions. To 5 ml of solution containing $0.5$–3 $\mu$moles of sulphate is added 1 ml of lanthanum chloride (prepared by dissolving $58.6$ g of $La_2O_3$ in 200 ml water, adding 250 ml of HCl and making up to 1 l with water) and 2 ml of $15\%$ (w/v) $BaCl_2$. The precipitate of $BaSO_4$ is centrifuged off and washed twice with 5 ml portions of water. It is then dissolved in 10 ml of EDTA reagent (10 g of disodium ethylenediaminetetraacetate in 500 ml water and 100 g of NaOH made up to 2 l) and the barium content of the solution determined by atomic absorption spectrophotometry. Roe *et al.* (1966) also described methods for the oxidation of organic compounds of sulphur to sulphate ions which were determined as described above.

## Reduction of sulphate to sulphide

The reduction of sulphate to sulphide is the basis of several sensitive analytical methods for sulphate (Luke, 1943, 1949; Roth, 1951; Bethge, 1956). In the following method of Johnson & Nishita (1952) sulphate is reduced in an apparatus illustrated in fig. 4.1 and the hydrogen sulphide estimated by the formation of methylene blue from *p*-aminodimethyl-aniline in the presence of ferric ions (equations 4.1 and 4.2).

$$2\,p\text{-aminodimethylaniline} + S^{2-} \rightarrow \text{leucomethylene blue} \qquad (4.1)$$

$$\text{leucomethylene blue} + Fe^{3+} \rightarrow \text{methylene blue} + Fe^{2+} \qquad (4.2)$$

The reducing mixture is prepared by slowly heating a mixture of 15 g of red phosphorus, 100 ml of hydriodic acid (S.G. $1.7$, methoxyl grade) and 75 ml of $90\%$ formic acid to $115°$ in a 250 ml flask through which a stream of nitrogen, introduced to the bottom of the flask, is passed. The heating is continued at $115$–$117°$ for $1$–$1\frac{1}{2}$ hr. The reagent is stable for 2–3 weeks but can be used for longer times with reduced sensitivity and precision.

A sulphur-free lubricant is prepared by boiling about 5 g of silicone stopcock grease with 10 ml of a $50\%$ (v/v) mixture of hydriodic acid and hypophosphorous acid for 45 min. At the end of the boiling period the acid is decanted off and the grease washed thoroughly with sulphur-free water. All reagents used in the digestion procedure should be made up in sulphur-free water prepared by distilling water containing alkaline permanganate from an all-glass still.

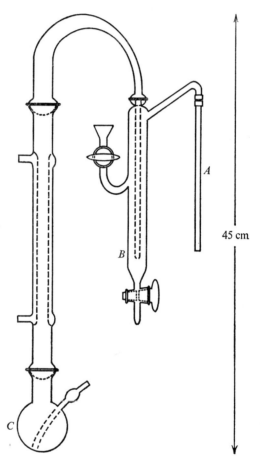

Fig. 4.1 Apparatus used in the quantitative reduction of sulphate of sulphide.

*Procedure*. All spherical joints of the apparatus are lightly greased with the sulphur-free lubricant. The gas washing column (*B*) is charged with 10 ml of a solution of 10% (w/v) $NaH_2PO_4$ and 10% (w/v) pyrogallol. The connecting tube (*A*) is inserted below the liquid level of a 100 ml glass-stoppered volumetric flask, containing 70 ml of sulphur-free water and

10 ml of a solution of 5% (w/v) zinc acetate dihydrate and 1·25% sodium acetate trihydrate. The sample to be analysed (containing less than 300 $\mu$g of sulphur in a volume of less than 2 ml) is introduced into the digestion flask (C). The digestion mixture is shaken to suspend the red phosphorus and 4 ml added to the digestion flask from a fast-delivery pipette. The flask is immediately fixed to the condenser and a stream of nitrogen introduced through the gas inlet tube in the digestion flask. The nitrogen should be washed by passage through a solution of 5–10% (w/v) $HgCl_2$ in 2% (w/v) $KMnO_4$ and the flow rate adjusted to 100–200 ml per min. The digestion mixture is then heated and maintained at boiling point for 1 hr. (A gas micro-burner is suitable for heating but the flask should be protected from direct flame by means of an asbestos pad.) After 1 hr the receiving flask, together with the connecting tube, is removed and 10 ml of 0·1% (w/v) $p$-aminodimethylaniline sulphate in 3·6 N-$H_2SO_4$ added rapidly from a rapid-delivery pipette. After mixing, 2 ml of 12·5% (w/v) ferric ammonium sulphate in 0·45 N-$H_2SO_4$ is added and the contents of the flask mixed thoroughly and made to 100 ml with water. The optical density at 670 m$\mu$ is measured after 10 min.

A linear relationship between optical density and sulphide concentration is obtained over the range 1–50 $\mu$g of sulphur per 100 ml. Methylene blue solutions equivalent to 50–300 $\mu$g of sulphur may be measured after suitable dilution with water containing the same concentrations of $p$-aminodimethylaniline and ferric ammonium sulphate as present in the standard reaction mixture. The method does not distinguish between sulphate, sulphite, thiosulphate, polythionates and sulphide but a number of organic compounds such as cystine, cysteine, methionine, glutathione and taurine are not reduced to hydrogen sulphide. Nitrate interferes when more than 6 mg is present in the sample; under these conditions sulphate should be separated as $BaSO_4$ prior to digestion.

### 4.1.3   The determination of sulphate esters

No specific technique exists for the determination of sulphate esters and the only general method depends upon their hydrolysis to give $SO_4^{2-}$ ions which may then be determined by one of the methods described above, or gravimetrically as $BaSO_4$ if sufficient material is available. The particular choice of hydrolysis conditions must obviously depend upon the nature of the ester under investigation, and upon whether the remainder of the molecule is required in an unaltered state: complete hydrolysis may well be incompatible with the latter requirement. As far as most simple sulphate esters are concerned, treatment with 1 N-HCl at 100° for 1 hr will give quantitative hydrolysis. Polysaccharide sulphates, in particular heparin, are more resistant to hydrolysis and require quite drastic conditions such as treatment with 1 N-HCl in sealed tubes at 110° for 5 hr (Dodgson & Price, 1962) or with 30% formic acid in sealed tubes at 100° for 8 hr (Muir & Jacobs, 1967): the former method was followed by the nephelometric determination of the liberated $SO_4^{2-}$ ions and the latter by their precipitation with 4-chloro-4'-aminodiphenyl or the reduction of sulphate to sulphide.

## Methylene blue method

A useful method for the determination of some sulphate esters is that developed by Roy (1956 *a*) which depends upon the solubility in chloroform of their methylene blue salts. A generally applicable technique is as follows. To 1 ml of a solution containing about 0·02 $\mu$mole of sulphate ester is added 1 ml of methylene blue reagent [0·025% methylene blue chloride (zinc free) in 1% (w/v) $H_2SO_4$ containing 5% (w/v) $Na_2SO_4$]. To this is added 5 ml of chloroform and after vigorous shaking for 30 sec the organic phase is separated by centrifuging, dried over $Na_2SO_4$, and its absorbance at 650 m$\mu$ determined against the appropriate reagent blank. The methylene blue reagent is quite stable when kept cool and in the dark but it rapidly alters when exposed to sunlight. It should also be noted that not all samples of methylene blue give a satisfactory reagent: some give very high blank values. Again careful attention is necessary to the maintenance of constant conditions because not only do high concentrations of inorganic salts alter the partition of the methylene blue salt between the aqueous and organic phase, but also, for example, chloride increases the blank value. The method cannot be used in the presence of certain large anions such as trichloroacetate, permanganate, or phosphotungstate.

Further, many sulphate esters, particularly steroid sulphates, are rather firmly bound to protein and the latter must be removed before the method is applied. Deproteinization can be brought about by heating the solution in a boiling-water bath (Roy, 1956 *a*) or preferably by treatment with ethanol (Roy, 1956 *b*). A suitable technique is to add 5 ml of ethanol to 1 ml of the protein-containing solution and to remove any precipitated protein by centrifuging. A 5 ml sample of the supernatant is transferred to a test-tube and, after the addition of an anti-bumping granule, taken to dryness in a boiling-water bath. The solid residue is cooled and dissolved in 2 ml of methylene blue reagent (previously diluted with an equal volume of water). The extraction with chloroform etc. is then performed as described above.

Under carefully controlled conditions this method is undoubtedly useful for the determination of sulphate esters which are derived from relatively non-polar alcohols or phenols. It is not suitable for the determination of simple alkyl sulphates, aminophenyl sulphates or carbohydrate sulphates because these give methylene blue salts which are poorly soluble in chloroform. The method can also be used for the determination of certain aryl sulphamates, such as 2-naphthyl sulphamate. In general terms the sensitivity of the method is inversely related to the solubility in water of the parent hydroxyl-containing compound and the amounts of sulphate ester (in the 1 ml of aqueous solution taken for analysis) which are required to produce an absorbance of 0·10 at 650 m$\mu$ when determined by the method described above are as follows:

| | |
|---|---|
| 2-Naphthyl sulphamate | 30·7 n-moles |
| Phenyl sulphate | 20·5 n-moles |
| 2-Naphthyl sulphate | 6·6 n-moles |
| 1-Naphthyl sulphate | 6·3 n-moles |
| p-Nitrophenyl sulphate | 6·3 n-moles |
| Androstenolone sulphate | 6·1 n-moles |

One of the advantages of this procedure is that all sulphate esters having related structures, for example most steroid sulphates, have very similar chromogenicities, a feature which is obviously of considerable value. Under the above conditions there is no interference by phosphate esters or by glucuronides, compounds which may sometimes be associated with sulphate esters in biological materials.

### 4.1.4   The determination of sulphatophosphates

The only specific methods for the determination of sulphatophosphates are enzymatic, and even these are only available for adenylyl sulphate and 3′-phosphoadenylyl sulphate. Otherwise such compounds can only be determined as a 'nucleotide bound' sulphate which readily gives rise to $SO_4^{2-}$ ions on gentle hydrolysis. The sulphatophosphates are readily separable from the more conventional nucleotides because of their lower pK values which give them a much higher mobility on paper electrophoresis (Robbins & Lipmann, 1957) and a much stronger affinity for Dowex 1 than simple nucleotides. They are only eluted from the latter by 1 M-NaCl (Brunngraber, 1958) or by 5 N-formic acid–0·3 N-ammonium formate (Robbins & Lipmann, 1957). Once the sulphatophosphates have been separated from other nucleotides they can be determined by measuring the amount of $SO_4^{2-}$ ions liberated by a brief hydrolysis in 0·1 N-HCl (see § 2.7) or by measuring their absorption at 260 m$\mu$ which is due entirely to the purine or pyrimidine moiety. Accurate values for the extinction coefficients of APS and of PAPS do not seem to have been determined although Baddiley *et al.* (1957) report an $\epsilon_{259}$ of 15,200 for APS. Both Robbins & Lipmann (1957) and Cherniak & Davidson (1964) use a value 14,500 for the $\epsilon_{260}$ of PAPS but this appears to be an assumed value rather than an experimentally determined one.

Alone among the common sulphatophosphates, adenylyl sulphate can be determined by its conversion to ATP through the action of ATP-sulphurylase, as described by Robbins & Lipmann (1957), while 3'-phosphoadenylyl sulphate can be determined by the transfer of its sulphuryl group to a suitable acceptor through the action of a sulphotransferase. The most suitable enzyme is phenol sulphotransferase (see § 6.1) with 2-naphthol as the acceptor. A satisfactory method is the following: the reaction mixture (volume 1 ml) contains 0·2 M-acetate buffer, pH 5·6, 0·1 mM-2-naphthol and approximately 0·02 mM-PAPS together with a suitably large quantity of partially purified phenol sulphotransferase (Banerjee & Roy, 1966; see § 6.1). The requisite amount of enzyme must be determined by experiment. After incubation at 37° for 2 hr the amount of 2-naphthyl sulphate formed is determined by the methylene blue method described above. When a partially purified enzyme preparation is used the amount of protein present (less than 1 mg/ml) is small and deproteinization of the reaction mixture is unnecessary: the methylene blue procedure can therefore be carried out directly.

### 4.1.5 Sulphide

Many gravimetric and iodometric procedures for the analysis of sulphide have been reported but most are of limited application in biological experiments. One of the most sensitive and specific methods involves the formation of methylene blue from *p*-aminodimethylaniline and has already been described on p. 65 in connection with the estimation of sulphate. The sample containing sulphide is acidified under a stream of oxygen-free nitrogen and the hydrogen sulphide trapped in a zinc acetate solution. Methylene blue is generated as described in the method for sulphate.

## Reaction with N-ethylmaleimide

Ellis (1964 a) described a method for the determination of sulphide based on the fact that sulphide forms an adduct with $N$-ethylmaleimide at pH 8 which gives a red colour on the addition of alkali (Margolis & Mandl, 1958; see also Broekhuysen, 1958). The sample is dissolved in 0·1 M phosphate or 0·1 M-tris at pH 8 and 0·5 ml of this solution, containing 10–250 n-moles of sulphide, is added to 0·5 ml. of 0·2 M-$N$-ethylmaleimide and incubated at 37° for 10 min: 1·5 ml. of 2 M-$Na_2CO_3$ is added and the incubation continued for a further 10 min at 37°. The optical density of 520 m$\mu$ is measured immediately against water: 0·1 $\mu$mole of sulphide gives an optical density of 0·42 at 520 m$\mu$ with a 1 cm light path.

At a given sulphide concentration the colour intensity is linearly proportional to the amount of $N$-ethylmaleimide added: the same batch of reagent, therefore, should be used for all samples, including standards, in any particular analytical run. $N$-Ethylmaleimide hydrolyses slowly in water and should be freshly dissolved before use. Slight changes in colour intensity also occur when the incubation times are increased or decreased; in addition the colour fades slowly at 20°. A standardized procedure should, therefore, be used and the absorbance determined immediately following the incubation with $Na_2CO_3$. Some interference is caused by thiosulphate and thiols; at equivalent concentrations the following relative absorbancies at 520 m$\mu$ are obtained: sulphide 100; thiosulphate 8·8; cysteine 6·6; glutathione 11·4. Optical densities (1 cm light path) of less than 0·02 are given by 10 $\mu$moles of sulphate, sulphite, cysteic acid, $S$-sulphocysteine and methionine; 10 $\mu$moles of dithionite give an optical density of 0·18.

## Preparation of standard sulphide solutions

Large crystals of $Na_2S.9H_2O$ (about 1·2 g) are rinsed with distilled water and then dissolved in one litre of cool, boiled, distilled water in a tightly stoppered flask. About 100 ml of the solution is pipetted into a flask containing 250 ml of water, 20 ml of 0·1 N-iodine and 25 ml 0·1 N-HCl. The excess iodine is titrated with standard thiosulphate using starch as an indicator. The iodine equivalent of the sulphide solution is calculated (1 ml of 0·1 N-iodine $\equiv$ 1·6 mg S) and dilutions made in boiled, cool distilled water to give the required sulphide concentrations. The stock and standard solutions are unstable and should be prepared immediately before using.

### 4.1.6 Elemental sulphur

The precise measurement of elemental sulphur in biological systems is difficult and few attempts have been made to carry out such assays. In the majority of cases insoluble sulphur, analysed as sulphate after oxidation, has been taken as a measure of 'elemental sulphur' but, particularly where intact cell preparations are used, the sulphur of insoluble cell constituents contributes to the value obtained. This test could be made more specific by prior extraction of the elemental sulphur with organic solvents such as $CS_2$ or benzene.

A number of methods for the estimation of elemental sulphur in inorganic and organic materials have been reported. Those based on the reaction of sulphur with sulphite to give thiosulphate (e.g. Morris, Lacombe & Lane, 1948) or with cyanide to give thiocyanate (e.g. Minatoya, Aoe & Nagai, 1935; Bartlett & Skoog, 1954) could, perhaps, be adapted to biological systems.

### 4.1.7 Sulphite

Iodometric methods are most commonly used for the analysis of sulphite. Direct titration of sulphite with iodine, however, usually gives low results due to air oxidation and loss of $SO_2$ by volatilization. The most satisfactory procedure involves the addition of a measured quantity of the sample to be analysed to an excess of iodine solution, in 5% acetic acid, followed by back titration of the remaining iodine with standard thiosulphate, with starch as an indicator. Sulphide, thiosulphate and other reducing agents interfere. Sulphite combines with formaldehyde to form formaldehyde sulphoxylate which is not attacked by iodine. If interfering reducing substances are not in great excess, therefore, the difference in the iodine titres in the presence and absence of formaldehyde gives a measure of sulphite.

#### Reaction with fuchsin formaldehyde

A sensitive method for sulphite is based on the formation of a red colour by the action of sulphur dioxide on fuchsin in the presence of formaldehyde and a strong mineral acid (Steigmann, 1942; Grant, 1947). Thiols, sulphide and thiosulphate interfere but the two former compounds can be removed by zinc or mercury ions. The following modification of the method is that of Quentin & Pachmayr (1961) as reported by Trüper & Schlegel (1964). The sample, containing up to about 3 $\mu$moles of sulphite is mixed with 20 ml of 2% (w/v) zinc acetate in a 100 ml volumetric flask and diluted to about 80 ml with water. Then 10 ml of 0·04% (w/v) basic fuchsin in 12·5% (w/v) $H_2SO_4$ are added and the flask shaken. After 10 min 1 ml of 32% (w/v) formaldehyde is added and the mixture brought to 100 ml with water. Exactly 10 min after the addition of formaldehyde the optical density at 570 m$\mu$ is measured.

Sulphite has also been measured by the decolorization of basic fuchsin in alkali (e.g. Kaji & McElroy, 1959).

## 4.1.8 Thiosulphate

Thiosulphate is classically determined by iodometric titration in 5% acetic acid using starch as an indicator. If volumes are kept small, about 2 $\mu$moles of thiosulphate can be measured accurately. The addition of formaldehyde (0·05 ml of 40% (w/v) per ml of titration mixture) eliminates interference from sulphite. This method is simple and rapid and is satisfactory for many purposes, particularly for the analysis of chromatographic column eluates and in experiments using intact bacterial cells. In experiments where animal tissues or bacterial extracts are used, however, interference from endogenous iodine-reacting substances may render such titrations impracticable.

### *Reaction with cyanide*

Sörbo (1957 *a*) described a colorimetric procedure for the estimation of thiosulphate based on the formation of thiocyanate when thiosulphate is mixed with cyanide in the presence of catalytic amounts of copper ions. The sample to be analysed is brought to the blue colour of thymolphthaleine (about pH 10) with NaOH. To 4·2 ml of the sample, containing 0·05–1·5 $\mu$moles of thiosulphate, 0·5 ml of 0·1 M-KCN and, after mixing, 0·3 ml 0·1 M-CuCl$_2$ are added. The CuCl$_2$ must be well mixed with the sample immediately after addition. A precipitate forms slowly but dissolves on the addition of 0·5 ml of ferric nitrate reagent (100 g Fe(NO$_3$)$_3$.9H$_2$O; 200 ml of 65% nitric acid and distilled water to 1000 ml). The optical density at 460 m$\mu$ is then determined and is corrected for a 'blank' obtained by adding to another aliquot of the sample, first the ferric nitrate reagent followed by KCN and CuCl$_2$: 1 $\mu$mole of thiosulphate gives a corrected optical density of about 0·58 in a 1 cm cell.

Positive interference is given by preformed thiocyanate, polythionates, sulphide and other thiol compounds. Thiocyanate interference is included in the blank value and does not cause difficulty if the concentration of thiocyanate is not too high. Tetrathionate and higher polythionates react spontaneously with cyanide to form thiocyanate and thiosulphate: they must therefore be determined separately and a correction applied (see below). Thiols and sulphide may be precipitated by cadmium ions which interfere only slightly with the assay.

### Reaction with p-benzoquinone

Thiosulphate reacts with p-benzoquinone in acid solutions to form a yellow-coloured quinone dithiosulphuric acid (Rzymkowski, 1925; Brauer & Staude, 1953; Sandved & Holte, 1940). This reaction has been utilized by Schöön (1959) for a colorimetric procedure for thiosulphate which is not affected by polythionates or by sulphite (when formaldehyde is present). In this method 10 ml of 1·7 M-acetic acid and 10 ml of mM-p-benzoquinone are added to 10 ml of the thiosulphate solution and the optical density at 400 m$\mu$ measured after 2 hr. The method is suitable for about 0·3–3 $\mu$moles of thiosulphate per 10 ml of sample but the relationship between thiosulphate concentration and optical density is not strictly rectilinear over this range and standard curves must be constructed. The optical density for a given concentration of thiosulphate appears to vary a little with different batches of p-benzoquinone.

### 4.1.9  Dithionate

No specific methods exist for the estimation of dithionate and it is generally assayed as sulphate after its separation and oxidation. Vasudeva Murthy (1953) has described a method in which all oxidizable sulphur compounds except dithionate are oxidized with alkaline permanganate (10–15 ml of saturated $KMnO_4$ and 10–15 ml of 10% NaOH). Excess permanganate is then destroyed by adding 2% manganous sulphate. After removal of the precipitated manganese dioxide the dithionate remaining in the solution is oxidized with excess potassium dichromate by boiling for 1 hr in the presence of 10 N-$H_2SO_4$. Excess dichromate is then determined iodometrically and the dithionate content calculated by the difference from a reagent blank determination.

### 4.1.10  Polythionates

A number of methods for estimating polythionates have been reported. Most of these involve the titration of thiosulphate formed by reactions of polythionates with sulphite, sulphide, cyanide or hydroxyl ions (see for example Kurtenacker & Goldbach, 1927; Goehring, Feldmann &

Helbing, 1949) and are not very satisfactory for the analyses of small amounts of polythionates. Methods based on the measurement of the acid produced when polythionates are heated with $HgCl_2$ were described by Starkey (1935 $a$) and by Jay (1953).

## Reaction with cyanide

Recently colorimetric methods for the estimation of tetrathionate and higher polythionates have been reported: these are based on the spontaneous formation of thiocyanate from polythionates and cyanide (Nietzel & DeSesa, 1955; Urban, 1961; Koh, 1965; Koh & Iwasaki, 1965, 1966 $a$) (equation 4.3).

$$S_nO_6^{2-} + (n-3)CN^- + OH^- \rightarrow S_2O_3^{2-} + HSO_4^- + (n-3)SCN^- \quad (4.3)$$

The method of Koh is described here. To 10 ml of the solution to be analysed are added 4·0 ml of 0·2 M-phosphate buffer, pH 7·4 and 2·6 ml of 0·05 M-NaCN: the final pH of the mixture should be between 8·0 and 8·8. The mixture is incubated at 40° for 30 min and then 3·0 ml of 1·5 M-ferric nitrate in 4 N-perchloric acid is added. The mixture is diluted to 25 ml and the optical density at 460 m$\mu$ read against distilled water (a blank containing all reagents other than polythionate would be preferable [see Koh & Iwasaki, 1966 $b$]). Calibration curves are prepared using standard KSCN solutions.

The method estimates 0·5–6 $\mu$equivalents of thiocyanate; one equivalent of thiocyanate is formed from tetrathionate, two from pentathionate and three from hexathionate. Thiosulphate itself does not interfere but those ions which interfere in the determination of thiosulphate do (see p. 71). The sensitivity of the method may be increased by reducing the volumes of the solutions and also by converting the thiosulphate formed in the reaction to thiocyanate and sulphate in the presence of copper ions (Koh & Iwasaki, 1966 $b$). In the latter case, however, preformed thiosulphate will cause positive interference.

Trithionate reacts with cyanide according to equation 4.4.

$$S_3O_6^{2-} + CN^- + OH^- \rightarrow SCN^- + HSO_3^- + SO_4^{2-} \quad (4.4)$$

This reaction is very much slower than those between cyanide and higher polythionates (Foss, 1945) but it is sufficiently rapid at 40° to cause interference in the method described above for the latter. Kelly, Chambers & Trudinger (1969) have found that Koh's method for higher polythionates works equally well at 5 to 10° at which temperature the reaction between cyanide and trithionate is negligible. The use of low temperatures is recommended, therefore, for the estimation of tetrathionate and higher polythionates when trithionate is also present. Kelly, Chambers & Trudinger (1969) have also found that trithionate can be estimated essentially by Koh's method using 5 ml of 0·1 M-KCN and heating at 95 to 100° for 45 min: the mixture is then cooled before the addition of the ferric nitrate reagent.

### Estimation of polythionates in acid solutions

The cyanolytic method is not suitable for the estimation of polythionates in strong acid solutions, for example eluates from ion exchange columns. Under these conditions oxidimetric titration with standard iodate is suitable (Andrews, 1903; Jamieson, 1915). The reaction for tri-, tetra-, penta- and hexathionate are given by equations 4.5–4.8).

$$S_3O_6^{2-} + 2IO_3^- + 2Cl^- \rightarrow 3SO_4^{2-} + 2ICl \tag{4.5}$$

$$2S_4O_6^{2-} + 7IO_3^- + 7Cl^- + 2H^+ \rightarrow 8SO_4^{2-} + 7ICl + H_2O \tag{4.6}$$

$$S_5O_6^{2-} + 5IO_3^- + 5Cl^- + 2H^+ \rightarrow 5SO_4^{2-} + 5ICl + H_2O \tag{4.7}$$

$$2S_6O_6^{2-} + 13IO_3^- + 13Cl^- + 6H^+ \rightarrow 12SO_4^{2-} + 13ICl + 3H_2O \tag{4.8}$$

The solution to be analysed is made 6 N with respect to HCl and cooled in ice water in a glass stoppered flask. The volume should be sufficiently large so that the concentration of HCl does not fall below 3 N during the titration. Chloroform or carbon tetrachloride (5 ml) is added and the mixture titrated with standard KIO₃. After each addition of KIO₃ the mixture is shaken. When the reductant is in excess, free iodine is liberated which imparts a violet colour to the chloroform or carbon tetrachloride layer. At the end-point when KIO₃ is in slight excess the iodine is oxidized to iodine monochloride and the violet colour is discharged. The method is suitable for about 1 μmole and upwards of trithionate and is increasingly more sensitive for the higher polythionates (see equations 4.5–4.8).

Iodate is a general oxidizing agent for many inorganic and organic reductants. This method therefore is only suitable for the estimation of polythionates once they have been separated from other oxidizable compounds.

### 4.1.11 Analysis of mixtures of inorganic sulphur compounds

Iodometric and colorimetric procedures for the analysis of complex mixtures of inorganic sulphur compounds have been described in the literature (see, for example, Starkey, 1935 a; Kurtenacker, 1938;

Goehring, Feldmann & Helbing, 1949; Koh & Iwasaki, 1966 b). These procedures rely on the use of two or more chemical reactions and simultaneous equations to discriminate between the components of the mixtures. They are suitable for simple mixtures containing mM amounts of material but are subject to considerable errors and uncertainties when small amounts of material are to be analysed and when more than two components are present in the mixtures. For more precise results mixtures should first be separated by chromatography or electrophoresis.

Kelly, Chambers & Trudinger (1969) have developed a method which is suitable for the micro-analysis of mixtures of tetrathionate, trithionate and thiosulphate. The method is based on the observations that when Koh's (1965) procedure for polythionates is used (see p. 73) only tetra-thionate reacts to form thiocyanate at 10°, both tetrathionate and thio-sulphate react at 10° in the presence of $0.01 M$-CuSO$_4$ while all three com-pounds are converted to thiocyanate by heating the mixture at 100° for 45 min and adding CuSO$_4$ after cooling.

## 4.2 Special techniques for the analysis of $^{35}$S-labelled compounds

Sulphur has three radioisotopes (table 4.1) but of these only sulphur-35 is sufficiently stable to be useful in biological experiments. The particle energy of $^{35}$S is in the same order as that of $^{14}$C and the conventional scintillation, Geiger and proportional techniques used for the latter are equally applicable to $^{35}$S. When planchette counting is used the samples should, if possible, be prepared by drying solutions of the material to be analysed so as to give infinitely thin layers. Alternatively the material may be oxidized and the $^{35}$S isolated as BaSO$_4$ which is then mounted on the planchette. Difficulties may be experienced, however, due to the tendency of BaSO$_4$ to adhere to the apparatus (filter funnels, centrifuge tubes, etc.) used in its isolation. Sulphide and sulphite solutions should not be dried directly on to planchettes owing to the possibility of losses

TABLE 4.1  *Properties of the radio isotopes of sulphur*

| Nucleide | Half life | Type of decay | Particle energy (meV) | γ-ray energy (meV) |
|---|---|---|---|---|
| $^{31}$S | 2·4 sec | $\beta^-$ | 4·5 | . |
| $^{35}$S | 87·1 days | $\beta^-$ | 0·167 | . |
| $^{37}$S | 5·0 min | $\beta^-$ and $\gamma$ | 1·6 (90%) and 4·3 (10%) | 2·6 |

of $H_2S$ or $SO_2$ respectively from the samples; likewise thiosulphate and polythionates may undergo some decomposition to $SO_2$, particularly if heat is applied during the drying process. Solutions containing reduced inorganic sulphur compounds should, therefore, be oxidized to convert the sulphur to sulphate. This is readily accomplished by saturating the solution with bromine which is then removed on a boiling-water bath in the fumehood. A more vigorous oxidation procedure, which is also suitable for a number of organic sulphur compounds, consists of boiling the sample with a mixture of 5 ml of concentrated HCl and 25 ml of bromine-saturated concentrated $HNO_3$ on a sand bath and evaporating to dryness. The residue is dissolved in water, filtered if necessary, and plated on planchettes for the assay of $^{35}S$.

### 4.2.1 The determination of $^{35}S$-labelled inorganic and ester sulphate

The measurement of $^{35}SO_4^{2-}$ has been much used by Dodgson's group in their extensive studies of the metabolism of sulphate esters. Their techniques were developed primarily for use with urine but they can easily be adapted to handle other body fluids or tissue extracts. The $^{35}SO_4^{2-}$ ions are precipitated as barium sulphate which is more convenient to handle than the benzidine sulphate used by Hawkins & Young (1954) but note must be taken of the tendency of barium sulphate to adhere to glass.

The method given below is that described by Lloyd (1961). Inorganic sulphate is determined as follows: to a 15 ml sample of the solution in a modified 50 ml centrifuge tube (Walkenstein & Knebel, 1957) are added 3 ml of 0·15 $M$-$K_2SO_4$ followed by 5 ml of 4 N-HCl and 4 ml of 10% (w/v) $BaCl_2$. After standing for 6 hr at 2° the precipitate of $Ba^{35}SO_4$ is centrifuged off, washed three times with 40 ml portions of water and finally with 40 ml of acetone before drying, in the centrifuge tube, in an oven at 110°. Combined inorganic plus ester sulphate is similarly determined. To a 15 ml sample of the solution is added 5 ml of 4 N-HCl and the mixture heated on a boiling-water bath for 4 hr: after cooling, 3 ml of 0·15 $M$-$K_2SO_4$ followed by 4 ml of 10% (w/v) $BaCl_2$ are added and the remainder of the procedure is carried out as described above.

The $Ba^{35}SO_4$ is plated at 'infinite thickness' on a planchette (area 1 cm²) and compressed with a machined steel plunger before counting with a conventional end-window Geiger–Muller tube. Corrections for background, decay and coincidence are made as usual.

### 4.2.2 Determination of isotope distribution in thiosulphate and polythionates

In experiments involving the use of $^{35}$S-labelled compounds it is often necessary to determine the distribution of $^{35}$S between the different sulphur atoms of polysulphur compounds such as thiosulphate and polythionates. The following methods may be used for this purpose.

*Degradation of thiosulphate with silver nitrate*
(Brodskii & Eremenko, 1954; Kelly & Syrett, 1966 *a*)

This method depends upon the reaction between thiosulphate and silver ions to give silver sulphide derived from the sulphane group and sulphate from the sulphonate group (equation 4.9).

$$\overset{*}{S}.SO_3^{2-} + 2Ag^+ + OH^- \rightarrow Ag_2\overset{*}{S} + HSO_4^- \tag{4.9}$$

Sufficient carrier thiosulphate should be added to the solution to be analysed to make it 0·01 M or greater with respect to thiosulphate. An aliquot of the solution is heated at 95° for 45 min with a slight excess of $AgNO_3$. The precipitated $Ag_2S$ is collected by centrifuging and washed twice with water. The supernatant and washings are combined and analysed for $^{35}$S (from the sulphonate group). For the determination of $^{35}$S from the sulphane group the $Ag_2S$ is dissolved in 2% (w/v) KCN, or oxidized to sulphate with $HCl-HNO_3-Br_2$ (see above); any precipitated AgCl or AgBr is removed by filtration prior to analysis of $^{35}$S.

*Degradation of polythionates with cyanide*
(Eremenko & Brodskii, 1955; Trudinger, 1961 *b*)

In the cyanolytic degradation of polythionates (see p. 73) thiocyanate and sulphate are formed from the sulphane and sulphonate groups respectively. The reaction in the case of tetrathionate and higher polythionates is given in equation 4.10 while that with trithionate is given in equation 4.11.

$$[_3OS.\overset{*}{S}_n.SO_3]^{2-} + (n-1)CN^- + OH^- \rightarrow \overset{*}{S}.SO_3^{2-} + (n-1)\overset{*}{S}CN^- + HSO_4^- \tag{4.10}$$

$$[_3OS.\overset{*}{S}.SO_3]^{2-} + CN^- + OH^- \rightarrow SO_3^{2-} + \overset{*}{S}CN^- + HSO_4^- \tag{4.11}$$

This provides the basis for the following method for discriminating between the two sulphur species in polythionates.

For tetrathionate and higher polythionates 1 ml of approximately 0·01 M-polythionate is made alkaline to phenolphthalein and mixed with 1 ml of 0·1 M-KCN: after 5 min at room temperature the products are separated by chromatography on Dowex $2 \times 8$ (acetate). Sulphate is eluted with 2 M-ammonium acetate, pH 5, thiosulphate with 5 M-ammonium acetate, pH 5, and thiocyanate with 2 N-HNO$_3$. Approximately 100 ml of each eluant is required for a $12 \times 1$ cm column; 4–5 ml fractions are collected. The elution is monitored by determining the $^{35}$S content of an aliquot of each fraction. The tubes corresponding to each radioactive peak are then combined and the total $^{35}$S determined.

For trithionate a similar procedure is followed except that cyanolysis is carried out at 90 to 100° for 45 min and the sulphite then oxidized by the addition of a slight excess of iodine to the cooled solution.

## Degradation of polythionates with mercuric chloride

(van der Heijde & Aten, 1952; Trudinger, 1961 b; Kelly & Syrett, 1966a).

Polythionates react with mercuric chloride according to equation 4.12.

$$2[_3OS.\overset{*}{S}_n.SO_3]^{2-} \quad + \quad 3HgCl_2 \quad + \quad 4OH^-$$

$$HgCl_2.2Hg\overset{*}{S} \quad + \quad 2(n-1)\overset{*}{S} \quad + \quad 4SO_4^{2-} \quad + \quad 4HCl$$

$$(4.12)$$

This reaction provides the basis of the determination of isotope distribution in polythionates.

To 1 ml of the solution to be analysed are added 0·5 ml of 40% formaldehyde followed by 0·5 ml of 5% (w/v) HgCl$_2$ in 2% (w/v) sodium acetate (formaldehyde is added to trap small amounts of sulphite which are produced by a secondary reaction). After 1 hr at room temperature the precipitate is collected by centrifuging and washed twice with 1 ml portions of 0·01 M-Na$_2$SO$_4$. The combined supernatants are mixed with 20 mg of unlabelled polythionate and the above procedure is repeated. The supernatants, containing sulphate and sulphite from the sulphonate groups, are combined and analysed for $^{35}$S. The precipitates of HgCl$_2$, HgS and sulphur are oxidized with concentrated HNO$_3$, saturated with bromine, and made to standard volume. Aliquots are plated on glass planchettes for determination of $^{35}$S derived from the sulphane groups.

## 4.3 Chromatographic and electrophoretic methods

### 4.3.1 Paper chromatography of inorganic sulphur compounds

Paper chromatographic methods for the separation and identification of inorganic sulphur compounds have been successfully employed in a number of studies on sulphur metabolism. Conventional techniques are used and some useful solvents are listed in table 4.2.

TABLE 4.2   *The $R_f$ values of some inorganic sulphur-anions in paper chromatography*

| Anion | Solvent | | | | | | | | | |
|---|---|---|---|---|---|---|---|---|---|---|
| | 1 | 2 | 3 | 4 | 5 | 6 | 7 | 8 | 9 | 10 |
| $S^{2-}$ | . | . | . | . | . | 0·46 | 0·38 | . | . | . |
| $SO_3^{2-}$ | . | . | . | . | . | 0·64 | 0·27 | . | . | . |
| $SO_4^{2-}$ | 0·18 | . | 0·02 | . | . | 0·76 | 0·38 | . | 0·11 | . |
| $S_2O_6^{2-}$ | . | . | . | . | . | 0·56 | 0·52 | . | 0·25 | . |
| $S_2O_3^{2-}$ | 0·23 | . | 0·03 | 0·43 | 0·37 | 0·58 | 0·40 | 0·26 | 0·17 | . |
| $S_3O_6^{2-}$ | 0·45 | 0·20 | 1·00 | 0·52 | 0·60 | 0·82 | 0·58 | 0·50 | 0·30 | 0·06 |
| $S_4O_6^{2-}$ | 0·59 | 0·30 | 1·43 | 0·53 | 0·67 | 0·88 | 0·63 | 0·66 | 0·34 | 0·12 |
| $S_5O_6^{2-}$ | . | 0·48 | . | . | 0·73 | 0·72 | 0·55 | 0·67 | 0·36 | 0·21 |
| $S_6O_6^{2-}$ | . | 0·60 | . | . | . | . | . | 0·77 | . | 0·28 |

Solvent systems and conditions

1  Pyridine-acetic acid-$n$-butanol-water (20:6:30:24). Descending chromatography on Whatman No. 1 paper (Trudinger, 1965).

2  $n$-Butanol-acetic acid-acetoacetic ester-water(10:2:1:7) (Scoffone & Carini, 1955).

3  $n$-Butanol-acetone-water (2:2:1). Descending chromatography on Whatman No. 1 paper (Skarżyński & Szczepkowski, 1959; Kelly, 1967, unpublished results).

4  $n$-Butanol-methanol-water (1:1:1) (Radiochemical Centre Data Sheet, DS 3552).

5  Pyridine-$n$-propanol-water (7:10:10). Descending chromatography on Whatman No. 1 paper (Trudinger, 1964 c). Note that Bighi & Trabanelli (1955 a, b) report the following $R_f$ values for this solvent system: $S_3O_6^{2-}$, 0·59; $S_4O_6^{2-}$, 0·36; $S_5O_6^{2-}$ 0·72; $S_6O_6^{2-}$, 0·54.

6  Ethanol-pyridine-water-30% (w/v) $NH_4OH$ (30:30:40:2·5). Ascending chromatography on Toyo No. 15 paper (Okuzumi & Imai, 1965).

7  Glycerol-$n$-butanol-acetone-water-30% (w/v) $NH_4OH$ (5:35:30:30:2·5). Ascending chromatography on Toyo No. 15 paper (Okuzumi & Imai, 1965).

8  $n$-Butanol-methylcellosolve (35:65) (Wood, 1954).

9  Isopropanol-water (75:25). Ascending chromatography on Arches No. 302 paper (Garnier & Duval, 1959).

10  $n$-Butanol-acetic acid-water (4:1:5). Ascending chromatography at 20° (Bighi, Trabanelli & Pancaldi, 1958).

79

## 'Rear phase' chromatography

Pollard, McOmie & Jones (1955) introduced a technique called 'rear phase chromatography' for the separation of inorganic sulphur compounds. Trithionate and tetrathionate form double zones in solvent systems containing potassium acetate due to the existence of two phase-regions on the chromatogram, a leading acid phase free from potassium ions and a rear phase, alkaline to methyl orange and containing potassium ions (Pollard, 1954; Pollard, McOmie & Stevens, 1951). Trithionate and tetrathionate give one zone in each phase-region. To overcome this effect Pollard et al. (1955) pre-eluted chromatograms with the solvent and applied the samples to be analysed after the phase boundary was well passed the starting line.

The chromatogram strips are folded at the bottom to prevent the solvent from running off the end of the paper and are then mounted in a chromatography tank which has been modified so that the first 4 cm of the paper at the top of the chromatogram is horizontal. The lid of the tank has a series of 1 mm holes (normally covered by glass slides) situated above the starting line, approximately mid-way along the horizontal portion of the paper, through which the samples are applied. The strips are equilibrated with solvent vapour for about 12 hr and then the solvent is poured into the troughs and allowed to run until the phase boundary is well passed the starting line (7–24 hr). The solvent used by Pollard et al. (1955) was t-butanol-acetone-water-potassium acetate (3:13:5: 0·1 g). Provided that it does not interfere with the subsequent detection of the chromatographed compounds, 1% (w/v) phenol red may be added to the solvent to indicate the position of the phase boundary.

At the end of the 'pre-elution' period, samples are applied to the starting line through the holes in the tank lid from capillaries which are

TABLE 4.3   *Rear phase chromatography: movement of anions*
(*Pollard, Nickless & Glover, 1964a*)

| Anion | $R_{S_4O_6^{2-}}$ | Anion | $R_{S_4O_6^{2-}}$ |
|---|---|---|---|
| $SO_4^{2-}$ | 0·02 | $SO_3^{2-}$ | 0·77 |
| $S^{2-}$ | 0·04 | $S_2O_8^{2-}$ | 0·89 |
| $S_2O_3^{2-}$ | 0·11 | $S_4O_6^{2-}$ | 1·00 |
| $S_2O_6^{2-}$ | 0·38 | $S_5O_6^{2-}$ | 1·25 |
| $S_3O_6^{2-}$ | 0·67 | $S_6O_6^{2-}$ | 1·43 |
| Sodium formaldehyde sulphoxylate | 0·71 | | |

held in contact with the paper for approximately 3 sec (the volume applied should be less than 5 $\mu$l). Development is then continued for 10–15 hr. The movements of a number of sulphur compounds relative to tetrathionate are shown in table 4.3.

### 4.3.2 Chromatography of sulphate esters

Solvents suitable for the paper chromatography of steroid sulphates have been described by, among others, Schneider & Lewbart (1956) and for the thin-layer chromatography of steroid sulphates by Crépy, Judas & Lachese (1964). Thin-layer chromatography of several types of sulphate ester was considered in detail by Wusteman, Dodgson, Lloyd, Rose & Tudball (1964). Tables 4.4 and 4.5 summarize some $R_f$ values for sulphate esters in both types of chromatography. In general, alkaline solvent systems are preferable to acid ones because of the greater danger of hydrolysis in the latter.

TABLE 4.4  *The $R_f$ values of some representative steroid sulphates in paper chromatography*

| Parent steroid | Position of sulphate group | Solvent system | | | | |
|---|---|---|---|---|---|---|
| | | 1 | 2 | 3 | 4 | 5* |
| 3$\alpha$-hydroxy-5$\alpha$-androstan-17-one | 3 | . | . | . | 0·70 | 29 |
| 3$\beta$-hydroxy-5$\alpha$-androstan-17-one | 3 | . | . | . | 0·70 | 22 |
| 3$\alpha$-hydroxy-5$\beta$-androstan-17-one | 3 | 0·36 | 0·39 | . | 0·70 | 24 |
| 3$\beta$-hydroxy-5$\beta$-androstan-17-one | 3 | . | 0·37 | . | 0·70 | 27 |
| 3$\beta$-hydroxyandrost-5-en-17-one | 3 | 0·33 | 0·37 | 0·77 | 0·70 | 17 |
| 3$\beta$-hydroxyandrost-5-ene-7,17-dione | 3 | . | . | 0·59 | 0·40 | 1 |
| 3$\beta$,7$\alpha$-dihydroxyandrost-5-en-17-one | 3 | . | . | 0·30 | . | . |
| | 3,7 | . | . | 0·12 | . | . |
| 3$\alpha$-hydroxy-5$\alpha$-androstan-11,17-dione | 3 | . | . | . | 0·48 | 3 |
| 17$\beta$-hydroxyandrost-4-ene-3-one | 17 | 0·16 | 0·19 | . | 0·70 | 7 |
| 3$\beta$-hydroxypregn-5-en-20-one | 3 | 0·37 | 0·43 | . | . | . |
| 21-hydroxypregn-4-ene-3,20-dione | 21 | . | 0·18 | . | . | . |

\* The figures for this solvent system are the distances (cm) migrated in 36 hr.

1  0·2% NH$_4$OH-ethyl acetate-*n*-butanol (200:175:25), Schneider & Lewbart (1956).

2  *n*-butyl acetate-*n*-butanol-10% formic acid (80:20:100), Lewbart & Schneider (1955).

3  *n*-butanol-toluene-3% NH$_4$OH (1:1:2), Starka, Sulcova & Silink (1962).

4  iso-amyl alcohol-0·88 NH$_4$OH-water (55:27:18), Baulieu (1962).

5  iso-propyl ether-ligroin-t-butanol-NH$_4$OH (5:2:3:1:9), Baulieu (1962).

TABLE 4.5   *The $R_f$ values of some sulphate esters in thin-layer chromatography*

| Parent compound | Position of sulphate group | Solvent system and support | | | | | | |
|---|---|---|---|---|---|---|---|---|
| | | 1 | 2 | 3 | 4 | 5 | 6 | 7 |
| 3-hydroxyoestra-1,3,5-trien-17-one | 3 | 0·54 | 0·71 | . | 0·03 | 0·07 | 0·05 | 0·08 |
| 3,17β-dihydroxyoestra- | 3 | 0·48 | 0·60 | . | . | . | . | . |
| 1,3,5-triene | 17 | 0·46 | 0·60 | . | . | . | . | . |
| | 3,17 | 0·16 | 0·27 | . | . | . | . | . |
| 3,16α,17β-trihydroxyoestra- | | | | | | | | |
| 1,3,5-triene | 3 | 0·35 | 0·48 | . | . | . | . | . |
| 3α-hydroxy-5β-androstan-17-one | 3 | . | . | . | 0·15 | 0·22 | 0·18 | 0·29 |
| 3β-hydroxyandrost-5-en-17-one | 3 | 0·49 | 0·66 | . | 0·09 | 0·14 | 0·14 | 0·23 |
| 3β-hydroxypregn-5-en-20-one | 3 | . | . | . | . | 0·27 | . | 0·34 |
| 3β,17α-dihydroxypregn- | | | | | | | | |
| 5-en-20-one | 3 | . | . | . | . | 0·09 | . | 0·12 |
| 17α,21-hydroxypregn- | | | | | | | | |
| 4-ene-11,20-dione | 21 | 0·44 | 0·58 | . | . | . | . | . |
| 3β-hydroxycholest-5-ene | 3 | 0·58 | 0·67 | . | . | . | . | . |
| methanol | . | . | . | 0·31 | . | . | . | . |
| ethanol | . | . | . | 0·38 | . | . | . | . |
| n-hexanol | . | . | . | 0·56 | . | . | . | . |
| cyclohexanol | . | . | . | 0·53 | . | . | . | . |
| phenol | . | . | 0·46 | . | 0·45 | . | . | . |
| p-nitrophenol | . | . | 0·57 | . | 0·52 | . | . | . |

1   Kieselgel: benzene-ethyl methyl ketone-ethanol-
    water (3:3:3:1)

2   Kieselgel: isopropanol-chloroform-methanol-    Wusteman, Dodgson, Lloyd,
    10 N-NH₄OH (10:10:5:2)                              Rose & Tudball (1964).

3   Kieselgel: isopropanol-chloroform-methanol-water
    (10:10:5:2)

4   DEAE-Cellulose: 0·5 M-acetate buffer, pH 4·75

5   DEAE-Cellulose: 1·0 M-acetate buffer, pH 4·75    Oertel, Tornero & Groot,
6   Ecteola-Cellulose: 0·5 M-acetate buffer, pH 4·75     (1964).
7   Ecteola-Cellulose: 1·0 M-acetate buffer, pH 4·75

### 4.3.3   Paper electrophoresis of inorganic sulphur compounds

Paper electrophoretic methods have not been extensively used in studies on the metabolism of inorganic sulphur compounds although quite useful separations of some classes of sulphur compounds can be achieved (table 4.6). In general high voltages and short duration times result in sharper bands and more reproducible results. Cooling may be facilitated by immersing the paper strips in a volatile organic fluid such as chloro-benzene (Wood, 1954) or carbon tetrachloride (Trudinger, 1961 b).

TABLE 4.6  *The mobilities of inorganic sulphur compounds in paper electrophoresis*

| Buffer | Paper | Tempera-ture | Applied voltage (V/cm) | Mobility of $S_2O_3^{2-}$ ($\mu$/sec/V/cm) | Mobilities relative to $S_2O_3^{2-} = 1$ | | | | | | | | | | | Ref. |
|---|---|---|---|---|---|---|---|---|---|---|---|---|---|---|---|---|
| | | | | | $S^{2-}$ | $SO_3^{2-}$ | $SO_4^{2-}$ | $S_2O_5^{2-}$ | $S_2O_8^{2-}$ | $S_3O_6^{2-}$ | $S_4O_6^{2-}$ | $S_5O_6^{2-}$ | $S_6O_6^{2-}$ | $SCN^-$ | |
| 0·1 M-$(NH_4)_2CO_3$ | Whatman 3 MM | 12–14° | 100·0 | 2·87 | 1·0 | 0·77 | 0·91 | 0·77 | 1·00 | · | · | · | · | 0·88 | 1 |
| 0·05 M-potassium hydrogen phthalate | * | 18–19° | 5·2 | 3·81 | · | · | · | · | · | 0·88–0·92 | 0·78–0·83 | 0·65–0·73 | 0·56–0·64 | · | 2 |
| 0·05 M-sodium citrate, pH 5·2 | Whatman No. 1 | ** | 19·6 | 3·65 | · | · | · | · | · | 0·95 | 0·85 | 0·77 | 0·68 | | 3 |
| 2% $(NH_4)_2CO_3$ | Arches No. 302 | * | † | * | · | 0·81 | 0·91 | 0·93 | · | | | | | 0·74 | 4 |
| 0·1 N-NaOH | Arches No. 304 | * | †† | ** | · | 0·70–0·78 | 0·82–0·91 | · | 0·96 | | | | | 0·81–0·87 | 5 |

\* Not recorded.
\** Paper immersed in a tank of CCl₄ at room temperature.
† 150 V for 1 hr, length of paper not recorded.
†† 240 V for 1 hr, length of paper not recorded.

1 Gross (1957).
2 Wood (1954).
3 Trudinger (1961b).
4 Lederer (1957).
5 Grassini & Lederer (1959).

83

### 4.3.4   Detection of ions on paper

*Inorganic ions*

A number of spray reagents for the detection of inorganic ions on paper are listed in table 4.7. The sensitivity of the silver method (no. 4 of table 4.7) for thiosulphate and polythionates is increased by carefully heating the sprayed papers in an oven at 110°. (Note, overheating causes blackening of the paper.) About $0.02\,\mu$mole of thionate over $1\,cm^2$ area is detected; somewhat greater sensitivity is achieved by viewing the spots under ultraviolet light. The spots are preserved by washing the papers successively in distilled water, 10% (w/v) $Na_2S_2O_3.5H_2O$ and water.

*Sulphate esters*

The methylene blue method (see p. 66) has been adapted by Crépy for the detection of the sulphate esters of steroids (and presumably also of other compounds) on chromatograms. Paper chromatograms are thoroughly dried, dipped in an acid solution of methylene blue (0·0125% (w/v) in 1% (w/v) $H_2SO_4$ containing 5% (w/v) $Na_2SO_4$), dried between strips of filter paper and then immersed in a dish of chloroform. On standing for about 20 min, and preferably with a second wash in chloroform, the areas containing sulphate esters appear as white spots on a blue background (Crépy & Judas, 1960). Amounts of about 10 n-moles of steroid sulphate can be detected in a spot 3 cm in diameter. Thin layer plates are sprayed with a solution of methylene blue (12 mg/100 ml of 0·025 N-$H_2SO_4$) when the areas containing steroid sulphates rapidly appear as coloured spots on a blue background (Crépy *et al.* 1964). Different steroids give rise to slightly differently coloured spots: androstenolone sulphate, for instance, gives a pink spot while androstenediol 3-sulphate gives a violet spot. If such plates are subsequently 'chromatographed' with chloroform then the areas containing the sulphate esters are again obvious as white spots on a blue background. Amounts as small as 2 n-moles of steroid sulphate can be detected.

Steroid sulphates have also been detected on paper chromatograms by their solvolysis in HCl-dioxan vapour at room temperature to give $SO_4^{2-}$ ions which can be detected by a modification of the rhodizonic acid test described above (Schneider & Lewbart, 1956). This method should obviously be applicable to the detection of any sulphate ester provided that a suitable method of hydrolysis can be found.

TABLE 4.7 *Spray reagents for the detection of inorganic sulphur compounds (Servigne & Duval, 1957; Baddiley et al. 1957; Garnier & Duval, 1959; Pollard, Nickless & Burton, 1962; Pollard, Nickless & Glover, 1964a; Okuzumi & Imai, 1965; Bowen & Cook, 1966)*

| Compound | Colour produced by reagent | | | | | | |
|---|---|---|---|---|---|---|---|
| | 1 | 2 | 3 | 4 | 5 | 6 | 7 |
| $S^{2-}$ | · | · | brown-black | · | black | · | yellow (yellow in UV) |
| $SO_3^{2-}$ | blue on brown | red | · | · | grey-black | · | light brown (green in UV) |
| $SO_4^{2-}$ | blue on brown | · | · | white on pink | · | · | green (dark blue in UV) |
| $S_2O_3^{2-}$ | · | · | brown-black | · | black | white on blue | action of UV forms dark red spot when viewed in visible light |
| $S_2O_6^{2-}$ | blue on brown | · | · | · | · | white on blue | green (dark blue in UV) |
| $S_3O_6^{2-}$ | blue on brown | · | black | · | black | white on blue | · |
| $S_4O_6^{2-}$ | blue on brown | · | brown-black | · | yellow | white on blue | · |
| $S_5O_6^{2-}$ | blue on brown | · | brown-black | · | · | white on blue | · |

Reagents

1  1% (w/v) benzidine in absolute ethanol followed by 0·001 M-$KMnO_4$ (limit of detection 8–10 µg).
2  5% (w/v) $ZnSO_4$ + 5% (w/v) sodium nitroprusside freshly prepared.
3  0·1 N-$AgNO_3$, aqueous or ammoniacal solution (see text).
4  0·15% (w/v) $BaCl_2$ followed by 0·25% (w/v) rhodizonic acid.
5  5% (w/v) $HgNO_3$.
6  Acetone-30% (w/v) HCl-30% $H_2O_2$ (75:4:3:0·17 to 100 ml with water). Dry at 50° for 40–50 min and spray with 3% (w/v) KI in 1% (w/v) starch.
7  1 part of 10% (w/v) $AgNO_3$ + 5 parts 0·2% (w/v) sodium fluoresceinate in absolute ethanol.

85

## 4.3.5   Ion exchange chromatography of sulphur compounds

Procedures for the separation of mixtures of inorganic sulphur compounds by ion-exchange chromatography are summarized in table 4.8. Conventional techniques are used except that the elutions should be performed at low temperature (below 10°) to minimize decomposition of

TABLE 4.8   *Separation of mixtures of inorganic sulphur compounds by ion exchange chromatography*

| Resin | Eluant | Compounds eluted | References |
|---|---|---|---|
| Diaion SA100 (100–200 mesh). Nitrate form, 4·9 g in 0·7 cm diam. column | (A) 0·1 M-NH₄NO₃ adjusted to pH 9·7 with NH₄OH+30% (v/v) acetone | $SO_3^{2-}$ (25–45 ml)* $S^{2-}$ (55–75 ml)* | (1) |
| | (B) 0·1 M-NaNO₃ | $SO_4^{2-}$ (35–60 ml)* | |
| | (C) 1 M-NaNO₃ | $S_2O_3^{2-}$ (5–20 ml)* | |
| Dowex 1×2 (50–100 mesh). Chloride form, 2·16 g in 0·9 cm diam. column | (A) 1 N-HCl | $S_2O_6^{2-}$ (35–60 ml)* | (2) |
| | (B) 3 N-HCl | $S_3O_6^{2-}$ (25–60 ml)* | |
| | (C) 6 N-HCl | $S_4O_6^{2-}$ (5–45 ml)* | |
| | (D) 9 N-HCl | $S_5O_6^{2-}$ (10–45 ml)* | |
| Deacidite FF (50–100 mesh). 20×0·75 cm column | (A) 2 M-potassium phthalate (250 ml) | $SO_3^{2-}$, $S_2O_3^{2-}$ (incomplete separation) | (3) |
| | (B) sodium acetate-HCl buffer pH 4 (250 ml) | | |
| | (C) 3 N-HCl (250 ml) | $S_3O_6^{2-}$ | |
| | (D) 6 N-HCl (250 ml) | $S_4O_6^{2-}$ | |
| | (E) 9 N-HCl (250 ml) | $S_5O_6^{2-}$ | |
| Dowex 1×2 (200–400 mesh). Acetate form, 12×1 cm diam. column | (A) 0·5 M-ammonium acetate pH 5 (100 ml) | $S^{2-}$, $SO_3^{2-}$ (5–90 ml)* (no separation) | (4) |
| | (B) 1 M-ammonium acetate pH 5 (100 ml) | $SO_4^{2-}$ (50–100 ml)* | |
| | (C) 2 M-ammonium acetate pH 5 (100 ml) | $S_2O_3^{2-}$ (20–80 ml)* | |
| | (D) Water (100 ml) | | |
| | (E) 3 N-HCl (100 ml) | $S_3O_6^{2-}$ (40–70 ml)* | |
| | (F) 6 N-HCl (100 ml) | $S_4O_6^{2-}$ (0–80 ml)* | |
| | (G) 9 N-HCl (100 ml) | $S_5O_6^{2-}$ (0–50 ml)* | |
| | (H) Conc. HNO₃ saturated with Br₂ (100 ml) | Higher polythionates, elemental sulphur† | |

*   Approximate elution volumes.
†   Oxidized and eluted as sulphate.

(1)   Iguchi (1958 a).
(2)   Iguchi (1958 b).
(3)   Pollard, Nickless & Glover (1964 b).
(4)   Trudinger (1964 b).

the polythionates. The fourth method in table 4.8 utilizes volatile ammonium acetate solutions to elute one- and two-sulphur compounds and has the advantage that the eluant can be removed by evaporation *in vacuo* at low temperatures thus facilitating analysis of small amounts of material. Owing to the volatility of sulphide and sulphite at pH 4 the fraction containing these substances should be eluted directly into iodine solution; sulphide and sulphite are oxidized to sulphur and sulphate respectively. The sulphur is separated by centrifuging, oxidized with bromine water and estimated as sulphate.

The methods described in table 4.8 have the disadvantage that the polythionates are eluted in strong acid solutions. They are therefore not suitable for preparative work on polythionates nor in cases where subsequent chemical tests on the isolated polythionates are necessary. Trudinger (1964 *b*) described the separation of mixtures of inorganic sulphur compounds by chromatography on Ecteola-cellulose using ammonium acetate solutions as eluants. With this technique polythionates can be recovered with minimum decomposition. The elution pattern depends, however, upon the composition of the mixture to be analysed and varies between different batches of Ecteola. The method is thus unsuitable for routine analytical work.

## 4.4 Direct spectroscopy of sulphur compounds

### 4.4.1 Inorganic compounds

Thiosulphate, polythionates, polysulphides and elemental sulphur (in organic solutions) absorb strongly in the ultra-violet region (see for example, Lorenz & Samuel, 1931; Ley & König, 1938; Baer & Carmack,

TABLE 4.9   *The detection of inorganic sulphur compounds by infra-red spectrophotometry (Garnier & Duval, 1959)*

| Compounds | Absorption bands (cm$^{-1}$) | Detection limit ($\mu$g/0·01 ml) |
|---|---|---|
| $SO_4^{2-}$ | 1103 | 50 |
| $S_2O_3^{2-}$ | 996 | 89 |
| | 1117 | 179 |
| $S_2O_6^{2-}$ | 991 | . |
| | 1233 | 50 |
| $S_3O_6^{2-}$ | 1099 | 50 |
| $S_4O_6^{2-}$ | 1221 | 50 |
| $S_5O_6^{2-}$ | 993 | 60 |

1949; Awtrey & Connick, 1951). This absorption is, however, not sufficiently specific to be useful for the estimation of these compounds in most situations, although the analysis of simple mixtures of poly-thionates by ultraviolet spectroscopy has been reported by Schmidt & Sand (1964 a). The identification of inorganic compounds on paper chromatograms by infra-red spectroscopy has been reported by Garnier & Duval (1959); their spectral data are shown in table 4.9. The infra-red spectra of polythionates has also been studied by Schmidt & Sand (1964 b).

### 4.4.2 Sulphate esters

Direct determination of sulphate esters is not possible by spectrophoto-metry in the visible or ultraviolet regions but the $-O \cdot SO_3^-$ group does show several absorption bands in the infra-red. Witmer & Austin (1960) have used that at $8 \cdot 02 \mu$ (1247 cm$^{-1}$), attributable to a vibrational mode of the S—O bond, for the quantitative determination of cerebroside sulphates. The method depended upon the quantitative separation of these lipids from the tissue and their incorporation into a KBr disk, the absorption of which was measured under carefully controlled condi-tions. Presumably this method could be a general one for the determina-tion of sulphate esters but it does not appear so far to have had much use for this purpose. A considerable improvement in the technique would presumably result if the spectrophotometry were carried out using a solution of the ester rather than a KBr disk: unfortunately suitable solvents might be hard to find although aqueous ethanol or methanol could be useful in many cases.

Infra-red spectrophotometry has been of greater use for structural studies of the sulphate esters (Lloyd, Dodgson, Price & Rose, 1961; Lloyd & Dodgson, 1961; Lloyd, Tudball & Dodgson, 1961) than for their quantitative analysis and it probably has been of most value in the field of carbohydrate sulphates. Matthews (1958), for instance, showed that chondroitin 4-sulphate could be distinguished from chondroitin 6-sulphate by infra-red spectrophotometry, the former having absorption bands at 928 cm$^{-1}$, 852 cm$^{-1}$ and 725 cm$^{-1}$ while the latter had bands at 1000 cm$^{-1}$, 820 cm$^{-1}$ and 775 cm$^{-1}$. These figures reflect one of the most useful features of this technique, the possibility of distinguishing between isomeric carbohydrate sulphates: sulphate esters derived from a primary equatorial hydroxyl group show an absorption band at 820 cm$^{-1}$ whereas those derived from a secondary equatorial hydroxyl group have the

TABLE 4.10 *Polarographic half-wave potentials of inorganic sulphur compounds*

| Compound | Conditions | Diffusion current | $E_{\frac{1}{2}}$ v. SCE† (volts) | References |
|---|---|---|---|---|
| sulphide | (a) 1 mM-$S^{2-}$, 0·1 N-NaOH | Anodic | −0·76* | 1,2 |
| | (b) 0·1 M-$Na_2HPO_4$+5% of aqueous 36% (w/v) HCHO | | −0·35 | 3 |
| elemental sulphur | (a) 25% (w/v) pyridine, 50% (w/v) ethanol | Cathodic | −0·63 | 4 |
| | (b) 0·044 M-sodium acetate; 0·44 M-acetic acid | | −0·50 | 5 |
| | (c) 90% ethanol, 0·05 N-$H_2SO_4$ | | −0·20 | 11 |
| | (d) 0·1 M-acetic acid, 0·1 M-sodium acetate in 1:1, $CH_3OH$:$C_6H_6$, pH 6·3 | | −0·58 | 12 |
| sulphite | (a) 1 mM-$SO_3^{2-}$, 0·1 M-$KNO_3$ | Anodic | −0·007* | 2 |
| | (b) 0·1 N-$HNO_3$ | | −0·38 | 2 |
| | (c) 0·04 M-acetate, pH 3·5, 25° | | −0·52 | 6 |
| thiocyanate | (a) 1 mM-$SCN^-$, 0·1 M-$KNO_3$ | Anodic | +0·18* | 2 |
| thiosulphate | (a) 1 mM-$S_2O_3^{2-}$, 0·1 M-$KNO_3$ | Anodic | −0·145* | 2,3 |
| | (b) 0·04 M-acetate, pH 3·5, 25° | | −0·15 | 6 |
| | (c) M-$(NH_4)_2HPO_4$, 17–22° | | −0·18 | 7 |
| trithionate | (a) M-$(NH_4)_2HPO_4$, 17–22° | Cathodic | −1·32* | 7 |
| | (b) 0·5% K.Na tartrate, 0·1 N-$BaCl_2$, 0·01% gelatin | | −1·4 | 3 |
| | (c) M-KCl, 50% ethanol | | −1·21 | 9 |
| | | | −1·34 | 10 |
| tetrathionate | (a) M-$(NH_4)_2HPO_4$, 2 mM-$S_4O_6^{2-}$ | Cathodic | −0·256* | 8 |
| | (b) N-HCl | | −0·28 | 7 |
| | (c) M-KCl, 0·01% gelatin, 50% ethanol | | −0·285 | 10 |
| | (d) 0·5% K.Na tartrate, 0·1 N-$BaCl_2$, 0·01% gelatin | | −0·300 | 10 |
| | | | −0·15 | 9 |
| pentathionate | (a) M-$(NH_4)_2HPO_4$, 17–22° | Cathodic | −0·67 | 7 |
| | (b) 2N-HCl, 50% ethanol | | −0·211 | 10 |
| | (c) 0·5% K.Na tartrate, 0·1 N-$BaCl_2$, 0·01% gelatin | | −0·27 | 9 |
| hexathionate | (a) 0·5% K.Na tartrate, 0·1 N-$BaCl_2$, 0·01% gelatin | Cathodic | −0·66 | 9 |

\* Slightly dependent upon concentration.

† SCE = Saturated calomel electrode.

1 Revenda, 1934.
2 Kolthoff & Miller, 1941.
3 Furness, 1950.
4 Poulton & Tarrant, 1951.
5 Hall, 1950.
6 De Ley & van Poucke, 1961.
7 Pankhurst, 1964.
8 Furness & Davies, 1952.
9 Cavallaro, Bighi, Pancaldi & Trabanelli, 1958.
10 Subrahmanya, 1955.
11 Gerber & Shusharina, 1950.
12 Hall, 1953.

corresponding band at 832 cm$^{-1}$. This type of analysis is of great value in the study of complex polysaccharide sulphates.

## 4.5 Polarography of inorganic sulphur compounds

Most reduced forms of sulphur yield well-defined diffusion currents with the dropping mercury electrode. Depolarization may be due either to reduction of the compound at the electrode as in the case of sulphite, sulphur and polythionates or to the formation of slightly soluble or complex compounds of mercury as is the case with thiocyanate, sulphide and thiosulphate. The reaction with thiosulphate for example is described by equation 4.13.

$$2S_2O_3^{2-} + Hg \rightarrow Hg(S_2O_3)_2^{2-} + 2e^- \qquad (4.13)$$

A general discussion of the depolarization reactions of sulphur compounds is given by Kolthoff & Lingane (1952).

Polarographic methods have been used by Green & Westley (1961) in an elegant study of the mechanism of rhodanese action (see chapter 8) and by De Ley & van Poucke (1961) and Pankhurst (1964) to identify the products of thiosulphate metabolism by thiobacilli (chapter 9). Bighi & Trabanelli (1955 $a, b$) used polarographic methods in conjunction with paper chromatography to separate and identify polythionates. Half-wave potentials of some inorganic sulphur compounds are listed in table 4.10.

# 5

## THE ACTIVATION OF SULPHATE IONS

The key intermediates in the metabolism of the inorganic compounds of sulphur are the two sulphate-containing nucleotides adenylyl sulphate and 3'-phosphoadenylyl sulphate which have already been described in § 2.7. The latter compound is the long-sought 'active sulphate', the existence of which had been suspected for many years and was essentially proven by the pioneer work of Bernstein & McGilvery (1952 b) and of De Meio, Wizerkaniuk & Fabiani (1953) on the synthesis of aryl sulphates by liver preparations. However, it was not until the work of Hilz & Lipmann (1955) and of Robbins & Lipmann (1956 a) that 'active sulphate' was isolated and its structure finally elucidated.

### 5.1    The biosynthesis of PAPS

This metabolic sequence was clarified almost entirely by the work of Wilson & Bandurski (1956) and of Robbins & Lipmann (1956 b, 1958 a) who showed that two enzymes, ATP-sulphurylase and APS-kinase, were involved, these two enzymes together making up the sulphate activating system. A third enzyme, ADP-sulphurylase, is present in some micro-organisms but the role, if any, of this enzyme in the formation of PAPS is obscure.

### 5.1.1    ATP-sulphurylase

The reaction catalysed by this enzyme can be represented by equation 5.1.

$$ATP + SO_4^{2-} \rightleftharpoons APS + PP_i \qquad (5.1)$$

It is generally regarded as a nucleophilic displacement by $SO_4^{2-}$ ions on the inner phosphorus atom of ATP with the elimination of pyrophosphate. As Kosower (1962) has pointed out, however, the low nucleophilicity of the $SO_4^{2-}$ ion and the repulsion to be expected between this and the negatively charged ATP, even allowing for the presence of a $Mg^{2+}$ ion chelated to the nucleotide, makes such a route improbable and he has suggested as an alternative mechanism the formation of an enzyme-AMP complex which subsequently reacts with $SO_4^{2-}$ ions to give APS. These two mechanisms should readily be distinguishable by detailed kinetic studies but unfortunately these have not yet been carried out.

The equilibrium for the reaction lies far in favour of ATP, the apparent equilibrium constant being of the order of $10^{-8}$ (Robbins & Lipmann, 1958 b; Akagi & Campbell, 1962 b) so that the standard free energy change of the reaction is approximately $+11,000$ calories. This implies that the sulphate group potential in APS must be of the order of 19,000 calories compared with a phosphate group potential in ATP of about 8,000 calories. With such an unfavourable equilibrium APS will only accumulate in the system if the pyrophosphate is continuously removed by, for instance, the action of pyrophosphatase.

The assay of ATP-sulphurylase is not without difficulty. Obviously the most direct method would be the determination of the amount of APS produced on incubation of the enzyme with ATP and $SO_4^{2-}$ ions but, as already pointed out, the unfavourable equilibrium of the reaction precludes the accumulation of significant amounts of this compound. The sulphurylase reaction must therefore be coupled with the APS-phosphokinase reaction (see § 5.1.2) to form PAPS which does accumulate. Sulphurylase activity can then be determined by measuring the incorporation of $^{35}SO_4^{2-}$ into PAPS. Other direct methods of assay are based on the measurement of the disappearance of pyrophosphate or the appearance of ATP which occurs when APS is incubated with pyrophosphate and ATP-sulphurylase: the only limitation of this method is the relative inaccessibility of APS. The simplest assay has been developed from the very important studies of Wilson & Bandurski (1958 b) on the effect of $SeO_4^{2-}$ ions, and other similar group VI anions, on the activity of ATP-sulphurylase. Essentially the method is based on the production of pyrophosphate, or better the orthophosphate derived therefrom, when ATP-sulphurylase is incubated with ATP and, for instance, $MoO_4^{2-}$ ions. The rationale of this assay is described below, but it must be kept in mind that very careful controls are required before the method can be used with any confidence because errors can easily be introduced through the action of an ATP-ase, or through a molybdate-catalysed fission of ATP. Another potential source of error is the displacement by $MoO_4^{2-}$ ions of phosphate ions bound to proteins (Wheldrake & Pasternak, 1965). Perhaps because of the difficulty of determining ATP-sulphurylase there is remarkably little direct information available on the occurrence of the enzyme in nature.

One preparation of ATP-sulphurylase which has been obtained in a reasonably purified state is that in bakers' yeast, the final product from which had a specific activity some thousand-fold greater than that of the initial extract and gave a single sharp peak in the ultracentrifuge (Robbins & Lipmann, 1958 b). The value for the sedimentation coefficient was not reported and no evidence for homogeneity with respect to sedimentation coefficient was presented. Nor indeed was the peak shown to be due to the enzyme and not to some extraneous protein. The enzyme has been detected in several other micro-organisms but has been partially purified only from *Desulfotomaculum nigrificans* and from *Desulfovibrio desulfuricans* (Akagi & Campbell, 1962 b). It has been shown to occur in wild-type *Neurospora* (Hilz & Lipmann, 1955) but it is lacking in certain sulphite-less mutants thereof (Ragland, 1959). In higher plants it has been detected in spinach chloroplasts (Asahi, 1964). In animal tissues ATP-sulphurylase is presumed to be of widespread distribution but there is direct evidence for its occurrence only in rat liver and colon (Sundaresan, 1966). Despite the commonly held view that the sulphate-activating system, and hence ATP-sulphurylase, occurs only in the soluble fraction of the cell, Sundaresan has shown that at least in rat liver and colon some ATP-sulphurylase is associated with the particulate fraction.

The ATP-sulphurylases of micro-organisms are apparently all rather stable proteins which require $Mg^{2+}$ ions as cofactors but which are not affected by SH reagents. The enzyme from the thermophile *D. nigrificans* is thermostable and can withstand heating at 60° for 1 hr under conditions where the sulphurylase from *Desulfovibrio* is completely inactivated in 3 min. The latter enzyme is, however, protected to some extent by the presence of its substrate, ATP. Little is known of the kinetic properties of the enzyme because of the difficulties caused by the unfavourable equilibrium which essentially precludes measurement of the initial rate of the forward reaction. What little information is available is for the reverse reaction, the pyrophosphorolysis of APS, but even here the saturating levels of the two substrates, APS and pyrophosphate ions, are too low to be measured. The reaction has a rather broad pH optimum between 7·5 and 9·0 (Robbins & Lipmann, 1958 b). Robbins (1962) has stated that in the overall production of PAPS by yeast extracts the $K_m$ for $SO_4^{2-}$ ions is 5 mM: this presumably represents fairly closely the $K_m$ for $SO_4^{2-}$ ions in the forward reaction of ATP-sulphurylase.

As far as the ATP-sulphurylases of animal tissues are concerned little is known of their properties but they appear to be considerably less stable than the microbial enzymes, probably being SH enzymes, although they also are to some extent heat-stable because De Meio, in his classical separation of the sulphate-activating enzymes, destroyed the sulpho-transferases by heating to 52° for 15 min (De Meio, Wizerkaniuk & Schreibman, 1955). Attempts to obtain the ATP-sulphurylases of animal tissues have in the past been quite unsuccessful but recently Levi & Wolf (1969) have described the extensive purification of this enzyme from rat liver, a purification of some thousand-fold from the original homogenate having been achieved. A partial purification (200-fold) of the enzyme from sheep liver has also been obtained (Panikkar & Bachhawat, 1968). Levi & Wolf (1969) have shown that, despite earlier claims, the ATP-sulphurylase of rat liver is not stabilized by sulphate ions, although it is by phosphate ions, and undoubtedly one of the reasons for the previous failures to purify this enzyme is its rapid inactivation in tris-HCl buffers, about 85% inactivation occurring overnight. The molecular weight of the enzyme is quite high, a value of about 900,000 having been obtained by gel filtration. Perhaps the most interesting observation by Levi & Wolf was that their most highly purified preparations were devoid of any radioactivity when they were prepared from rats treated with [$^{14}$C]vitamin A after having been rendered deficient in this vitamin. This seems to quite definitely exclude any role for vitamin A (or for any derivative thereof containing carbon atoms 6 and 7 of the vitamin) in the actual mechanism of the reaction catalysed by ATP-sulphurylase (see § 5.1.4).

Some preliminary observations on the kinetics of liver ATP-sulphurylase have been made by Levi & Wolf (1969) and their results are summarized in table 5.1 together with the scanty, but apparently very different, data for sheep liver and for yeast ATP-sulphurylases. These figures must be treated with some reserve, however, because in no case was it shown that

TABLE 5.1 *The $K_m$ values for the substrates of purified ATP-sulphurylases. It should be stressed that the significance of these values in terms of dissociation constants has not been shown*

| Substrate | $K_m$ (mM) for enzyme from | | |
| --- | --- | --- | --- |
| | Rat liver | Sheep liver | Yeast |
| ATP | 1·6 | . | . |
| SO$_4^{2-}$ | 0·1 | . | 5 |
| APS | 0·25 | 2·0 | * |
| PP$_i$ | 0·037 | 1·7 | * |

\* Too low to be measured by conventional methods.

the $K_m$ of any one substrate was independent of the concentration of the other. This is a particularly important omission because Levi & Wolf have proposed that their studies of the exchange-reactions catalysed by ATP-sulphurylase show that the reaction has a ping-pong mechanism (Cleland, 1963) which can be represented as follows:

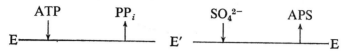

If this indeed be so then the $K_m$ values of the different substrates are not independent and the values given in table 5.1 can have no absolute significance. It is to be hoped that more detailed kinetic studies of this enzyme will soon be forthcoming so that the above mechanism, with its rather characteristic type of kinetics, can be confirmed or denied. Incidentally, such a mechanism is essentially that predicted by Kosower (1962).

The only specificity studies have been made with the enzymes from yeast (Robbins & Lipmann, 1958 a; Wilson & Bandurski, 1958 b) and from *D. nigrificans* or *Desulfovibrio* (Akagi & Campbell, 1962 b): all three sulphurylases are reported as being absolutely specific for ATP. As far as the second reactant, the inorganic ion, is concerned the important work of Wilson & Bandurski has already been mentioned. Briefly they showed that when yeast ATP-sulphurylase is incubated with ATP and a group VI anion other than the $SO_4^{2-}$ ion—such as the $SeO_4^{2-}$ or $MoO_4^{2-}$ ion—then the reaction went to completion in that all the ATP was converted to AMP and the inorganic ion remained unaltered in concentration. It was suggested that these results could be explained if an unstable anhydride link, analogous to the sulphatophosphate link of APS, were formed between the inner phosphorus atom of ATP and the group VI anion. Traces of such a nucleotide were in fact detected by the electrophoresis of reaction mixtures containing $^{95}SeO_4^{2-}$ ions although the compound was not isolated. This type of reaction was further studied by measuring the incorporation of labelled pyrophosphate into ATP when this was incubated with ATP-sulphurylase and a group VI anion. Only with $SO_4^{2-}$ and $SeO_4^{2-}$ ions did incorporation of pyrophosphate occur. Three slightly different types of reaction can therefore be catalysed by ATP-sulphurylase:

1. With $SO_4^{2-}$ ions pyrophosphate exchange occurs and APS accumulates in amounts predicted from the exchange reaction. The reaction can be represented as

$$ATP + SO_4^{2-} \rightleftharpoons APS + PP_i \qquad (5.2)$$

2. With $SeO_4^{2-}$ ions pyrophosphate exchange occurs but only traces of the anhydride can be detected, in amount far less than that expected from the pyrophosphate exchange. The reaction proceeds slowly to completion and can be represented as

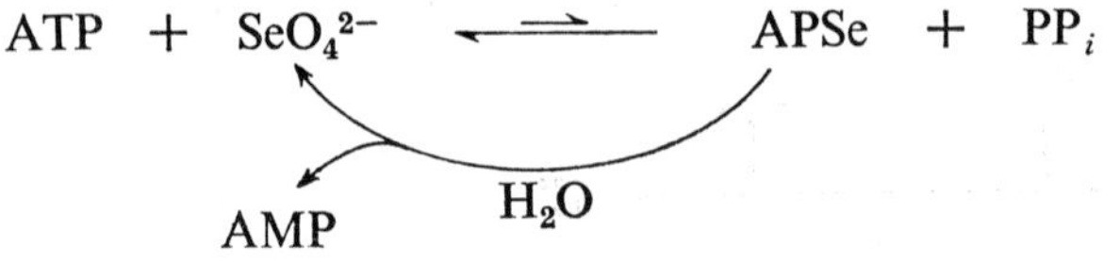

$$ATP \; + \; SeO_4^{2-} \; \longleftrightarrow \; APSe \; + \; PP_i \tag{5.3}$$

$$AMP \qquad H_2O$$

3. With $SO_3^{2-}$, $CrO_4^{2-}$, $WO_4^{2-}$ or $MoO_4^{2-}$ ions no pyrophosphate exchange occurs and no anhydride can be detected. The reaction proceeds rapidly to completion and can be written as

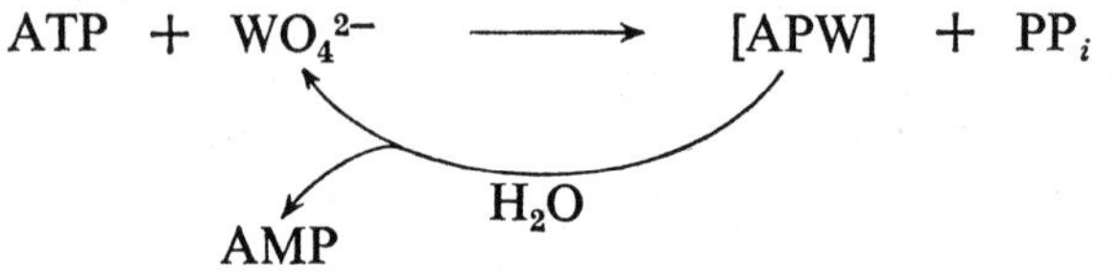

$$ATP \; + \; WO_4^{2-} \; \longrightarrow \; [APW] \; + \; PP_i \tag{5.4}$$

$$AMP \qquad H_2O$$

In reaction 5.4 the 'anhydride', if it exists at all, must have a half-life too short to allow pyrophosphate exchange to occur. The only alternative explanation would seem to be that the reaction is in fact a direct anion-catalysed fission of ATP, reminiscent of the arsenate-catalysed decomposition of some phosphate esters. Whatever be the exact mechanism of the reaction it certainly has provided a very useful means of determining the activity of ATP-sulphurylase not only in yeast but also in animal tissues. The limitations of the method must, however, always be kept in mind.

Once again there is little information on the kinetics of reaction 5.4. Other nucleoside triphosphates (UTP, CTP and GTP) could not substitute for ATP in the reaction and both APS and $SO_4^{2-}$ ions were inhibitors of the molybdolysis, 20 mM $SO_4^{2-}$ ions inhibiting by 67% the reaction with 10 mM $MoO_4^{2-}$ ions (Wilson & Bandurski, 1958 $b$). Robbins (1962) has reported that the molybdolysis of ATP catalysed by the ATP-sulphurylase of yeast attains half its maximum velocity in 0·5 mM-$MoO_4^{2-}$ ions and that the velocity of the reaction slowly increases between pH 7 and pH 9. For the same reaction extracts of *D. nigrificans* gave an optimum pH of about 7 but with preparation from *D. desulfuricans* the rate of the reaction increased up to pH 8 and then remained constant over the pH range studied (up to pH 9·5).

This lack of kinetic data for any of the reactions catalysed by ATP-sulphurylase is unfortunate because it prevents any conclusions being drawn about the detailed mechanism of the reactions. It is to be hoped that this lack will soon be filled.

### 5.1.2   APS-kinase

This enzyme catalyses the reaction shown in equation 5.5. It is an essentially irreversible reaction with a standard free energy change of $-5,000$

$$APS + ATP \rightarrow PAPS + ADP \qquad (5.5)$$

calories. The only enzyme of this type which has been purified to any extent is that from yeast (Robbins & Lipmann, 1958 a): it required the presence of $Mg^{2+}$ ions as cofactor, had a pH optimum between 8·5 and 9·0 and was completely saturated with APS at the lowest concentration of APS tested, 5 $\mu$M. Above this concentration substrate inhibition was pronounced. This ability to function at low substrate concentrations, and the irreversible nature of the reaction, makes APS-kinase a most efficient means of trapping the very small amounts of APS produced through the action of ATP-sulphurylase, allowing PAPS to accumulate up to concentrations of about 1 mM in reaction mixtures (Brunngraber, 1958). Adams & Rienits (1961) have reported that the APS-kinase of chick embryo cartilage is rather powerfully inhibited by galactosamine, but not by glucosamine nor by the two corresponding N-acetyl sugars.

The enzyme does not appear to be absolutely specific, certainly in so far as all the nucleoside triphosphates which have been tested have been found to act as phosphate donors (Robbins & Lipmann, 1958 a): unfortunately the presence of nucleosidediphosphate kinase in the enzyme preparations was not excluded. Less is known of the specificity with respect to the phosphate acceptor but there is indirect evidence that here too the specificity may not be absolute. Wilson & Bandurski (1958 b) showed that when APS-kinase was incubated with ATP-sulphurylase, ATP and $SeO_4^{2-}$ ions there was an increased production of phosphate ions, suggesting that the small amounts of APSe produced by the ATP-sulphurylase were being effectively removed through the formation of PAPSe, so driving the sulphurylase reaction in the forward direction to form pyrophosphate, and then orthophosphate. However, as neither PAPSe nor its expected degradation product adenosine 3',5'-diphosphate could be detected this interpretation of the finding must remain in some doubt.

The distribution of the enzyme is relatively unstudied. It has been isolated only from yeast but it must be presumed to occur in all those animal tissues which can form PAPS and so to have rather a wide distribution. It should be noted, however, that it need not always accompany ATP-sulphurylase: the latter enzyme occurs, for instance, in *Desulfovibrio desulfuricans* whereas APS-kinase does not (Wheldrake & Pasternak, 1965).

### 5.1.3   The sulphate-activating system

The above enzymes, ATP-sulphurylase and APS-kinase, together form the sulphate-activating system which is of very widespread distribution in nature and which catalyses the overall reaction represented by equation 5.6.

$$2ATP + SO_4^{2-} \rightarrow ADP + PAPS + PP_i \qquad (5.6)$$

Although the component enzymes have been isolated, or even identified, in only a few cases it seems justifiable to assume that they occur in all tissues which form PAPS, and only a little less justifiable to assume that they occur in all tissues which can form sulphate esters in the presence of ATP and $SO_4^{2-}$ ions because the only known route of this reaction is through PAPS (see chapter 6). Accepting such evidence, the sulphate-activating system must be present in most mammalian tissues including adrenal, brain, cartilage, cornea, heart muscle and valves, intestinal mucosa, kidney, liver, lung, mast cells, muscle, ovary, pancreas, placenta, retina, skin, spleen and several types of tumor. It also occurs in hen oviduct (Suzuki & Strominger, 1960) and in frog liver (Bridgwater & Ryan, 1957). Among the invertebrates, it has been shown to occur in snails (Goldberg & Delbruck, 1959; Yoshida & Egami, 1965) and some sea urchins (Creange & Szego, 1967). In the plant kingdom it occurs in higher plants (Nissen & Benson, 1961), in *Euglena gracilis* (Abraham & Bachhawat, 1963; Davies, Mercer & Goodwin, 1966), and in marine algae (Goldberg & Delbruck, 1959). In micro-organisms the system seems to be of quite general occurrence.

In mammals little is known of the quantitative distribution of the system, or of variations in its activity under physiological or pathological conditions. Wengle (1963) and Carroll & Spencer (1965 *a*) have reported that the activity of the PAPS-forming system in the livers of foetal and of new-born rats is much lower than in the livers of adult animals, a finding which Carroll & Spencer (1965 *b*) relate to the low vitamin A content of foetal rat liver, as is discussed later (§ 5.1.4). Balasubramanian

& Bachhawat (1961) have claimed that the ability of the rat brain to form PAPS reached a maximum at twelve days after birth, at the time when myelination is at a maximum: it must be admitted, however, that their published figures do not show this very convincingly. Nevertheless there is other evidence for a heightened activity of the sulphate-activating system in growing tissues. Gerlach (1963), for instance, has clearly shown that the heart muscle of young rats has a high level of sulphate activation which drops rather sharply when the body weights of the animals attain between about 50 g and 150 g. Sasaki (1967) has likewise shown that in granulation tissue produced in response to wounding the level of sulphate activation is highest about six days after the initial stimulus.

When the sulphate-activating system is assayed *in vitro* the activity of the preparation may be influenced by the behaviour of other enzyme systems therein. For example, Spolter, Rice, Yamada & Marx (1967) have shown that the addition of $NAD^+$ stimulates the formation of PAPS by a mitochondria-free enzyme preparation from a mouse mastocytoma. This is probably the result of the $NAD^+$ stimulating glycolysis and so giving rise to relatively high concentrations of ATP which are known to favour the formation of PAPS from APS (Suzuki & Strominger, 1960). In a typical experiment without added $NAD^+$ the ratio PAPS/APS was 3, in the presence of $NAD^+$ it was 15. Finally, an unexplained observation, must be mentioned. Barker, Cruickshank & Webb (1965) claimed that the PAPS formed by rat skin or by guinea pig skin *in vitro* was non-dialysable and associated with amino acids: they therefore suggested that in this preparation the PAPS was bound to a peptide chain. Perhaps it should be recalled that Panikkar & Bachhawat (1968) have claimed that the APS formed by their preparation of ATP-sulphurylase from sheep liver was partially protein-bound. They implied that the APS was bound to ATP-sulphurylase and that in this form it was a substrate for APS-kinase. Their own evidence is, however, not in accord with this view. In particular, chromatography on Sephadex showed that the 'bound' APS was eluted after the void volume from a column of Sephadex G-75 yet the enzyme itself was eluted at the void volume from Sephadex G-200; the APS must therefore either be bound to a protein smaller than ATP-sulphurylase or the latter must disaggregate during the reaction catalysed by it. It is to be hoped that further investigations will be made both of this phenomenon and of the claim (Barker *et al.* 1965) that rat skin contains a sulphated cytidine nucleotide.

There is now some understanding of the control of the sulphate-activating system in micro-organisms. Pasternak (1962) has shown that in *E. coli* the formation of PAPS is repressed by cysteine, and in *Bacillus subtilis* by both cysteine and glutathione. Both enzymes of the activating system appear to be repressed simultaneously (Wheldrake & Pasternak, 1965) and there is an inverse relationship between the intracellular concentration of cysteine and the specific activity of the sulphate-activating system (Wheldrake, 1967). Such a control mechanism is by no means universal because it does not exist in *Desulfovibrio* in which ATP-sulphurylase is not repressed by either cysteine or by sulphite.

### 5.1.4 Vitamin A and sulphate activation

It is now apparent that a deficiency of vitamin A causes alterations in the metabolism of $SO_4^{2-}$ ions in the mammal, but the exact nature of these effects and the way in which they arise are far from clear (Roels, 1967). The initial observation was that of Wolf & Varandani (1960) who showed that the ability of colonic segments from rats to incorporate $^{35}SO_4^{2-}$ ions and [$^{14}$C]glucose into mucopolysaccharides *in vitro* was greatly depressed during vitamin A deficiency. The activity of preparations from such deficient rats could be restored to normal by the addition of retinol, retinal or retinoic acid *in vitro*. Further study of preparations of pig colonic mucosa localized the defect which occurred in vitamin A deficiency to somewhere in the metabolism of $SO_4^{2-}$ ions (Wolf, Varandani & Johnson, 1961) and this was further restricted to the formation of PAPS (Varandani, Wolf & Johnson, 1960). In tissue preparations from animals deficient in vitamin A the rate of synthesis of PAPS was very low but could be restored to normal by the addition of retinol *in vitro*.

This work was essentially confirmed by the results of Subba Rao, Sastry & Ganguly (1963) and Subba Rao & Ganguly (1964) who showed a decreased ability to form PAPS in the liver and colon of rats deficient in vitamin A. Again this activity could be restored to normal by the addition of retinol (2·5 $\mu$g/ml) or of retinoic acid (10 $\mu$g/ml) *in vitro*. Unfortunately this relatively simple picture was somewhat complicated by the results of Pasternak, Humphries & Pirie (1963). These authors could detect no change from normal in the ability to form PAPS of colonic mucosa or corneal epithelium from rats or rabbits deficient in vitamin A. The activation of sulphate in rabbit corneal stroma was possibly decreased in vitamin A deficiency, but no reactivaction was given by the addition of vitamin A *in vitro*. Subba Rao & Ganguly (1964) have pointed out that the criteria of vitamin A deficiency used by

100

Pasternak—namely changes in the cornea and the disappearance of vitamin A from the liver—are not adequate and that the experimental animals were perhaps in fact not deficient. On the other hand, Hall & Straatsma (1966) have shown that in rats deficient in vitamin A by the criteria of Subba Rao & Ganguly the synthesis of PAPS in the liver is decreased whereas that in the retina is actually increased. To add to the confusion, Mukherji & Bachhawat (1967) claim that a deficiency of vitamin A does not depress the activity of the sulphate-activating system in rat liver, colon or brain but does do so in epiphyseal cartilage!

The low activity of the sulphate-activating system in preparations from the livers of foetal rats has been ascribed by Carroll & Spencer (1965 b) to a deficiency therein of a factor derived from vitamin A and they claim a restoration of the system to the normal adult values by the addition of retinol or retinoic acid *in vitro* in concentrations of about 10 $\mu$g/ml. Unlike all other workers in this field, they have also obtained an activation, by some 25%, of the sulphate-activating system in preparations from adult rat liver by the addition of vitamin A *in vitro*. In preparations from foetal guinea pig liver the sulphate-activating activity is only about 20% of the adult value but this is not altered by the addition of vitamin A *in vitro*. Spencer ascribes these different results in the two species to differing levels of vitamin A in the foetal tissues but gives no figures to support this contention.

The effect of vitamin A deficiency on the formation of PAPS has now apparently been further localized by the studies of Sundaresan (1966) who has shown that the ATP-sulphurylase activity, measured by all three of the techniques discussed above (§ 5.1.1), was severely depressed in the liver and colon of vitamin A deficient rats. The activity could not be restored to normal by the addition of retinol or of retinoic acid *in vitro*. On the other hand, the activity of the preparation from deficient animals could be restored to normal by the addition of an acidic lipid-soluble factor prepared from extracts of the liver of normal animals by extraction with butanol at pH 5. This factor was shown, by chromatography, to be not identical with retinol or retinoic acid. Nevertheless, when an ATP-sulphurylase preparation isolated from rats which had been dosed with [14C]vitamin A was chromatographed on DEAE-cellulose there was a radioactive peak corresponding to the peak of sulphurylase activity. This work certainly suggested that the activity of ATP-sulphurylase was dependent upon the presence of vitamin A or a metabolite thereof but, as Sundaresan pointed out, it was not possible to say that this dependency implied a co-factor requirement. The more recent work of Levi & Wolf (1969) has shown that the ATP-sulphurylase of rat liver can be obtained free from vitamin A, or any derivative of vitamin A still retaining carbon atoms 6 and 7, so that some other explanation must be sought for the undoubted influence of vitamin A on the activity of ATP-sulphurylase in animal tissues. Perhaps some relatively non-specific stabilizing effects are involved because it is striking that the enzyme is closely associated with a derivative of vitamin A during the early stages of its purification.

It is obvious that in these studies of the effect of vitamin A there are many discrepancies yet to be explained. Firstly there is the failure of Pasternak to detect any decreased ability to form PAPS during vitamin A deficiency: this is perhaps, but not necessarily, explicable by the experimental animals not being truly deficient in the vitamin. However, there are other reports which at least partially confirm this finding. Secondly, there is the discrepancy between the results of Sundaresan and of Subba Rao with respect to the reversal of the effects of vitamin A deficiency by the addition of retinol or retinoic acid *in vitro*: Sundaresan obtains no such reversal whereas Subba Rao does. One explanation of this discrepancy could be that the enzyme preparations of Sundaresan are slightly more purified than those of Subba Rao so that the former may be lacking in an enzyme necessary for the conversion of vitamin A into the unknown acidic 'factor'. An alternative explanation has been offered by Subba Rao & Ganguly (1966) themselves in so far as they have shown that reversal by retinol *in vitro* of the effects of vitamin A deficiency is only possible during the early stages of the deficiency: in the later stages the effects are irreversible. Now that Levi & Wolf (1969) have shown that vitamin A is not directly involved in the sulphurylase reaction it seems clear that much more detailed studies will be required to resolve these discrepancies.

Obviously highly pertinent is recent work which shows that the level of ATP-sulphurylase in rat tissues is highly dependent upon the nutritional state of the animal: starvation for only two days causes a considerable drop in the activity of the enzyme (Levi, Geller, Root & Wolf, 1968; Geison, Rogers & Johnson, 1968). Even in this work there are, however, discrepancies. The latter authors state that when such nutritional effects are minimized, the levels of ATP-sulphurylase in normal and in vitamin A deficient rats are identical within experimental error while the former group claims, on the contrary, that under similarly controlled conditions a deficiency of vitamin A does cause a drop in the activity of ATP-sulphurylase. This drop occurs in all tissues studied but is particularly obvious in the adrenal glands. Again further investigations of this problem are obviously required.

### 5.1.5 ADP-sulphurylase

Again this enzyme was first described by Robbins & Lipmann (1958 $a$) and partially purified by them from yeast. The reaction catalysed by the enzyme is shown in equation 5.7 but it must be stressed that the forward

$$ADP + SO_4^{2-} \rightleftharpoons APS + P_i \tag{5.7}$$

reaction, the formation of APS, has never been studied and the enzyme has been detected only through the reverse reaction. ADP-sulphurylase is easily differentiated from ATP-sulphurylase either by the actual separation of the two activities or by the fact that the activity of the former enzyme is not dependent upon the presence of $Mg^{2+}$. Further, ADP-sulphurylase does not catalyse a reaction between ADP and group VI anions (Peck, 1962 $a$), although it can catalyse the arsenolysis of APS.

The significance of the enzyme in metabolism is obscure, at least in the yeast from which it was isolated, but Peck (1960) has suggested that in *Thiobacillus thioparus* its function is in fact the formation of ADP from APS and phosphate ions and therefore, indirectly, the formation of ATP by this organism (see chapter 9).

Recently Grunberg-Manago, Campillo-Campbell, Dondon & Michelson (1966) have detected an enzyme in yeast extracts which catalyses an exchange between the terminal phosphate of ADP and orthophosphate ions.

$$ADP + {}^{32}P_i \rightleftharpoons AMP.{}^{32}P + P_i \tag{5.8}$$

The enzyme was quite non-specific with regard to the nucleotide, which could be either a ribotide or a deoxyribotide, and as well as catalysing an exchange with phosphate ions it also did so with $SO_4^{2-}$ ions. On these grounds it was claimed that the enzyme involved was in fact ADP-sulphurylase but such an identification requires further confirmation before it can be accepted.

### 5.2 The enzymatic degradation of APS and PAPS

This is a matter of considerable practical importance because of the quite general occurrence of enzymes capable of degrading PAPS in animal and plant tissues. These enzymes can cause considerable difficulties when crude tissue preparations are used to prepare PAPS, or as a source of enzymes capable of utilizing this nucleotide for synthetic reactions.

The enzymatic hydrolysis of PAPS was used by Robbins & Lipmann (1957) in their studies of the structure of this compound. In particular they used the purified 3'-nucleotidase of rye grass in their determination of the position of the phosphate group added during the reaction catalysed by APS-kinase: this nucleotidase hydrolysed PAPS to adenylyl sulphate, thus showing that the kinase had caused the phosphorylation of the 3' position of the ribose. Robbins & Lipmann also showed that the 5'-nucleotidase of bull semen could attack PAPS with the formation of adenosine, phosphate ions and $SO_4^{2-}$ ions. These same products were produced through the action of the venom of *Crotalus atrox* on APS (Baddiley, Buchanan & Letters, 1957): presumably this hydrolysis was caused by one or more of the many nucleotidases and phosphatases present in snake venom.

Liver contains a 3'-nucleotidase which will hydrolyse PAPS, presumably to APS although this product was not isolated (Brunngraber, 1958). This nucleotidase differs from other similar enzymes in requiring the presence of a charged group at the 5' position.

Another type of enzyme also attacks the sulphate-containing nucleotides. This is the so-called PAPS-sulphatase which splits the sulphato-phosphate bond. It is, however, by no means certain that the enzyme responsible is a sulphatase and information on the actual bond split would be most useful. As discussed in chapter 7, sulphatases split the O—S bond of their substrates and it would be useful to determine whether this also holds in the case of PAPS-sulphatase. If it should be the P—O bond which is split, then it would appear better to consider the enzyme as being a phosphatase rather than a sulphatase.

The activity attributed to PAPS-sulphatase, the hydrolysis of PAPS to adenosine 3'5'-diphosphate and $SO_4^{2-}$ ions, has been detected in hen oviduct (Suzuki & Strominger, 1960), in several mammalian tissues (Spencer, 1960 b; Balasubramanian & Bachhawat, 1962; Balasubramanian *et al.* 1967), in serum (Adams, 1964 c), in the mucous gland of the marine mollusc *Charonia lampas* (Yoshida & Egami, 1965) and in *Euglena gracilis* (Abraham & Bachhawat, 1964) but very little purification of the enzyme has been achieved. A convenient property of PAPS-sulphatase is its rather powerful inhibition by either phosphate ions or $F^-$ ions, a property which has been utilized when the action of PAPS-sulphatase had to be minimized (Suzuki & Strominger, 1960). According to Balasubramanian

& Bachhawat (1962) the PAPS-sulphatase of sheep brain is activated by $Co^{2+}$ or $Mn^{2+}$ ions and inhibited by SH-containing compounds or by ADP. Much further work is needed to establish the true nature of the enzyme responsible for the fission of the sulphatophosphate link and indeed to show whether both APS and PAPS are substrates for the enzyme. Despite this, there can be no doubt of the practical importance of the elimination of this enzyme in all studies involving the formation or utilization of PAPS in relatively crude tissue preparations.

### 5.3   Other sulphate-containing nucleotides

Only one other sulphate-containing nucleotide has been found in nature: this is a compound of 6-succinyladenosine 5'-sulphatophosphate and a peptide containing equal numbers of glutamic acid and serine residues, the nucleotide and the peptide being linked in an unknown manner (Tsuyuki & Idler, 1957). This most interesting compound was isolated from extracts of cod liver by chromatography in refrigerated columns: if the columns were not refrigerated then the compound apparently decomposed to 6-succinyladenylic acid. Treatment of the nucleotide with 0·01 N-HCl at room temperature for 1 hr removed the peptide but apparently left intact the remainder of the molecule, including the sulphatophosphate link. No further information on this compound has appeared since its original description, despite its very obvious interest.

Synthetic analogues of APS have been prepared chemically and their biochemical properties have been investigated. Ishimoto & Fujimoto (1961) prepared guanididyl sulphate, cytididyl sulphate and urididyl sulphate and showed that all three were substrates for the APS-reductase of *Desulfovibrio* (see chapter 10). More recently Yount *et al.* (1966) have prepared adenosine 5'-sulphatopyrophosphate, an analogue of ATP, and have shown that it is not a substrate for myosin, actomyosin nor myosin ATP-ase. Nor can it serve as an energy source for the contraction of myosin fibres *in vitro.*

# 6

## THE SULPHOTRANSFERASES

The transfer of the sulphuryl group from PAPS to a suitable acceptor is catalysed by one of a group of enzymes, the sulphotransferases (sometimes incorrectly known as the sulphokinases), with the formation of the appropriate sulphate ester and adenosine 3′,5′-diphosphate (PAP) as represented in equation 6.1.

$$PAPS + R.OH \rightleftharpoons R.OSO_3^- + PAP \qquad (6.1)$$

This reaction is quite irreversible except in the particular case of the nitrophenyl sulphates which is discussed below. Few of the sulphotransferases have been purified to any extent and there is still considerable doubt about the number of different types of those enzymes. It is generally assumed that they are rather specific with respect to their sulphate acceptors but this has by no means been proven because they are rather difficult enzymes to study, being unstable and far from easy to separate from one another.

A number of different types of method are available for the determination of the sulphotransferases. The most generally applicable are those based on the use of [$^{35}$S]PAPS: the simplest of these are chromatographic (Vestermark & Boström, 1959 a; Spencer, 1960 b) but the most valuable is undoubtedly that of Wengle (1964 a) which depends on the fact that the barium salts of PAPS and of $SO_4^{2-}$ ions are insoluble whereas the barium salts of sulphate esters are, in general, soluble. Another fairly general and rather useful method is based on the solubility in chloroform of the methylene blue salts of certain sulphate esters (Roy, 1956 a: see p. 66). As already pointed out, this method cannot be used to determine the sulphate esters of simple alcohols or of carbohydrates but it has proved most valuable in the study of the synthesis of aryl sulphates and of steroid sulphates. Other methods are available for the determination of specific compounds such as p-nitrophenyl sulphate (De Meio, 1952; Gregory & Lipmann, 1957), m-aminophenyl sulphate (Bernstein & McGilvery, 1952 a) or serotonin sulphate (Hidaka, Nagatsu & Yaka, 1967), but their use is obviously restricted.

In the present discussion it is not proposed to consider in detail the enormous amount of information which is available on the formation of sulphate esters—especially of steroids—by unfractionated tissue preparations incubated with ATP and $SO_4^{2-}$ ions. Although such reactions undoubtedly involve sulphotransferases, in most cases nothing is known of the properties of the enzymes responsible for the syntheses.

Even with the sulphotransferases which have to some extent been purified, accurate kinetic data are often not available because very frequently the assays have not been carried out with known concentrations of PAPS but rather the latter has been generated, under arbitrary conditions, by a sulphate-activating system present during the actual assay of the sulphotransferase. This procedure may be permissible for specificity studies but it is quite valueless for any kinetic work. For instance, pH effects must obviously be a resultant of the individual responses of the sulphate-activating enzymes and the sulphotransferases. Or again, the strong inhibition of the sulphotransferases by PAP which is often found in such assays need reflect only the vanishingly small concentrations of PAPS, with which PAP competes, likely to be present in these conditions.

Slightly more justifiable are procedures in which the PAPS is generated by a sulphate-activating system prior to the addition of the sulphotransferase but even these are not without their disadvantages. Not only is the actual concentration of the PAPS unknown but inhibitory nucleotides such as PAP and ADP are likely to be present in the reaction mixture. Certainly the technique of stopping the activation reaction by the addition of EDTA is an extremely dangerous one because many sulphotransferases require the presence of $Mg^{2+}$ ions before they can exhibit their full activity. Now that well-tried methods are available for the preparation of PAPS it is to be hoped that all future studies of the sulphotransferases will be made using known concentrations of the purified nucleotide as substrate. Even when this is done a further complication must be borne in mind: this is the apparently ubiquitous presence in crude—or even appreciably purified—tissue preparations of enzymes capable of degrading PAPS (see § 5.2). For example, of the three sulphotransferases prepared from guinea pig liver by Banerjee & Roy (1966) only one, the phenol sulphotransferase, was completely free from PAPS-degrading enzymes.

Table 6.1 summarizes some of the properties of the few sulphotransferases which have been purified. Only one other general property need be mentioned here: this is their quite powerful inhibition by PAP and by ADP, a property which must be kept in mind when using crude preparations of either PAPS or the sulphotransferases.

TABLE 6.1   *The general properties of some partially purified sulphotransferases*

| | Phenol sulphotransferase (Liver) | Oestrone sulphotransferase (Adrenal) | Oestrone sulphotransferase (Liver) | Androstenolone sulphotransferase (Liver) | Choline sulphotransferase (Aspergillus) |
|---|---|---|---|---|---|
| optimum pH | 5·6 | 8·0 | 6·0 | 7·5 | 8·0 |
| $Mg^{2+}$ requirement | − | + | + | + | − |
| $K_m$ (mM) for PAPS | 0·036 | 0·07 | . | 0·04 | 0·022 |
| $K_m$ (mM) for PAP | 0·02 | . | . | . | . |
| SH enzyme | Yes | Yes | ? | ? | Yes |

## 6.1   Phenol sulphotransferases

The phenol sulphotransferases are probably among the most studied of this group of enzymes but even here remarkably little information is available on the properties of the purified enzymes. Certainly the phenol sulphotransferases are among the most widespread of the group and there is direct evidence for their occurrence in many ox and human tissues (Holcenberg & Rosen, 1965; Boström & Wengle, 1967). Their distribution in other animals has not been directly studied but if the ability to excrete 'ester sulphate' be accepted as indirect evidence for the occurrence of phenol sulphotransferases then these enzymes must be present in all mammals—including several species of whale (Schmidt-Nielsen & Holmsen, 1921)—although undoubtedly in very different amounts in different species. For instance, Stekol (1936) long ago claimed that pigs could only very inefficiently form the sulphate ester of isobarbituric acid which is rapidly sulphurylated by many other mammals. Birds, reptiles and amphibia can also form aryl sulphates but fish apparently cannot (Smith, J.N., 1964, 1968; Maickel, Jondorf & Brodie, 1958). In invertebrates, insects and arachnids are known to be able to form aryl sulphates (Smith, J. N. 1964, 1968; Hitchcock & Smith, 1964; Darby, Heenan & Smith, 1966) and molluscs of the genus *Murex* have long been known to contain the substituted indoxyl sulphates which are precursors of Tyrian Purple. The occurrence of aryl sulphates, and so presumably of phenol sulphotransferases, in micro-organisms is rare but Ruelius & Gauhe (1950) isolated a hydroxynaphthoquinone sulphate, fusarubin sulphate, from the culture medium of *Fusarium solani*. There appears to be no report of the occurrence of phenol sulphotransferases in higher plants but it is possible that enzymes of this type are responsible for the formation of rhamnizin 3-sulphate, a flavonol sulphate ester, which occurs in *Polygonum hydropiper* (Horhammer & Hansel, 1953). Similar flavonoid sulphates have apparently also been found in two species of eel-grass by Nissen & Benson (1964) and although these compounds are not strictly aryl sulphates it seems possible that the flavonols may have sufficient aromatic character to allow them to be substrates for phenol sulphotransferase. In algae, a phenol sulphotransferase must presumably be responsible for the formation of 2,3-dibromobenzylalcohol 4,5-disulphate (formula *6.4* below) which has been found in *Polysiphonia lanosa* (Hodgkin, Craigie & McInnes, 1966).

The reaction catalysed by phenol sulphotransferase is that shown in equation 6.1, where R.OH is a phenol, and it may or may not be reversible depending upon the 'sulphate group potential' of the aryl sulphate: if this potential is high, as in the case of *p*-nitrophenyl sulphate or the dinitrophenyl sulphates, then the reaction is reversible and with suitable precautions the reverse reaction can be used for the assay of phenol sulphotransferase (Gregory & Lipmann, 1957). With simple aryl sulphates, such as phenyl sulphate or the naphthyl sulphates, the reaction is essentially irreversible.

It is unfortunate that *p*-nitrophenol has been used as substrate in so many studies of the phenol sulphotransferases. This acceptor is to some extent atypical because the partial anhydride character of the product (p. 30) allows the transferase reaction to be freely reversible.

A second, but closely related, reaction is catalysed by crude preparations of phenol sulphotransferase from rabbit liver (Gregory & Lipmann, 1957; Brunngraber, 1958). This is the transfer of the sulphuryl group from *p*-nitrophenyl sulphate to some suitable phenolic acceptor, such as phenol itself, with adenosine 3′,5′-diphosphate acting as a cofactor. Potential acceptors of other types, such as steroids, are inactive in this system. It must be presumed that PAPS is formed as an intermediary and that the reaction can be represented as in equation 6.2.

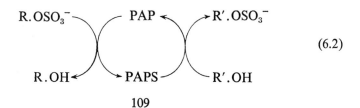

$$R.OSO_3^- \qquad PAP \qquad R'.OSO_3^-$$

$$R.OH \qquad PAPS \qquad R'.OH \tag{6.2}$$

The PAPS must remain enzyme-bound otherwise transfer to non-phenolic acceptors would have been expected because the enzyme preparation certainly contained other types of sulphotransferase. The concentration of adenosine 3',5'-diphosphate required for the reaction is very low: saturation of the crude phenol sulphotransferase from rabbit liver was reached at 2 $\mu$M adenosine diphosphate and concentrations greater than 10 $\mu$M inhibited quite powerfully (Brunngraber, 1958). This transfer reaction obviously provides a rather simple means of determining either phenol sulphotransferase or adenosine 3',5'-diphosphate but it should be noted, however, that it has not been shown to be a general reaction catalysed by all phenol sulphotransferases. It has not been possible, for instance, to detect it using the purified phenol sulphotransferase of guinea-pig liver (Banerjee & Roy, 1968).

How many individual phenol sulphotransferases occur in mammalian tissues, in particular in liver, is not known. To judge by the considerable variations in the reported properties of the phenol sulphotransferases there may well be more than one. The only direct observations bearing on this problem are those of Banerjee & Roy (1966) who, by chromatography on DEAE-Sephadex, have separated three fractions showing phenol sulphotransferase activity from extracts of guinea-pig liver. One of these appears to be a true phenol sulphotransferase in that the enzyme will not catalyse the transfer of the sulphuryl group from PAPS to any other type of acceptor which has been investigated. The other phenol sulphotransferases are, on the other hand, closely associated with steroid sulphotransferase and arylamine sulphotransferase. It is not clear whether all three of these activities are due to the one enzyme: indirect evidence suggests that they are, but much further work is needed to prove this contention.

The true phenol sulphotransferase isolated and partly purified from guinea-pig liver is the only example of the class which has been investigated in any detail from the standpoint of reaction kinetics (Banerjee & Roy, 1968). The enzyme has a pH optimum of 5·6 in acetate buffers (at which pH it is quite unstable), it is an SH enzyme which is apparently very sensitive to metals or to oxidation, its activity is not dependent upon the addition of exogenous $Mg^{2+}$ ions and it is not inhibited by EDTA. It has a molecular weight of about 65,000. The reaction catalysed by the enzyme is, in the nomenclature of Cleland (1963), a rapid equilibrium random bi bi reaction with one dead-end ternary complex of enzyme-adenosine diphosphate-nitrophenol. This reaction can be represented as follows:

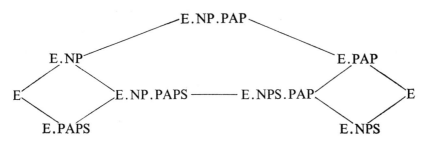

The values of the various kinetic constants for this reaction are summarized in table 6.2.

TABLE 6.2 *The dissociation constants of the enzyme-substrate complexes formed by the phenol sulphotransferase of guinea-pig liver at pH 5·6 and at 37°. The values are given with their 95% confidence limits*

| Substrate | $K_s$ (mM) |
|---|---|
| $p$-nitrophenol | $0·070 \pm 0·012$ |
| PAPS | $0·036 \pm 0·010$ |
| $p$-nitrophenyl sulphate | $0·068 \pm 0·008$ |
| PAP | $0·024 \pm 0·002$ |

No information is available on the detailed mechanism of the phenol sulphotransferase reaction, or indeed of any sulphotransferase reaction. Recently Weidman, Mayers, Zaborsky & Kaiser (1967) showed that the oxidation of hydroquinone monosulphate in the presence of methanol leads to the formation of methyl sulphate and they suggested that this system might serve as a model for the enzymatic transfer of the sulphuryl group. There is no evidence, however, that such enzymatic reactions are oxidative in nature. It should perhaps be recalled that Ford & Ruoff (1965) have previously shown that the liberation of $SO_4^{2-}$ ions from 5,6-isopropylidene-L-ascorbic acid 3-sulphate can be an oxidative process. Mayers & Kaiser (1968) have also suggested that a compound related to the $SO_3$ adduct of imidazole (an analogue of 1-phosphoimidazole) might be involved in the sulphotransferase reaction but there is again no evidence for this, apart from the fact that such adducts are powerful sulphating agents in simple chemical systems.

The specificity of the phenol sulphotransferases has been little studied but considering the very wide range of aryl sulphates which have been found naturally—for example, tyrosine $O$-sulphate (*6.1*) in mammalian fibrinogens (Doolittle & Blombäck, 1964), bufothionine (*6.2*) in toad

111

venom (Wieland & Vocke, 1930), ommatin D (*6.3*) in insects (Butenandt, Biekert, Koga & Traub, 1960) and 2,3-dibromobenzyl alcohol 4,5-disulphate (*6.4*) in the marine alga *Polysiphonia lanosa* (Hodgkin, Craigie & McInnes, 1966)—they must either be rather non-specific or there must exist a very large number of specific phenol sulphotransferases.

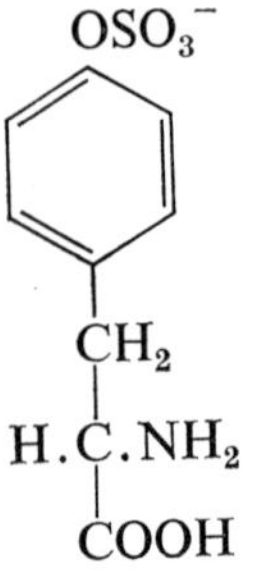

$OSO_3^-$

$CH_2$

$H.\overset{|}{C}.NH_2$

$COOH$

(*6.1*)

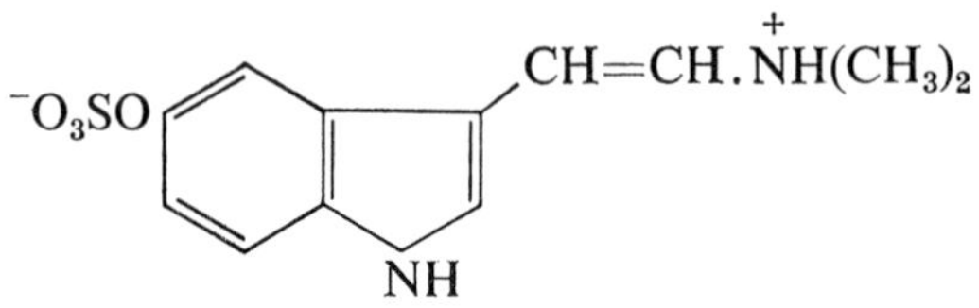

$^-O_3SO$ ... $CH{=}CH.\overset{+}{N}H(CH_3)_2$

$NH$

(*6.2*)

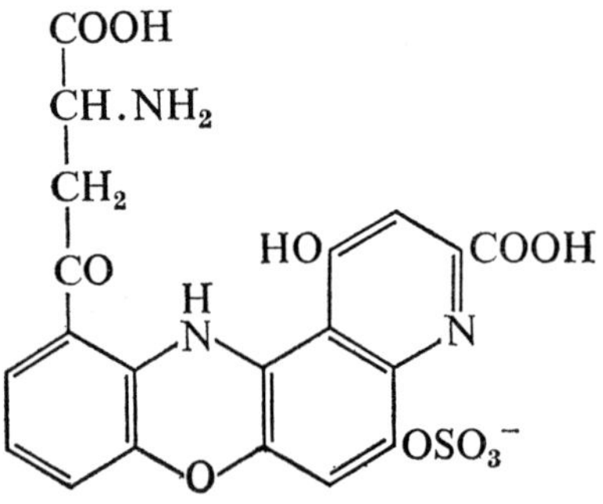

$COOH$

$\overset{|}{C}H.NH_2$

$\overset{|}{C}H_2$

$\overset{|}{C}O$

$HO$ ... $COOH$

$H$

$N$ ... $N$

$O$ ... $OSO_3^-$

(*6.3*)

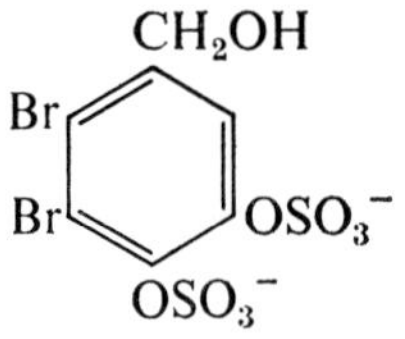

$CH_2OH$

$Br$

$Br$ ... $OSO_3^-$

$OSO_3^-$

(*6.4*)

A similar conclusion can be drawn from studies of the detoxication of phenols *in vivo* and of the action of partially purified sulphotransferases *in vitro*, all of which have shown that the majority of phenols can be sulphurylated, be they purely artificial like *p*-nitrophenol or more physiological such as adrenalin (Boström & Wengle, 1964), serotonin (Chadwick & Wilkinson, 1960) or the thyroid hormones (Cohn, 1965).

The only detailed, but nevertheless limited, studies of the specificity of a purified phenol sulphotransferase are those of Banerjee & Roy (1968) using the enzyme from guinea pig liver. The results are summarized in table 6.3. It is clear that rather a wide range of phenols can act as substrates, only oestrone being inactive in this respect. Oestrone also does not inhibit the formation of nitrophenyl sulphate, suggesting that the alicyclic rings of the steroid have prevented its combination with the enzyme, presumably because of their rigid, non-planar structure.

TABLE 6.3 *The specificity of the phenol sulphotransferase of guinea-pig liver. The reaction was in all cases studied at pH 5·6 in acetate buffer at 37° with PAPS as the sulphate donor. The maximum velocities are expressed relative to that with p-nitrophenol*

| Substrate | $K_m$ (mM) | V |
|---|---|---|
| phenol | 2·5 | 0·30 |
| p-nitrophenol | 0·070 | 1·00 |
| 1-naphthol | 0·025 | 1·02 |
| 2-naphthol | 0·025 | 0·95 |
| 5,6,7,8-tetrahydro-2-naphthol | 0·056 | 0·63 |
| 4-nitro-1-naphthol | 0·018 | 0·39 |
| 2-phenanthrol | 0·017 | 0·51 |
| 15,16-dihydro-3-hydroxy-17-oxo-cyclopentena-[a]phenanthrene | 0·015 | 0·13 |
| equilin | 0·020 | 0·21 |
| equilenin | 0·015 | 0·10 |
| oestrone | . | 0·00 |

A problem of particular interest is that of the synthesis of tyrosine O-sulphate which occurs in the urine of many mammalian species (John, Rose, Wusteman & Dodgson, 1966) as well as being present in certain polypeptides. Tyrosine itself is not a substrate for phenol sulphotransferases (Nose & Lipmann, 1958; Grimes, 1959) although derivatives of tyrosine in which the carboxyl group is blocked and the amino group is free (and uncharged) can be sulphated (Segal & Mologne, 1959). Tyrosine methyl and ethyl esters are sulphated by unfractionated sulphotransferase preparations from rat liver and the products are hydrolysed to free tyrosine O-sulphate by an esterase present in the enzyme preparations (Jones & Dodgson, 1965). Tyrosylglycine and tyrosylalanine are also sulphated, to give a mixture of products, but glycyltyrosine is not (Jones, Scotland & Dodgson, 1966). Whether or not a specific sulphotransferase is involved in those reactions is not known but all three of the fractions showing phenol sulphotransferase activity from guinea-pig liver can sulphurylate tyrosine methyl ester. Whatever enzyme, or enzymes, is involved, these observations imply that the tyrosyl O-sulphate residues which occur in proteins cannot be derived from tyrosine O-sulphate produced by the direct sulphurylation of tyrosine.

113

The apparently restricted occurrence of tyrosine $O$-sulphate in poly-peptides is puzzling. It has so far been found only in the fibrinopeptides B of several mammals (but not of man) (Doolittle & Blombäck, 1964; Mross & Doolittle, 1967), in hog gastrin II (Gregory, Hardy, Jones, Kenner & Sheppard, 1964) and in the two hypotensive peptides, phyllo-kinin (Anastasi, Bertaccini & Erspamer, 1966) and caerulein (Anastasi, Erspamer & Endean, 1967) from the skins of the reptile *Phyllomedusa rohdei* and the frog *Hyla caerulea* respectively. The structures of these peptides are

$$SO_4^-$$
$$|$$
Ala-Asp-Asp-Tyr-(Asp,Glu,Pro,Leu,Asp,Val)-Asp-Ala-Arg
*Rabbit fibrinopeptide B*

$$SO_4^-$$
$$|$$
Pyr-Gly-Pro-Trp-Glu-Glu-Glu-Glu-Glu-Ala-Tyr-Gly-Trp-Met-Asp-Phe.NH$_2$
*Hog gastrin II*

$$SO_4^-$$
$$|$$
Pyr-Gln-Asp-Tyr-Thr-Gly-Trp-Met-Asp-Phe.NH$_2$
*Caerulein*

$$SO_4^-$$
$$|$$
Arg-Pro-Pro-Gly-Phe-Ser-Pro-Phe-Phe-Ile-Tyr
*Phyllokinin*

The relationship between gastrin II and caerulein is most striking: not only do both peptides have a blocked $N$-terminus of pyrrolidone carboxylic acid but the $C$-terminal end consists of the same penta-peptideamide in both cases. It should perhaps also be pointed out that phyllokinin is a bradikinin derivative, being bradikinyl-isoleucyl-tyrosyl $O$-sulphate.

This restricted occurrence of tyrosyl $O$-sulphate residues may of course be apparent rather than real because of the difficulty of detecting the residue in proteins, it being destroyed under the usual conditions of acid hydrolysis used in studies of protein structure. The ester is, on the other hand, stable to alkali and an analytical method based on this property has been devised by Jevons (1963) but so far it has done little to further our understanding of the distribution of tyrosyl $O$-sulphate residues. It is perhaps worth noting, however, that if all the tyrosine $O$-sulphate present in human urine be derived from tyrosyl O-sulphate residues in proteins, then without doubt proteins other than fibrinogen must be sources of the ester. At present the function of sulphate ester groups in proteins cannot even be guessed but it is obvious that their presence must completely alter the charge structure of the protein and it is not without interest that the desulphation of phyllokinin causes a considerable decrease in its biological activity (Anastasi *et al.* 1966).

## 6.2 Oestrone sulphotransferase

This enzyme, which is responsible for the formation of oestrone sulphate (*6.5*) and presumably of other oestrogen sulphates, was first detected in rat liver by Nose & Lipmann (1958) and it was later found in human and bovine adrenal gland (Sneddon & Marrian, 1963; Adams, 1964 *a*). Rather surprisingly it also occurs in embryonic chick, but not calf, cartilage (Adams, 1963). Oestrone sulphotransferase must also be presumed to occur in the gut of the sea urchin *Strongylocentrotus fransiscanus* which can form oestradiol 3-sulphate at a rather high rate (Creange & Szego, 1967).

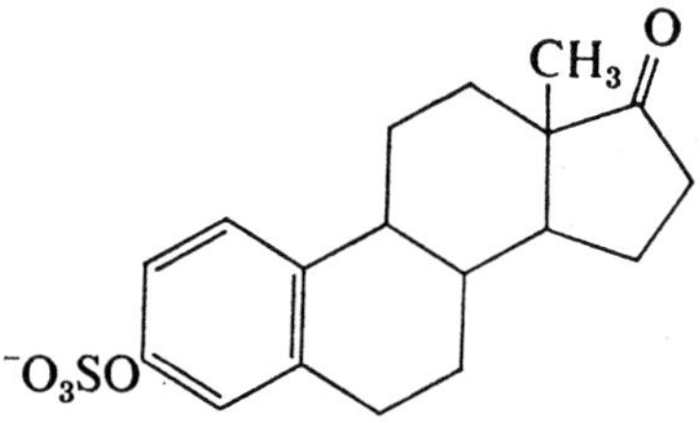

$(6.5)$

In adult mammalian tissues oestrone sulphotransferase does not appear to be widely distributed: in the ox only the adrenal and the liver are significant sources (Holcenberg & Rosen, 1965) and in the human only the adrenal, liver and jejunal mucosa are important (Boström & Wengle, 1967). In the human foetus, on the other hand, oestrone sulphotransferase is of rather widespread distribution, occurring in all the tissues examined except the cerebrum (Wengle, 1966). It has been suggested, therefore, that this enzyme plays a much more important role in the foetus than in the adult.

The only oestrone sulphotransferase which has so far been obtained free from other types of sulphotransferase activity is that from the ox adrenal gland (Adams & Poulos, 1967; Adams & Chulavatnatol, 1967): the final preparation of the enzyme could not utilize simple phenols, androstenolone, 17$\beta$-oestradiol 3-methyl ether nor 2-naphthylamine as substrates. This enzyme exists as a monomer-trimer (or monomer–tetramer) system, SH groups being involved in some way in facilitating the polymerization: if the isolation of the enzyme is carried out in the presence of mercaptoethanol only the polymeric form is obtained. The monomeric species of the enzyme (molecular weight 67,000) exhibits simple kinetics but the polymeric form shows much more complex behaviour which can be interpreted in terms of an interaction between the several catalytic sites of the polymer. Although detailed studies of the kinetics of the reaction have not yet been carried out, the results so far obtained are again consistent with the enzyme catalysing a rapid equilibrium random bi bi reaction. With the monomeric form of the enzyme the optimum pH was about 8, the $K_m$ for PAPS was 0·07 mM and that for the oestrone was 0·014 mM. The $K_m$ values for other oestrogens were the same order of magnitude as that for oestrone, as were the maximum velocities of the reactions. The enzyme was activated by $Mg^{2+}$ ions, and by certain other divalent cations, but an absolute requirement for such ions could not be shown because the enzyme was not inhibited by EDTA: the situation is therefore quite comparable to that found with the steroid sulphotransferases of guinea-pig liver (Banerjee & Roy, 1966).

Although the oestrone sulphotransferase of ox adrenal gland cannot utilize simple phenols as substrates it can so utilize the synthetic oestrogens stilboestrol and hexoestrol: dienoestrol, on the other hand, is not a substrate. The specificity of the only other oestrone sulphotransferase which has been partly purified, that from guinea-pig liver, is quite different because here the oestrone sulphotransferase activity is closely associated with both phenol sulphotransferase and arylamine sulphotransferase activities (Banerjee & Roy, 1966). All three types of activity appeared to be associated with the one enzyme but the possibility certainly existed that the different sulphotransferases formed an interacting system analogous to the polymerizing system of the oestrone sulphotransferase from ox adrenal gland.

One of the most interesting features of the enzyme from the adrenal is its apparent association with oestrone when isolated from the gland (Adams, 1967): at this stage it would be premature to comment further on this observation but if it can be confirmed with more highly purified preparations of the enzyme then some most interesting possibilities are opened up.

Considering only steroid substrates, there is no doubt that all three of the major oestrogens—oestrone, oestradiol and oestriol—are substrates for oestrone sulphotransferase, as also are 2-methoxyoestrone and 17-desoxyoestrone. More difficult to interpret is the report by Levitz, Katz & Twombly (1966) that a crude sulphotransferase preparation from guinea-pig liver can form the 3-sulphate of oestriol 16-glucuronide. If the sulphotransferase involved in this reaction is simply oestrone sulphotransferase then this enzyme shows a surprising lack of specificity: the differences between its usual substrate, a free steroid, and oestriol glucuronide are so great that it might well be thought that two distinct sulphotransferases must be involved. Such double conjugates are by no means artefacts and it is becoming clear that they are of widespread distribution in nature but their role in steroid metabolism is still obscure.

## 6.3   Steroid sulphotransferases

This group of enzymes has attracted much interest because of its obvious importance in the metabolism of steroids but it is nevertheless still unknown how many different steroid sulphotransferases exist. Nose & Lipmann (1958) were the first to point out that it was highly probable that there were several such enzymes, quite apart from oestrone sulphotransferase, in rat liver. Further evidence was provided by Banerjee & Roy (1966, 1967 a) in their investigation of the sulphotransferases of guinea-pig liver. From their work it was clear that there were at least two, if not three, steroid sulphotransferases—androstenolone sulphotransferase, testosterone sulphotransferase and desoxycorticosterone sulphotransferase with androstenolone (dehydroepiandrosterone, 3β-hydroxyandrost-5-en-17-one, 6.6), testosterone (6.7) and desoxycorticosterone (6.8) as respective typical substrates. There is indirect evidence, based essentially on the ratios of activities of the crude sulphotransferase preparations with different steroids as substrates, that other steroid sulphotransferases must exist in mammalian tissues. For instance, in rat (Roy, 1956 b) and rabbit (Nose & Lipmann, 1958) liver the two 3-hydroxy-5β-androstan-17-ones (aetiocholanolones) are sulphated much more slowly than androstenolone but in human liver (Boström & Wengle, 1964) and in human adrenal (Adams, 1964 a; Boström, Franksson & Wengle, 1964) the aetiocholanolones are sulphated as rapidly as androstenolone. Unless there are quite large species differences in the specificities of the enzymes, results such as these imply that there must exist an aetiocholanolone sulphotransferase.

Because of this doubt about the actual number of enzymes involved it is only possible to consider the steroid sulphotransferases in rather general terms. Like oestrone sulphotransferase, the steroid sulphotransferases appear to be of limited distribution in the adult, occurring in significant amounts in only the liver, the adrenal and the jejunal mucosa in the human (Boström & Wengle, 1967) and in the ox (Holcenberg & Rosen, 1965). Neither Adams (1964 b) nor Boström & Wengle (1967) could detect steroid sulphotransferases in the human ovary: that they must be present therein is, however, shown by the studies of Wallace & Silberman (1964) who used a rather sensitive method employing $^{14}$C- and $^{3}$H-labelled steroids. The level of steroid sulphotransferase in the ovary must nevertheless be much lower than in the above-mentioned tissues. Androstenolone sulphotransferase must also occur in the testes of the pig in which species androstenolone sulphate has been shown to be secreted into the spermatic vein (Baulieu, Fabre-Jung & Huis in't Veld, 1967).

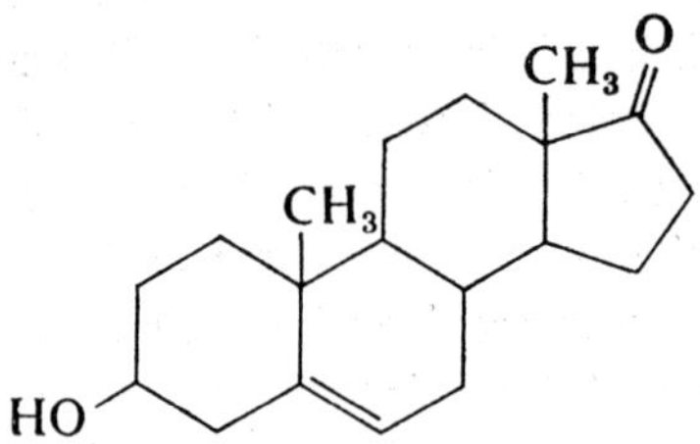

(6.6)

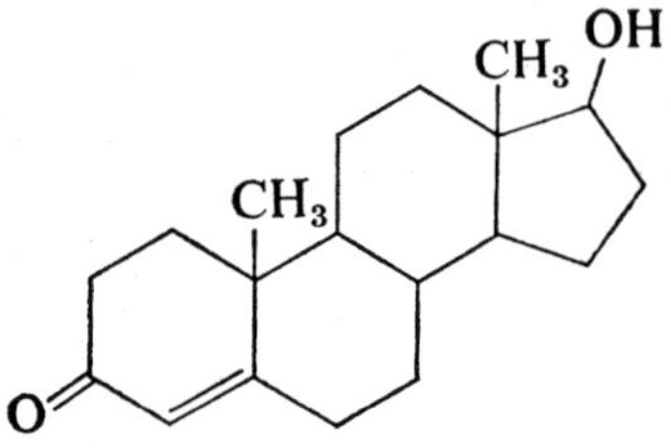

(6.7)

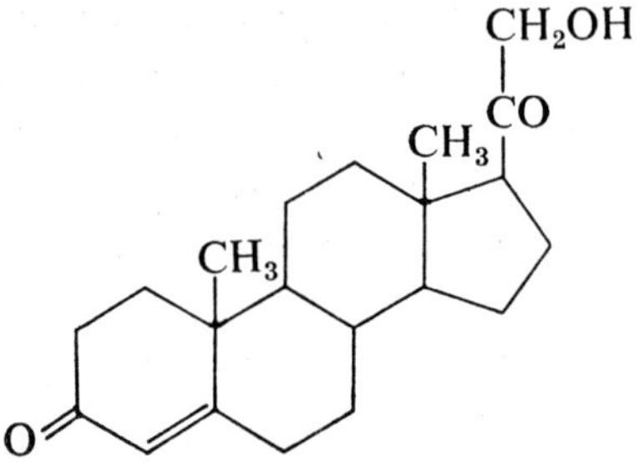

(6.8)

In the human foetus the steroid sulphotransferases are found in the same tissues as in the adult with the addition of the kidney; the level of activity is lower than in adult tissues with the exception of the adrenal gland, which is a very rich source of this enzyme (Wengle, 1964 b, 1966).

The steroid sulphotransferases appear to be SH enzymes and at least some of them require $Mg^{2+}$ ions before they can show their full activity. Their pH optima lie between 6 and 8, and the few $K_m$ values which have been determined are of the order of 0·02 mM steroid and 0·05 mM PAPS, comparable to the corresponding values for phenol sulphotransferase. Again, the data which are available would be consistent with the steroid sulphotransferase reaction being of the rapid equilibrium random bi bi type (Banerjee & Roy, 1967 $b$). These authors have also shown that the synthesis of cholesteryl sulphate, presumably by androstenolone sulpho-transferase, is competitively inhibited by androstenolone methyl ether, with a $K_i$ of 0·09 mM. Obviously this steroid cannot be a substrate for androstenolone sulphotransferase and it was suggested (Banerjee & Roy, 1967 $b$) that its combination with enzyme might be through the D ring, as occurred in the case of arylamine sulphotransferase (see § 6.5), so that a control function could be envisaged if cholesteryl sulphate were an obligatory intermediate in the formation of, say, androstenolone sulphate.

Adams & Edwards (1968) have recently made some most interesting observations on the androstenolone sulphotransferase of human adrenal gland. Their results are, however, difficult to interpret with certainty because no attempt was made to purify the enzyme. It was simply studied in a high-speed supernatant from a homogenate of the gland. The kinetics shown by the system were extremely complex and Adams & Edwards suggested that they were explicable if the enzyme existed in a number of polymeric forms of a monomer having a molecular weight of 65,000. Evidence for such a system was obtained by gel filtration and the different polymers were partially separated by sedimentation in a sucrose density gradient. The separated components slowly reformed an equilibrium mixture on standing for some three days but, most interestingly, the position of this equilibrium was greatly influenced by the presence of substrates or modifiers of the enzyme: PAPS favoured dissociation to the monomer while androstenolone, $Mg^{2+}$ and cysteine favoured associa-tion to polymeric forms. It is essential that these suggestive, but pre-liminary, results be extended by studies using purified preparations of the enzyme. Not only would possible complications caused by the presence of interfering enzymes (e.g. PAPS-degrading enzymes) be eliminated but the possibility that the 'polymers' were in fact complexes between androstenolone sulphotransferase and other proteins, perhaps other sulphotransferases, could definitely be excluded. As has been stated above, it is likely that such a system of interacting sulphotransferases exists in extracts of guinea-pig liver. Once again it appears that further progress is dependent upon the isolation of the enzyme in a more highly purified state.

Until purified preparations of steroid sulphotransferases are available

119

it is obviously impossible to consider their specificity in anything but the most general terms but it has long been known that steroids can be sulphated at the 3, 17 and 21 positions (Roy, 1956 $b$; Schneider & Lewbart, 1956; Nose & Lipmann, 1958). It is likely that separate sulpho-transferases are involved in the reaction at each of these positions and perhaps also for the reactions with the $5\alpha$ and $5\beta$ series of steroids. With polyhydroxy steroids, disulphates can be formed, especially of the $3\beta$, $17\beta$ and $3\beta,20\beta$ dihydroxy steroids (Wengle & Boström, 1963). These authors have pointed out some structural features which influence the behaviour of steroids as substrates for the steroid sulphotransferases. For instance, a double bond in the 4–5 position inhibits sulphation of a $3\beta$-hydroxy steroid; a methyl or ethyl group of the $17\alpha$ position inhibits sulphation of a $17\beta$-hydroxy steroid; and a $17\alpha$-hydroxy group inhibits sulphation at position 21. Many of these features can be correlated with the behaviour of steroids *in vivo*.

Once again the occurrence of double conjugates such as pregn-5-ene-$3\beta,20\alpha$-diol 20-(2'-acetamido-2'-deoxy-$\alpha$-D-glucoside) 3-sulphate in urine (Arcos & Lieberman, 1967) or glycolithocholic acid 3-sulphate and tauro-lithocholic acid 3-sulphate in bile (Palmer, 1967) makes the exact speci-ficity requirements of these enzymes a matter of considerable interest.

Other steroids of quite different types can also be sulphurylated by enzyme preparations from mammalian livers. For example, the sul-phurylation of various cardioactive genins, such a digitoxigenin, uzari-genin, sarmentogenin, etc., has been reported by Herrmann & Repke (1964). As these steroids contain $3\beta$-hydroxy groups perhaps andros-tenolone sulphotransferase is involved here but this has not been shown. More intriguing is the report of the sulphurylation of vitamin D by enzymes from mammalian liver (Higaki, Takahashi, Suzuki & Sahashi, 1965), and the subsequent isolation of vitamin D sulphate from both human and cows' milk (Sahashi, Suzuki, Higaki & Asano, 1967). Although vitamin D is derived from a $3\beta$-hydroxy steroid the vitamin itself cannot be regarded as such so that perhaps a further type of sulphotransferase is involved.

In the lower vertebrates other types of steroid sulphotransferases must certainly be present. Several of these animals synthesize sulphate-contain-ing bile salts, such as ranol sulphate (the 24-sulphate of $3\alpha,7\alpha,12\alpha,24\xi$, 26-pentahydroxy-$5\alpha$-27-norcholestane) in the bile of the frog *Rana temporaria* (Haslewood, 1964) or myxinol disulphate, the $3\beta,27$-disulphate of $3\beta,7\alpha,16\alpha,27$-tetrahydroxy-$5\alpha$-cholestane in hag fish bile (Haslewood,

1966). The biosynthesis of these primitive bile salts had hardly been investigated but it is known that enzyme preparations from frog liver can sulphurylate ranol, and the related alcohols scymnol and cholan-24-ol, when these steroids are incubated with the enzyme, ATP and $SO_4^{2-}$ ions (Bridgwater & Ryan, 1957). Presumably a quite distinct group of steroid sulphotransferases must be involved in the synthesis of these very characteristic types of steroid sulphate.

## 6.4 Choline sulphotransferase

Choline sulphotransferase, catalysing the transfer of the sulphuryl group from PAPS to choline with the formation of choline $O$-sulphate (6.9) is of much more restricted occurrence than the other sulphotransferases so far considered. It is found only in those higher fungi (Harada & Spencer, 1960; Spencer & Harada, 1960) which utilize choline $O$-sulphate as a store of sulphur. It was at one time claimed that choline sulphotransferase used APS as the donor of the sulphuryl group (Kaji & McElroy, 1958) but this was an error; it does in fact only utilize PAPS in a normal irreversible, sulphotransferase-catalysed reaction (Kaji & Gregory, 1959).

$$\underset{H_3C}{\overset{H_3C}{\diagdown}} \overset{CH_2.CH_2.OSO_3^-}{\underset{N}{\overset{+}{\diagup}}} \underset{CH_3}{\diagup} \qquad (6.9)$$

Choline sulphotransferase has been obtained from *Aspergillus nidulans* (Orsi & Spencer, 1964). The enzyme is rather unstable and extensive purification was not possible but it was shown to be an SH enzyme which did not require the addition of $Mg^{2+}$ ions for its full activity. This latter observation is contrary to the findings of Kaji & Gregory (1959). The $K_m$ for choline was 0·012 M and that for PAPS 0·022 mM, the value for each substrate being independent of the concentration of the other. This suggests that the choline sulphotransferase reaction might, like the phenol sulphotransferase and the oestrone sulphotransferase reactions, be a rapid equilibrium random bi bi reaction. A wide range of compounds structurally related to choline was tested as potential substrates for choline sulphotransferase but only choline, dimethylethylaminoethanol and dimethylaminoethanol were sulphurylated with $K_m$ values of 0·012, 0·020 and 0·025 M respectively. Several other analogues of choline, such as thiocholine, were inhibitors of choline sulphotransferase.

121

Choline sulphotransferase may have an important role to play in some of the higher plants: Benson & Atkinson (1967) have shown that of fifteen species of mangrove only those which excreted salt could form choline sulphate and they have suggested that in those species choline sulphate may play a role in salt transport.

## 6.5 Arylamine sulphotransferase

The present status of this enzyme is obscure. The activity ascribed to it was first found by Roy (1960 a) in extracts of rat and guinea-pig liver which catalysed the formation of 2-naphthyl sulphamate, a metabolite of 2-naphthylamine (Boyland, Manson & Orr, 1957). This reaction (6.3) was rather different from that catalysed by the other known sulphotransferases in that the acceptor molecule was an aromatic amine:

$$R.NH_2 + PAPS \rightarrow R.NH.SO_3^- + PAP \qquad (6.3)$$

Interest in this enzyme was stimulated by the fact that 17-oxosteroids had a pronounced effect on the reaction, acting as partially competitive inhibitors of the guinea-pig enzyme and partially competitive activators of the rat enzyme (Roy, 1961, 1962 a, 1964). Unfortunately it is no longer certain that arylamine sulphotransferase has a separate existence, at least in guinea-pig liver, because Banerjee & Roy (1966) have shown that both the androstenolone sulphotransferase and the oestrone sulphotransferase fractions from this organ show arylamine sulphotransferase activities which are inhibited by 3$\beta$-methoxyandrost-5-en-17-one, the model 17-oxosteroid used in the work. As with the phenol sulphotransferase activity also associated with these fractions, the arylamine sulphotransferase activity appears to be closely associated with the steroid sulphotransferases. There is no direct proof of their identity but the fact that there are two separable fractions exhibiting arylamine sulphotransferase activity makes the quantitative interpretation of the earlier kinetic studies of doubtful significance. Nevertheless, the finding of an association between steroid sulphotransferase activity and arylamine sulphotransferase activity makes it simpler to visualize competition between a steroid and an arylamine.

The distribution of this activity in nature is unknown but, as well as mammals, some birds (Smith, J.N. 1964, 1968) and arachnids (Hitchcock & Smith, 1964) can form aryl sulphamates.

## 6.6 Mucopolysaccharide sulphotransferases

This group is probably the largest and the most fundamental of the sulphotransferases but at present it is perhaps the least understood because none of the enzymes has been appreciably purified although they are apparently of common occurrence in those tissues which contain mucopolysaccharides—that is, in most tissues of the body. In mammals they have been found in brain (Balasubramanian & Bachhawat, 1964), cartilage (D'Abramo & Lipmann, 1957), the eye (Wortman, 1960, 1961) and serum (Adams, 1964 c). In the hen they are present in the isthmus of the oviduct (Suzuki & Strominger, 1960). They also occur in molluscs (Goldberg & Delbruck, 1959).

Little is known of the specificity of the enzyme, or more probably enzymes, and it is difficult to know whether there is a single non-specific mucopolysaccharide sulphotransferase or a large number of more specific enzymes. The latter situation seems much more probable. Some of the earlier results of specificity studies are summarized in table 6.4 which amply shows the rather wide spectrum of activity of this group of enzymes.

TABLE 6.4    *The acceptor specificity of some crude preparations of mucopolysaccharide sulphotransferases*

| Tissue | Potential acceptors | | Reference |
|--------|---------------------|--|-----------|
| | Utilized | Not utilized | |
| rabbit skin | dermatan sulphate | . | 1 |
| hen oviduct | heparitin sulphate | . | 2 |
| chick cartilage | chondroitin 4- and 6-sulphates | dermatan sulphate | 3 |
| human leimyosarcoma | dermatan sulphate | chondroitin 4- and 6-sulphates | 4 |
| human mammary carcinoma | chondroitin, chondroitin 4- and 6-sulphates, dermatan sulphate, heparitin sulphate | . | 5 |

1, Davidson & Riley (1960). 2, Suzuki, Trenn & Strominger (1961). 3, Adams (1960). 4, Hasegawa, Delbruck & Lipmann (1961). 5, Adams & Meaney (1961).

Detailed studies of the mucopolysaccharide sulphotransferases are few and one of the most significant contributions is undoubtedly that of Suzuki & Strominger (1960) who investigated these enzymes in the isthmus of the hen oviduct which contains not only the sulphotrans-

123

ferase(s) but also a sulphate-activating system, both of which can readily be obtained in a cell-free form. This system could transfer, at an optimum pH of 6·6, the [³⁵S]sulphuryl group from PAPS to a number of mucopolysaccharides, the relative rates of transfer being shown in table 6.5. It should be noted that the actual amount of sulphate, as distinct from radioactivity, transferred is minute, being measured in amounts of $\mu\mu$-moles so that chemical studies of these sulphotransferases are fraught with difficulty. The results clearly showed, however, that sulphation at position 4 of an $N$-acetylgalactosamine residue (as is free in chondroitin

TABLE 6.5    *The relative extents of sulphate transfer to different acceptors by the mucopolysaccharide sulphotransferases of hen oviduct (Suzuki & Strominger, 1960) and of chick embryo cartilage (Meezan & Davidson, 1967)*

| Acceptor | Oviduct | Cartilage |
|---|---|---|
| chondroitin 4-sulphate | 1·00 | 1·00 |
| chondroitin 6-sulphate | 1·8 | 0·46 |
| dermatan sulphate | 0·72 | 0·75 |
| heparitin sulphate | 0·45 | 1·3 |
| chondroitin (natural) | 1·8 | . |
| chondroitin (chemical) | 0·62 | 0·13 |
| hyaluronic acid | 0·00 | 0·08 |
| heparin | 0·00 | 0·26 |
| keratosulphate | 0·00 | 0·25 |

6-sulphate) occurred more rapidly than that at the corresponding position 6. Further studies of the system showed that simple di- and oligosaccharides could also act as acceptors for sulphotransferases present in the hen oviduct, although rather less efficiently than the polysaccharides themselves, and that both mono- and di-sulphated derivatives of $N$-acetylgalactosyl residues were formed. The polysaccharide sulphates formed when chondroitin or chondroitin 4-sulphate were the acceptors were investigated by hydrolysing them to oligosaccharides with testicular hyaluronidase: here also the results were consistent with the production of both mono- and disulphated derivatives of the $N$-acetylgalactosamine residues. Although these results clearly showed that mucopolysaccharide sulphotransferases existed in the hen oviduct, they gave little indication as to their physiological acceptors which could, apparently, be either oligosaccharides or polysaccharides. Further studies would be dependent on the availability of a more highly purified enzyme which will be difficult to obtain because of its instability.

More recently Meezan & Davidson (1967 $a, b$) have studied the muco-polysaccharide sulphotransferases of chick embryo cartilage, using not added PAPS as the donor but rather PAPS produced in the reaction mixture through the action of sulphate-activating enzymes. As already discussed, such a technique is open to criticism. This system could like-wise use many, but not all, mucopolysaccharides as acceptors and the relative rates with several of these compounds are again summarized in table 6.5. An endogenous acceptor occurred in chick cartilage and it was of interest that after the desulphation of this with methanolic-HCl it could no longer be sulphurylated: similarly chondroitin prepared by the chemical desulphation of chondroitin 4-sulphate was not an acceptor. Because of this, and the fact that hyaluronic acid was not an efficient acceptor, Meezan & Davidson have suggested that sulphation of a com-pletely non-sulphated polysaccharide probably does not occur: apart from the inherent unlikelihood of such a situation it should perhaps be pointed out that Suzuki & Strominger (1960) likewise found chemically prepared chondroitin to be a poor acceptor for the hen oviduct system but that 'natural' chondroitin, isolated from bovine cornea, was as effective an acceptor as was chondroitin 6-sulphate (table 6.5).

The nature of this endogenous acceptor is not clear but Meezan & Davidson (1967 $b$) have shown that it is probably related to the chon-droitin sulphates. Its ability to accept sulphate groups is little influenced by its molecular size but is strongly dependent on the charge on the molecule, as had been already discussed. Treatment of the acceptor with pronase or with alkali increases its ability to accept the sulphate group so that sulphation cannot be dependent upon the presence of protein: there is, however, a striking change in the type of sulphation following the removal of the protein. When the mucopolysaccharide is combined with protein the sulphate groups which are formed through the action of the enzyme are predominantly axial, as judged by their rates of hydrolysis (Rees, 1963), so that they are almost certainly localized at position 4 of the galactosamine: after removal of the protein, however, equatorial or, more probably, primary hydroxyl groups are sulphated. Meezan & Davidson (1967$b$) therefore make the interesting suggestion that although sulphation of the carbohydrate can occur whether or not this is bound to protein, the *type* of sulphation may be controlled by the presence of the latter. However, as mucopolysaccharides are built up on a protein core (Telser, Robinson & Dorfman, 1965) this situation is likely to be a purely artificial one.

An important problem in all this work is the complete lack of any demonstration of a net synthesis of sulphate ester groups: what has been demonstrated is an incorporation of $^{35}SO_4^{2-}$ ions into mucopolysaccharides but it would appear important to couple this with a demonstration of a net synthesis. Unfortunately the very small scale on which the transfer occurs will make such a demonstration difficult.

Perhaps related to—or even identical with—the above mucopolysaccharide sulphotransferases is the enzyme which has recently been described by Harada, Shimizu, Nakanishi & Suzuki (1967) as being present in the magnum of the hen oviduct (the albumen-secreting region). This enzyme catalyses the transfer of the sulphuryl group from PAPS to uridine diphosphate N-acetylgalactosamine 4-sulphate to give the corresponding 4,6-disulphate. Other simple compounds were also acceptors in this system and the appropriate data are summarized in table 6.6. Two interesting points arise from the kinetic study of this enzyme: first, the $K_m$ for the sugar is not independent of that for PAPS so that the kinetics of the reaction are more complex than those of the other sulphotransferases so far studied, and secondly, the optimum pH is at 4·8. This is much lower than the value usually reported—or assumed—for other mucopolysaccharide sulphotransferases and it would be most interesting to know whether the difference is due to the use of known concentrations of purified PAPS as the sulphate donor in this work or whether it simply shows the involvement of quite distinct enzymes in the two types of reaction.

TABLE 6.6   *The specificity of a sulphotransferase from hen oviduct which can utilize simple sugars or sugar derivatives as substrates*

| Acceptor | $K_m$ (mM) | V (relative) |
|---|---|---|
| UDP-N-acetylgalactosamine 4-sulphate | 0·05 | 1·00 |
| N-acetylgalactosamine 4-sulphate | 1·4 | 0·72 |
| N-acetylgalactosamine 1-phosphate 4-sulphate | 0·13 | 0·39 |
| $\Delta^{4,5}$-glucuronido-N-acetylgalactosamine 4-sulphate | 2·0 | 0·05 |

The mucopolysaccharide sulphotransferase of beef cornea is an interesting enzyme because it can apparently couple with the phenol sulphotransferase which is also present in corneal extracts and so in effect catalyse the transfer of the sulphuryl group from p-nitrophenyl sulphate of polysaccharide in the presence of adenosine 3′,5′-diphosphate (Wortman,

1961). This coupling is certainly rather surprising in view of the inability of the phenol sulphotransferase of liver to couple with any other type of sulphotransferase, but it is nevertheless a useful property in that it eliminates the necessity of using PAPS as a sulphate donor.

The enzymes considered so far have all been concerned with the formation of $O$-sulphate esters. Heparin, however, also contains $N$-sulphate groups. That a separate sulphotransferase is involved in the formation of the latter was shown by Korn (1959 $a, b$) who noted that while slices of a mast cell tumour incorporated $^{35}SO_4^{2-}$ ions almost equally into the $O$-sulphate and the $N$-sulphate groups of heparin, extracts of the tumour formed only $O$-sulphate groups. The formation of the $N$-sulphate groups has recently been studied in more detail by Eisenman, Balasubramanian & Marx (1967) and by Rice, Spolter, Tokes, Eisenman & Marx (1967), again using a mouse mast cell tumour as a source of the enzyme which was localized in a post-microsomal particulate fraction. With $N$-desulphoheparin as acceptor and PAPS as donor, the reaction had a pH optimum between 6·7 and 7·2, was dependent on added $Mg^{2+}$ ions and was not saturated at 0·04 mM PAPS. As might be expected this preparation could utilize other polysaccharides as acceptors, heparin sulphate, dermatan sulphate and chondroitin 4-sulphate all being sulphated although heparin itself was not.

Now that some preliminary information on these mucopolysaccharide sulphotransferases is available it is to be hoped that more detailed studies of them will be made because without a doubt they must be the most fundamental, and probably the most important, of the sulphotransferases of mammalian tissues.

## 6.7 Other sulphotransferases

Having regard to the many types of sulphate ester which are of natural occurrence, it must be obvious that many sulphotransferases remain to be studied. There is already evidence for the existence for certain other of these enzymes but for many of the sulphate esters—especially those of plant origin such as the mustard oil glycosides or the seaweed polysaccharides—there is no information whatever on the mechanism of the incorporation of sulphate groups into the molecule.

One of the groups of compounds for which PAPS has been shown to be sulphate donor is that of the cerebroside sulphates. The first observations here were those of Goldberg & Delbruck (1959) and of Goldberg (1960) who showed that the sulphuryl group of PAPS was incorporated into lipid fractions by enzyme preparations from rat liver or rat brain. The latter enzyme, unlike the previously known sulphotransferases, was found in a particulate fraction. The nature of the sulphate-containing lipid was not clarified, but it was unlikely to be cerebroside sulphate and was probably N-acetylsphingosine sulphate. More recently Balasubramanian & Bachhawat (1965 a, b) showed that PAPS was indeed the sulphuryl donor for the formation of cerebroside sulphates by a particulate fraction prepared from sheep brain. This system could, however, utilize only endogenous galactocerebrosides as acceptors. More important is the work of McKhann, Levy & Ho (1965) who have used solubilized preparations from a particulate fraction of rat brain and have shown the incorporation of $^{35}SO_4^{2-}$ ions into exogenous cerebrosides. This incorporation was dependent upon the presence of ATP and so presumably upon the formation of PAPS.

Another carbohydrate ester of interest, but of unknown function, in mammals is the neuramin lactose sulphate found in mammary gland and in milk. Barra & Caputto (1965) have shown that this, and lactose sulphate, are produced by extracts of rat mammary gland using PAPS as the sulphate donor.

Other sulphotransferases are presumably involved in the formation of the simple alkyl sulphates which are formed in vitro by preparations from rat liver (Vestermark & Boström, 1959 b; Carroll & Spencer, 1965 a,b). The synthesis of alkyl sulphates has also been shown to occur in extracts of embryonic chick liver (Spencer & Raftery, 1966), a not unexpected finding in view of the occurrence in the allantois of the hens egg of isopropyl sulphate and other simple alkyl sulphates (Yagi, 1964, 1966). It is perhaps of interest that the major metabolite of Dimetridazole (6.10) in turkeys is apparently the alkyl sulphate, (6.11) (Law, Mansfield, Muggleton and Parnell, 1963).

(6.10)                    (6.11)

In lower animals sulphotransferases must be involved in the formation of charonin sulphate, a complex polysaccharide whose composition is described later (§ 7.3.2), and other similar polysaccharide sulphates present in molluscs. Another sulphotransferase is likely to be involved in the biosynthesis of holuthurin A, a neurotoxin present in the sea cucumber *Actinopyga agassizi* in which the sulphate group is esterified with the complex carbohydrate moiety of a steroid glycoside. The toxicity of holuthurin A is, like that of phyllokinin, greatly decreased by desulphation (Friess, Durant, Chanley & Fash, 1967).

It is therefore clear that many other sulphotransferases await study but it is unfortunately also clear that any significant progress is dependent upon the development of methods for handling this family of closely related and apparently rather unstable enzymes.

## 6.8   Vitamin A and the sulphotransferases

It has been claimed that the activities of several types of sulphotransferase are influenced by the vitamin A status of the animal from which the preparations have been obtained: the evidence for this view is only indirect, but it is nevertheless suggestive. Subba Rao, Sastry & Ganguly (1963) and Subba Rao & Ganguly (1964) were the first to show that the level of phenol sulphotransferase activity was low in extracts from the livers of rats deficient in vitamin A. This activity could be restored to normal by the addition of retinol or of retinoic acid (in concentrations of about 5 $\mu$g/ml) *in vitro*. There were quantitative differences from the very similar effects of vitamin A on the sulphate-activating system: the latter was most strongly influenced *in vitro* by retinol whereas the transferase was more susceptible to retinoic acid.

Carroll & Spencer (1965 *a*) showed that in the livers of foetal rats the activities of several types of sulphotransferase were much lower than in the livers of adult animals. The activities increased quite rapidly after birth and reached adult values within about one week. Carroll & Spencer (1965 *b*) have suggested that this low sulphotransferase activity in foetal liver is due to a physiological deficiency of vitamin A because the addition of retinol or of retinoic acid (in concentrations of about 1 $\mu$g/ml) *in vitro* increased the activity to the adult value. They therefore postulated that in foetal rat liver the sulphotransferases were present in inactive forms which became active only after birth when vitamin A was derived from the milk.

This effect in foetal liver is not a general one. While it is obvious in the rat, it does not occur in the foetal guinea-pig (Spencer & Raftery, 1966) which may be replete with vitamin A, nor in the human foetus (Wengle, 1964 *b*) although the sulphotransferase activity of the latter is certainly generally lower than that of the adult.

At present, then, there is only circumstantial evidence that vitamin A is involved in the sulphotransferases and further investigations are certainly required. If vitamin A, or perhaps a metabolite thereof, is a participant in the transfer of the sulphuryl group as it may be in the activation of sulphate ions, then much remains to be discovered as to the exact way in which the vitamin functions. Once again, further progress would appear to be dependent on the availability of more highly purified sulphotransferases.

## 6.9    Other routes of formation of sulphate esters

Although there seems little doubt that the major route of formation of sulphate esters *in vivo*, as *in vitro*, involves the transfer of the sulphuryl group from PAPS, and so ultimately involves the utilization of $SO_4^{2-}$ ions, there have been some reports of other pathways. The first of these implies a direct utilization of the sulphur of cysteine, without the prior conversion of this to $SO_4^{2-}$ ions, and in recent years this has been propounded by Wellers (1960), despite the formidable chemical difficulties involved. Wellers has based his claims on some very detailed sulphur-balance studies in rats fed various artificial diets with and without the addition of *p*-nitrophenol or of indole as sulphate acceptors (Wellers & Boelle, 1960). This most carefully controlled work has shown quite clearly that *in vivo* the sulphur of dietary cysteine is used much more efficiently than the sulphur of dietary $SO_4^{2-}$ ions for the formation of ester sulphate. It is nevertheless difficult to accept Wellers' conclusion that this demonstrates the direct utilization of the sulphur of cysteine for the formation of sulphate esters: it seems much more likely that what has been quite conclusively demonstrated is the relatively inefficient absorption of $SO_4^{2-}$ ions from the gastrointestinal tract. In this connection it should perhaps be noted that the work of Bray *et al.* (1952) showed that, at least in the rabbit, the level of endogenous $SO_4^{2-}$ ions is normally rate-limiting in the formation of aryl sulphates *in vivo*.

Another postulated route for the formation of carbohydrate sulphates involves the normally hydrolytic aryl sulphatase: this is discussed below in chapter 7 (§ 7.3.3).

Finally it should be mentioned that Butenandt *et al.* (1960) have suggested that ommatin *D* (formula *6.3*, § 6.1) may act as a sulphate donor in an oxidative reaction which, considering only the functional part of the molecule, can be represented by equation 6.4 below. This reaction is of course identical in type to the oxidative transfer of the sulphate group from hydroquinone monosulphate to methanol (see

§ 6.1) (Weidman *et al.* 1967) and there is no more evidence for its occurrence *in vivo* than there was for that with hydroquinone sulphate.

$$+ R.OH \longrightarrow \quad + R.OSO_3^- + 2H^+ + 2e$$

$$(6.4)$$

In conclusion, therefore, it must be conceded that any evidence for the formation of sulphate esters by routes not involving a sulphotransferase is slight indeed.

## 6.10 The physiological role of the sulphotransferases

Although the sulphotransferases presumably fulfil *in vivo* their obvious function of forming sulphate esters, this gives no real answer to questions about their physiological role because the function of the sulphate esters themselves is by no means clear. Even if only the quantitatively more important types of naturally occurring ester (carbohydrate sulphates, choline sulphate ester, steroid sulphates, aryl sulphates and the mustard oil glycosides) are considered, the exact part which these play in the overall metabolism of the appropriate organism is far from clear.

Some carbohydrate sulphates, especially the high molecular weight mucopolysaccharide sulphates and sea-weed polysaccharides, appear to differ in their function from the other esters in that they seem to fulfill a purely structural function. However, this interpretation may well only reflect our ignorance. Nevertheless it must be stressed that many such compounds (e.g. the chrondroitin sulphates) have very high molecular weights, are highly charged and may be present in high concentration. Chondroitin sulphate, for instance, occurs to the extent of about 15% in cartilage. Under such conditions the osmotic pressures due to these compounds must be very high: it will be intrinsically high because of their presence in high concentrations but it will be increased, first by their highly non-ideal behaviour (see Ogston, 1966), a non-ideality which must itself be increased by the presence of other macromolecular components such as proteins, and secondly by large salt effects arising from their high charge. The resultant high osmotic pressure may well be of fundamental importance in maintaining the structure of tissues such as, for example, articular cartilage. Further, such highly charged macromolecules will show pronounced ion-binding and they will have many of the pro-

5-2

perties of ion-exchange resins, an aspect of their behaviour which has recently been considered in some detail by Marroudos (1968).

Of the sulphate esters of low molecular weight, choline sulphate ester functions as a store of sulphur in certain fungi and some steroid sulphates can (but do not necessarily do) act as metabolic intermediates. The aryl sulphates and mustard oil glycosides, on the other hand, have no obvious function unless the former be regarded—certainly erroneously—simply as detoxication products. That aryl sulphates and steroid sulphates could be involved in the transport of materials through lipoid membranes has been demonstrated by the most interesting model experiments of Keller & Blennemann (1961 b) who showed that methylene blue could be transported between two aqueous phases separated by chloroform if one of the former contained a system forming sulphate esters and the other a system hydrolysing them (a sulphatase). This transport depended upon the solubility in chloroform of the methylene blue salts of certain sulphate esters (see p. 66) and was of necessity accompanied by the movement of the phenol and of sulphate ions in the same direction as the methylene blue. Such a system may not be purely artificial because Burstein & Dorfman (1962, 1963) have shown that steroid sulphates are rendered soluble in toluene, and so presumably in lipids, by their combination with cationic phospholipids. Further, the sulphotransferases and the sulphatases certainly seem to be spatially separated in the intact cell.

Perhaps the only unifying principle lying behind the formation of sulphate esters of all types is the conversion of a slightly polar compound to a strong acid, completely ionized at all pH values compatible with life, with all the concomitant changes in general physical and chemical properties, particularly in solubility and in reactivity. One need consider only the great changes in chemical and physical properties which accompany the conversion of Sephadex to Sulphoethylsephadex to visualize the important consequences of the conversion of, say, chondroitin to chondroitin sulphate. The results of these changes in properties may then be utilized in quite different ways to give the different functions of the various types of sulphate ester. If, however, the prime function of the sulphate cycle (see chapter 1), and therefore of the sulphotransferases, is to form strong acids *in vivo* then perhaps the importance of sulphate ions becomes obvious. They could be replaced by no other anion present in biological fluids in significant amounts and, as has already been pointed out, sulphate ions must have been available to organisms from the earliest days of life on this earth.

# 7

## THE SULPHATASES

The sulphatases form a large group of hydrolytic enzymes which catalyse the hydrolysis of sulphate esters according to equation 7.1.

$$R.O.SO_3^- + H_2O \rightarrow R.OH + SO_4^{2-} + H^+ \qquad (7.1)$$

Although typical substrates for the sulphatases, the aryl sulphates, were recognized by Bauman in 1876, there was no systematic study of these enzymes until the appearance of a long series of papers by Neuberg and his co-workers in Berlin in the 1920s and by Soda and his group in Tokyo in the 1930s. This work lead to the recognition of several different sulphatases, each having different specificities, and the conclusions reached during this exploratory period were admirably summarized by Fromageot in his now classical review of 1938. Again there followed a period of quiescence, as can be judged from a later review by Fromageot (1950), and it was not until 1953 that the present active investigations began with the work of Dodgson & Spencer and of Roy. Since then interest in the sulphatases has grown rapidly.

It is now clear that there are many different types of sulphatase, each apparently being characterized by a fairly high specificity towards the organic part of the molecule and, as far as is known, an absolute specificity for the sulphate moiety. It must be stressed, however, that as yet few of the sulphatases have been obtained in a purified state and much further work is required before the relationships of the various groups are understood and their separate identities made certain.

Up to the present, work on the sulphatases has been concerned mainly with the isolation of the enzymes and with their specificities but interest has recently been growing in their possible metabolic roles and in the mechanisms of their control in micro-organisms. Unfortunately, the physiological functions of the sulphatases remain as much of a problem today as they were when Fromageot considered them in 1938 and it is undoubtedly in this area that the most important developments remain to be made in the not too distant future.

## 7.1 Arylsulphatases

The arylsulphatases catalyse the hydrolysis of aryl sulphates through the fission of the O—S bond (Spencer, 1958). They undoubtedly comprise the largest and most studied group of the sulphatases and the only one for which homogeneous enzyme preparations are available. Spectrophotometric methods of assay have been most used for the study of these enzymes and the commonest substrates are potassium *p*-nitrophenyl sulphate (*7.1*) and potassium 2-hydroxy-5-nitrophenyl sulphate (*7.2*, nitrocatechol sulphate) although others, such as *p*-acetylphenyl sulphate

$$NO_2 \qquad\qquad NO_2$$

$$OSO_3^- \qquad\qquad OSO_3^-$$
$$\qquad\qquad\qquad OH$$

$$(7.1) \qquad\qquad (7.2)$$

and phenolphthalein disulphate, have been used with no obvious advantage (Dodgson & Spencer, 1957 *a*). The same substrates can also be used in the pH-stat which allows the easy determination of initial reaction velocities and so has many advantages over the more commonly used spectrophotometric methods (Andersen, 1959 *a*; Nichol & Roy, 1964). It should be noted that the pK of the phenolic group of nitrocatechol sulphate is 6·4: as the pH optima of the various arylsulphatases cover the range of pH 5 to pH 8 this hydroxyl group may be ionized in the reaction mixtures used for the assay of some of these enzymes but unionized in others. Detailed studies of the influence of this ionization on arylsulphatase activity have not yet been made. A fluorimetric method which uses 4-methylumbelliferone sulphate has been described (Sherman & Stanfield, 1967) despite earlier criticisms of this substrate (Mead, Smith & Williams, 1955).

Arylsulphatases have been detected in almost all tissues of all animals and in many micro-organisms. They have also been found in the seeds of a few species of higher plants (Ney & Ammon, 1959) but claims of their occurrence in the vegetative tissues thereof are based almost entirely on not very satisfactory histochemical evidence (Olszewska & Gabara, 1964; Poux, 1966). Such a widespread distribution of the arylsulphatases might be taken to imply a rather fundamental role for these enzymes but certainly up to the present this has escaped detection.

The best known arylsulphatases are those occurring in the livers of the higher mammals and primarily on the basis of investigations of these it was suggested that they could be subdivided into two major types, type I and type II, containing sulphatase C and sulphatases A and B respectively. The evidence for this in the case of ox liver is summarized in table 7.1. However, it is now evident from more recent work on the

TABLE 7.1   *The general properties of the arylsulphatases of ox liver*

| | Sulphatase | | |
|---|---|---|---|
| | A | B | C |
| Substrate: nitrocatechol sulphate | | | |
| optimum pH | 5·0 | 5·6 | 7·5 |
| optimum substrate conc. (mM) | 3·0 | 15 | 2C |
| $K_m$ (mM) | 0·8 | 1·8 | 8·0 |
| Relative activity in: | | | |
| 0·025 M-$SO_4^{2-}$ | 0·29 | 0·63 | . |
| 0·1 M-$Cl^-$ | 1·21 | 0·78 | . |
| 0·025 M-$H_2PO_4^{2-}$ | 0·01 | 0·01 | . |
| Substrate: *p*-nitrophenyl sulphate | | | |
| optimum pH | 5·4 | 5·7 | 8·0 |
| optimum substrate conc. (mM) | 80 | 50 | 7·5 |
| $K_m$ (mM) | 23 | 4·4 | 2·0 |
| Relative activity in: | | | |
| 0·025 M-$SO_4^{2-}$ | 0·16 | 0·32 | 1·00 |
| 0·1 M-$Cl^-$ | 0·94 | 5·0 | 1·00 |
| 0·025 M-$H_2PO_4^{2-}$ | 0·03 | 0·01 | . |
| $V_{NCS}/V_{NPS}$ | 3·0 | 55 (0·6*) | 0·5 |

\*   Value for NPS determined in 0·1 M-KCl.

lower vertebrates and on micro-organisms that the distinction between the various enzymes is by no means as clear cut as is suggested by table 7.1 and it is likely that the data therein simply describe extreme examples of a wide and continuous range of types of arylsulphatase. Nevertheless, the distinction between the types I and II arylsulphatases and sulphatases A, B and C remains valid so long as only the arylsulphatases of the higher mammals are being considered and such a nomenclature is certainly useful, although its limitations must be recognized.

### 7.1.1   Arylsulphatases of eutherian mammals

*Sulphatase A.* The sulphatase A of ox liver was first detected by Roy (1953 *a,b*) using nitrocatechol sulphate as substrate: it was separated from the other arylsulphatases of liver and was characterized by the

135

highly anomalous kinetics of its reaction. A similar enzyme was subsequently found in human liver (Baum, Dodgson & Spencer, 1958) and this also exhibited the same anomalous kinetics (Baum & Dodgson, 1958). Both enzymes were localized in the lysosomes. Other sulphatases A have been isolated from human brain (Balasubramanian & Bachhawat, 1963), from ox brain (Błeszyński & Działoszyński, 1965) and probably also from pig intestine (Szafran & Szafran, 1964). The two enzymes from brain show kinetic anomalies similar to those of the liver enzymes.

The sulphatase A of ox liver has now been obtained in a homogeneous state, as judged by its behaviour in the ultracentrifuge and in moving boundary electrophoresis, in a yield of about 1 mg/kg liver which represents about 20% of the enzyme initially present (Nichol & Roy, 1964). Table 7.2 summarizes the physical properties of the enzyme which has a

TABLE 7.2   *The physical properties of sulphatases A and B of ox liver and of the arylsulphatase of* Aerobacter aerogenes

| | Sulphatase | | |
|---|---|---|---|
| | A | B | *Aerobacter* |
| specific activity* | 140 | 110 | 73 |
| $E_{280}^{1\%}$ | 7 | 14 | . |
| $E_{280}/E_{260}$ | 1·68 | . | 1·95 |
| p$I$ | 3·6 | 8·3 | . |
| $s_{20,w}^0$ | 6·5† | . | 3·5 |
| molecular weight | 107,000† | 45,000 | 40,700 |
| $f/f_0$ | 1·28 | . | 1·24 |

\*   $\mu$Moles substrate hydrolysed per minute per mg protein.
†   These values are for the monomer of sulphatase A at pH 7·5 in 0·1 $I$ buffers: at pH 5·0 in 0·1 $I$ buffers a tetramer is formed and the corresponding values are 14·2 and 411,000.

molecular weight of 107,000 at pH 7·5 in 0·1 ionic strength buffers. The amino acid composition of this enzyme has been determined (Nichol & Roy, 1965) and its most striking feature is the very high content of proline, some 90 residues per molecule: this very high value is only approached by some of the caseins and exceeded by the collagens. Also noteworthy is the presence of more than eight residues of glucosamine per molecule of protein: presumably other sugars are also present but these have not been identified. A relatively high content of glutamic and aspartic acids is reflected in the low isoelectric point of sulphatase A of 3·6 and in the high negative charge of about −30 at pH 7·4 (Nichol & Roy, 1966).

Sulphatase A forms a tetramer of molecular weight 410,000 when the pH is lowered to 5·0 in 0·1 ionic strength buffers. A detailed analysis of the sedimentation of sulphatase A showed that at protein concentrations of about 0·005 to 1% in solutions of ionic strength 0·1 only the monomer existed at pH values greater than 6·5 whereas at pH values below 5·5 only the tetramer existed. In the intermediate range of pH 5·5 to 6·5 the enzyme was present as an equilibrium mixture of monomer, dimer, trimer and tetramer, the dimer being the dominant species (Nichol & Roy, 1965). Studies of the effect of ionic strength, of temperature and of sodium dodecyl sulphate on the sedimentation of sulphatase A suggested that not only was the structure of the polymer dependent upon hydrophobic interactions but also the monomer itself was formed by the hydrophobic interaction of sub-units (Nichol & Roy, 1966).

The study of this interacting system in the ultracentrifuge required that the enzyme be present in concentrations of not less than 0·005% which is greater than that used in the study of the protein as an enzyme by a factor of a hundred or more. When lower, catalytic concentrations of the enzyme were examined by frontal analysis on Sephadex columns, it was shown that the monomer remained stable at pH 7·5 to protein concentrations of the order of $10^{-5}$ g/100 ml but that at such low concentrations the tetramer (at pH 5) was unstable and dissociated, presumably to the monomer (Nichol & Roy, 1965). It therefore follows that most, if not all, of the studies of the enzymatic properties of sulphatase A have been carried out under conditions where the monomer would be the dominant species although the presence of small amounts of polymeric forms can by no means be excluded.

As far as is known sulphatase A is a true arylsulphatase which attacks only aryl sulphates and related compounds such as indoxyl sulphate but recent work has suggested a possible action on cerebroside sulphates (see § 7.3.6). It will not attack chondroitin sulphate or any of a number of steroid sulphates: the claim (Keller & Blennemann, 1961 a) that sulphatase A hydrolysed cholesteryl sulphate was in error and the apparent reaction detected by these workers was simply a reflection of the rather powerful binding of cholesteryl sulphate to protein (Roy, 1963 a). Of the commonly used substrates nitrocatechol sulphate is the most readily hydrolysed and has the lowest $K_m$ but other compounds such as phenyl sulphate, m- and p-nitrophenyl sulphates, 1- and 2-naphthyl sulphates, 4-nitro-1-naphthyl sulphate and 2-phenanthryl sulphate are hydrolysed, although more slowly and with higher values of $K_m$ than nitrocatechol

137

sulphate. In general terms, the more polycyclic the phenol, the lower is the $K_m$ value for the sulphate ester. Accurate kinetic parameters cannot be given for the enzyme because of its highly anomalous kinetics, which are discussed below, but the optimum pH for the hydrolysis of the above substrates lies in the region of pH 5·0 to 5·5. Sulphate, sulphite, phosphate and fluoride ions are all quite powerful inhibitors at concentrations in the millimolar range, with sulphite ions being particularly effective.

Hydroxylamine, phenylhydrazine and similar carbonyl reagents powerfully inhibit sulphatase A, giving 50% inhibition at concentrations of 2 mM and 0·03 mM respectively. The inhibition is due to the reduction of traces of $Cu^{2+}$ ions in the reagents (or enzyme) to give $Cu^+$ ions which, like $Ag^+$ ions, inhibit sulphatase A, probably through a reaction with histidyl residues in the protein. The enzyme is also inhibited by bromine and more interestingly by N-acetylimidazole and by tetranitromethane, all of which would be consistent with the presence of a tyrosyl residue in the active centre (Jerfy & Roy, 1969). The sulphatase A of ox liver is inhibited by p-hydroxymercuribenzoate at pH 5·0 but not at pH 7·5, and at neither pH by N-ethylmaleimide, so that the previous suggestion (Roy, 1955) that this enzyme is an SH enzyme obviously requires further investigation. If it is such, then it is a remarkably stable one because during its preparation no precautions need be taken to exclude heavy metals or oxygen, either of which might be expected to produce a considerable degree of inactivation of a typical SH enzyme. It is of some interest that Szafran & Szafran (1964) have also shown a sensitivity of the presumed sulphatase A from pig intestine to mercurials at pH 5 but an insensitivity to N-ethylmaleimide.

As has already been pointed out, one of the most characteristic features of sulphatase A is its highly anomalous kinetics: these were first noted by Roy (1953 b, 1957 a) for the enzyme from ox liver and they have since been studied in considerable detail by Baum, Dodgson & Spencer (1958) and by Baum & Dodgson (1958) using the enzyme from human liver. The original observation was that the velocity of the enzymatic reaction, as measured by the extent of the hydrolysis of the substrate over one hour, was not rectilinearly related to the concentration of the enzyme and it was suggested that this phenomenon was caused by polymerization of the enzyme to a more active form. It was then shown that the velocity of the reaction varied in an extremely complex manner during its initial stages: typical progress curves for the hydrolysis of nitrocatechol sulphate are given in fig. 7.1. In general, the reaction has three stages: an initial stage in which the velocity falls rapidly from a high starting level to a relatively low value which remains almost constant throughout the second stage but which then increases again to a constant value in stage three, the velocity in the latter stage being considerably

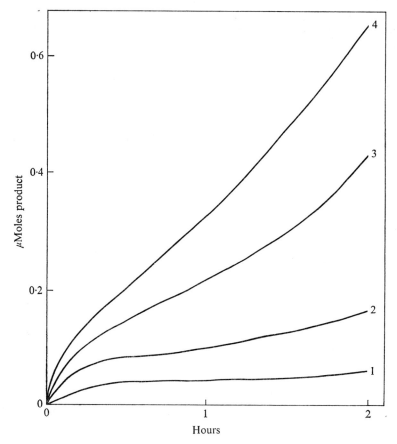

Fig. 7.1 Progress curves for the hydrolysis of nitrocatechol sulphate by increasing relative concentrations (1–4) of sulphatase A.

less than the initial velocity of the first stage. The lengths of these stages vary with the concentration of the enzyme present in the reaction mixture, as is obvious from fig. 7.1. In a very detailed study of these kinetics Baum & Dodgson (1958) showed that the reaction velocity was greatly influenced by the accumulation of the reaction products and that these, in particular the nitrocatechol and the sulphate ions liberated from the commonly used nitrocatechol sulphate, were responsible for the increase in velocity between the second and third stages. To account for the very complex kinetics, which were similar to those found with the enzyme from ox liver (Roy, 1957 *a*), Baum & Dodgson proposed that during the reaction the enzyme-substrate complex could form a catalytically inactive enzyme species which was subsequently reactivated by its combination with either of the products of the reaction, that is, with nitrocatechol or with sulphate ions. The scheme which they proposed was essentially the following of which the main feature was the formation of inactive enzyme species, E′ and E′S, which were converted to active forms either by combination with the reaction products, as in the following scheme, or by

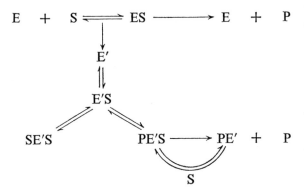

combination with certain competitive inhibitors (e.g. phosphate, pyrophosphate) to give inhibitor complexes analogous to PE′S. Obviously the kinetics of such a complex sequence of reactions would be extremely difficult to describe in terms of a steady-state—if indeed such could exist in the initial stages of the reaction—and Baum and Dodgson considered it in only qualitative terms which, nevertheless, sufficed to explain most of the anomalous kinetics of sulphatase A.

A rather similar reaction sequence was suggested by Andersen (1959 *a, b, c*) who followed the hydrolysis of nitrocatechol sulphate in a pH-stat, a technique which has the advantage of allowing the continuous recording of the progress of the reaction but, as used by Andersen, suffered from the disadvantage that the reaction mixture contained $0.15$ M-KCl to maintain a constant ionic strength during the reaction. The presence of Cl⁻ ions was unfortunate because these have a pronounced effect on the behaviour of sulphatase A. There was only one important difference between the findings of Andersen and those of Roy and of Baum and

Dodgson, namely, that sulphate ions did not reactivate the postulated inactive form of the enzyme although nitrocatechol did do so. The scheme proposed by Andersen can be represented as follows (where $P_1$ is nitrocatechol and $P_2$ sulphate ion).

$$E \; + \; S \rightleftharpoons ES \rightleftharpoons E \; + \; P_1 \; + \; P_2$$
$$\Big\Updownarrow$$
$$E'S$$
$$\Big\Updownarrow$$
$$P_1E'S \rightleftharpoons P_1E \; + \; P_1 \; + \; P_2$$
$$S$$

Obviously, this reaction sequence is basically similar to that of Baum and Dodgson and the only major point at issue is the inability of sulphate ions to reactivate the enzyme. This discrepancy could well be caused by the presence of high concentrations of $Cl^-$ ions, or perhaps by the absence of acetate ions, both of which influence the behaviour of sulphatase A. Another possible reason for the differences between the results is that during the preparation of the enzyme Andersen heated it to 65° for five minutes: although this apparently does not grossly alter the enzymatic activity it could perhaps alter some of the details of the reaction. It should also be noted that although the above reaction sequence apparently holds at 40°, at 25° the reactivation of the enzyme by nitrocatechol does not take place, so that only the reactions involving E, ES and E'S can occur. Once again, there is no proof of the correctness of the proposed reaction sequence and although Andersen treated his data in a more quantitative way than did Baum and Dodgson, it would seem that the problem of the anomalous kinetics of sulphatase A cannot yet be regarded as solved. Now that apparently homogeneous preparations of the enzyme are available the kinetics of the reaction would be well worth investigating by more refined techniques. In particular, attempts to isolate the different forms of the enzyme, as was previously tried by Baum and Dodgson, might give valuable information but it seems not improbable that if such forms do exist they may differ only in their conformation so that their isolation and characterization could well be difficult.

Whatever be the explanation of the anomalous kinetics, it is clear that their existence makes difficult any detailed study of the kinetics of sulphatase A.

*Sulphatase B.* No representative of this class of sulphatase has yet been obtained in a pure state despite the quite detailed studies of the sulphatase B of ox liver (Roy, 1954 *a*; Webb & Morrow, 1959, 1960), of human liver (Dodgson & Wynn, 1958) and of ox brain (Błeszyński & Działoszyński, 1965) which have simply served to show the fairly widespread distribution of this enzyme. The arylsulphatase isolated from calf aorta by Utermann, Lorenzen & Hilz (1964) is probably also a sulphatase B but confirmation of this is needed. Again these enzymes occur in the lysosomes and so are easily obtained in solution.

Wortman (1962) was the first to point out that mammalian tissues might contain as many as four arylsulphatases, only one of which seemed to be a sulphatase A, and more recently Błeszyński (1967) has suggested that ox brain may contain three sulphatases B. Allen & Roy (1968) have isolated from ox liver two sulphatases B (B$\alpha$ and B$\beta$) which apparently differ in their charge but not in their size: the homogeneity of the final preparation has not been shown but a purification of about thirty thousand-fold from the original extract of the liver has been achieved. So far no detailed study of the physical properties of these enzymes has been possible but their molecular weights (determined by chromatography on Sephadex) are about 45,000 and their isoelectric points about 8·3, both of which features clearly distinguish them from sulphatase A. The sulphatases B are also less stable than sulphatase A, particularly at pH values below neutrality, and care is required in their assay.

Sulphatase B was first detected in ox liver by its ability to hydrolyse nitrocatechol sulphate and it was shown that simpler substrates, such as *p*-nitrophenyl sulphate, were hydrolysed only extremely slowly by this enzyme or by the sulphatase B of human liver (Roy, 1954 *a*; Dodgson & Wynn, 1958). Webb and Morrow, however, pointed out that *p*-nitrophenyl sulphate was hydrolysed by sulphatase B, at a rate comparable to nitrocatechol sulphate, when Cl⁻ ions were present in the reaction mixture. These ions, and other monovalent anions, activated the breakdown of the enzyme-substrate complex formed between sulphatase B and *p*-nitrophenyl sulphate without altering the value of the $K_m$ for the substrate. No comparable effect was obtained with nitrocatechol sulphate and the hydrolysis of this by sulphatase B is, in fact, slightly inhibited by Cl⁻ ions. The two purified components, B$\alpha$ and B$\beta$, of the crude enzyme used by Webb and Morrow show exactly the same behaviour. The reason for the different properties of the two enzyme-substrate complexes has not been investigated and further developments would appear to require the availability of a pure enzyme in amounts sufficient to permit physical studies to be made on it.

Sulphatase B somewhat resembles sulphatase A in, for example, its inhibition by sulphate, sulphite, phosphate or fluoride ions but, as already shown in table 7.1, there are many quantitative differences. The kinetics of the crude enzyme from human liver have been investigated in some detail by Dodgson & Wynn (1958) and they have shown that the enzyme contains an ionizing group with a pK of 5·7 which is involved in the binding of the substrate: the group was not identified. The enzyme-substrate complex contained two ionizing groupings with pK values of 5·2 and approximately 7 respectively. From these results there is no doubt that the binding of the substrate is independent of its subsequent reaction at the catalytic site of the enzyme. Comparable data are not yet available for the hydrolysis of *p*-nitrophenyl sulphate in the presence of $Cl^-$ ions but such information would be most valuable and might throw considerable light on the mechanism of the reaction. Obviously, however, any future work must be carried out with more clearly defined enzyme preparations.

*Sulphatase C.* This enzyme, unlike sulphatases A and B, is very firmly bound to the microsomal fraction of liver and is extremely difficult to obtain in true solution. Only the sulphatase C of rat liver has so far been solubilized, by treatment of the isolated microsomes with crude pancreatic enzymes in the presence of non-ionic detergents (Dodgson, Rose & Spencer, 1957) and all attempts to obtain soluble preparations of the ox (Roy, 1956 *c*) or human (Dodgson, Spencer & Wynn, 1956) enzymes have been unsuccessful.

Recently Działoszynski, Gniot-Szulżycka & Barancewicz (1966) claimed to have solubilized a sulphatase C from an acetone powder of human placenta by autolysis at pH 9·8 for one week: unfortunately these authors have not provided an adequate characterization of this enzyme which could simply be a sulphatase A or B, both of which occur in placenta (Gniot-Szulżycka & Działoszynski, 1966).

Quite apart from its insolubility, sulphatase C differs strikingly from sulphatases A and B, as is shown by the data in table 7.1, the most obvious differences being the insensitivity of sulphatase C to sulphate or phosphate inhibition and the relative rates of hydrolysis of nitrocatechol and *p*-nitrophenyl sulphates. The data for sulphatase C must be treated with some reserve, however, because of the uncertainties introduced by the use of an insoluble enzyme and it would seem that any further progress must await the development of methods for its preparation in a soluble form.

An important point which is often forgotten is that all preparations of sulphatase C from mammalian liver also exhibit a steroid sulphatase activity (see § 7.2.1) and the relationship between these activities has certainly not been clarified although they appear to be due to separate enzymes (Roy, 1957 b; French & Warren, 1967).

### 7.1.2 Arylsulphatases of other vertebrates

None of these enzymes has been purified and therefore there is little unambiguous information on their nature. Microsomal arylsulphatases, analogous to sulphatase C of eutherian mammals, are apparently sporadic in occurrence and have been detected in the livers of only three of the seven species studied (Roy, 1963b). It was tentatively claimed (Roy, 1963 b) that the livers of the lower vertebrates contained only a single arylsulphatase of lysosomal origin. This had the electrophoretic properties of a sulphatase B although it differed from a typical sulphatase B of a eutherian mammal in that p-nitrophenyl sulphate was quite readily hydrolysed by it even in the absence of Cl⁻ ions. This generalization may be invalid because preliminary work on the arylsulphatases of the liver of the red kangaroo (*Macropus rufus*) has shown the presence of two soluble arylsulphatases which resemble, but are by no means identical with, sulphatases A and B of ox liver. The study by Działoszyński, Kuik & Leźnicki (1966) of the arylsulphatase of the liver of the bream (*Abramis brama*) also suggests that more than one such enzyme may be present in this species although the authors do not stress this interpretation. More detailed studies of these enzymes are awaited.

### 7.1.3 Arylsulphatases of the invertebrates

Again there has been little detailed investigation of the arylsulphatases of the invertebrates although such enzymes are certainly widespread. Only the arylsulphatases of a few molluscs, the European limpet *Patella vulgata*, the Roman snail *Helix pomatia* and the Japanese marine gastropod *Charonia lampas*, have been purified to a slight extent. Some of the properties of these enzymes are summarized in table 7.3 but these results must be viewed with some reserve because it is probable that some molluscan tissues contain more than one arylsulphatase. Wortman & Schneider (1960) have claimed the separation, by chromatography on

DEAE-cellulose, of three arylsulphatases from *P. vulgata* and Suzuki, Takahashi & Egami (1959) have separated two arylsulphatases, differing in their electrophoretic mobility and pH optima, from the mucous gland, but not from the hepatopancreas, of *C. lampas*.

TABLE 7.3  *The general properties of some arylsulphatases from the digestive glands of molluscs. Two values are given for the parameters of the enzyme from* H. pomatia *because of the very large variation of the optimum pH with substrate concentration*

| | Helix pomatia | | Patella vulgata | Charonia lampas |
|---|---|---|---|---|
| Substrate: nitrocatechol sulphate | | | | |
| optimum pH | 7·4 | 6·6 | 5·3 | 6·2 |
| optimum substrate conc. (mM) | 15 | 0·5 | 10 | 12 |
| $K_m$ (mM) | 0·73 | 0·09 | 0·7 | 1·9 |
| Relative activity in: | | | | |
| 0·025 M-$SO_4^{2-}$ | 0·95 | 0·89 | 0·95* | 0·92 |
| 0·025 M-$Cl^-$ | 1·05 | 1·00 | 1·18† | . |
| 0·025 M-$H_2PO_4^{2-}$ | 0·26 | 0·18 | 0·16* | 0·77‡ |
| Substrate: *p*-nitrophenyl sulphate | | | | |
| optimum pH | 7·5 | 6·3 | 5·6 | 6·1 |
| optimum substrate conc. (mM) | 15 | 0·5 | . | . |
| $K_m$ (mM) | 5·1 | 0·27 | 6·6 | . |
| Relative activity in: | | | | |
| 0·025 M-$SO_4^{2-}$ | 0·86 | 0·61 | . | . |
| 0·025 M-$Cl^-$ | 0·94 | 1·00 | . | . |
| 0·025 M-$H_2PO_4^{2-}$ | 0·24 | 0·20 | . | . |

\* 0·005 M    † 0·05 M    ‡ 0·001 M

The enzyme of this group which has been most thoroughly studied is undoubtedly that of *H. pomatia*: this has been investigated by Jarrige (1963) and more especially by Dodgson & Powell (1959). Neither group found any evidence for the occurrence of more than one arylsulphatase in their preparations. It is a less stable enzyme than most arylsulphatases and is extensively denatured by even gently agitating the reaction mixture in which it is being assayed. Strangely, the inactivation is much less if the substrate (nitrocatechol sulphate) is absent. An interesting property of the enzyme, and one which it seems to share with no related sulphatase, is the striking shift in its pH optimum with variations in the substrate concentration. An increase in the concentration of *p*-nitrophenyl sulphate from 0·5 to 15 mM shifts the pH optimum from 6·0 to 7·3 and an effect of similar magnitude occurs with nitrocatechol sulphate. The mechanism

145

of this phenomenon remains unknown: it is of course well known to be shown by other hydrolases but most of the explanations which have been put forward in these cases are not applicable in the present instance (Dodgson & Powell, 1959).

These molluscan arylsulphatases obviously have some analogies with the type II arylsulphatases of the eutherian mammals—for instance, their inhibition by sulphate or phosphate ions and their solubility—but they are not directly comparable to them. Their isoelectric points, as judged by their behaviour on paper electrophoresis, must lie between pH 6 and pH 7 so that they fall between those of sulphatases A and B and they certainly do not show any of the anomalous kinetics of sulphatase A.

### 7.1.4   Arylsulphatases of micro-organisms

The most studied examples of these enzymes are those of *Aspergillus oryzae* (the source of Taka-diastase) (Abbott, 1947; Robinson, Smith, Spencer & Williams, 1952; Young, 1958; Ammon & Ney, 1959), of *Alcaligenes metalcaligenes* (Dodgson, Spencer & Williams, 1955) and of *Aerobacter aerogenes* (Harada & Kono, 1954). The general properties of these microbial arylsulphatases are given in table 7.4. It has usually been assumed that only a single arylsulphatase is produced by any individual micro-organism but Cherayil & Van Kley (1961, 1962, 1963) were the first to cast doubt on this assumption by demonstrating that *A. oryzae*, the classical microbial source of arylsulphatase, contains three aryl-sulphatases, separable by chromatography or by electrophoresis, in proportions which vary according to the conditions under which the organism was grown. More detailed reports of those findings would be most valuable. Harada (1964) has shown that *Pseudomonas aeruginosa* must contain at least two arylsulphatases but he has not separated them.

*A. aerogenes*, on the other hand, apparently does contain only one arylsulphatase (Rammler, Grado & Fowler, 1964) and this is the only microbial arylsulphatase to have so far been obtained in a homogeneous state (Fowler & Rammler, 1964). The physical properties of this enzyme, which is obviously quite different from sulphatases A and B, have been summarized in table 7.2. The specificity of the enzyme is peculiar: although it can readily hydrolyse *p*-nitrophenyl sulphate and phenyl sulphate, it cannot hydrolyse 1-naphthyl sulphate nor phenolphthalein disulphate. Another interesting feature is its inhibition by cyanide: this occurs only in the presence of the substrate but is irreversible so that it cannot be strictly uncompetitive.

146

TABLE 7.4 *The general properties of some arylsulphatases isolated from micro-organisms*

| | *Aspergillus oryzae* | *Aerobacter aerogenes* | *Alcaligenes metalcaligenes* | *Proteus rettgeri* | *Proteus vulgaris* |
|---|---|---|---|---|---|
| Substrate: nitrocatechol sulphate | | | | | |
| optimum pH | 5·9 | . | 7·8 | 6·7 | 5·7 |
| optimum substrate conc. (mM) | 3·0 | . | 15 | 65 | 25 |
| $K_m$ (mM) | 0·35 | . | 0·13 | . | 5 |
| Relative activity in: | | | | | |
| 0·025 M-SO$_4^{2-}$ | . | . | 1·00 | 0·70* | 0·85 |
| 0·025 M-Cl$^-$ | . | . | 1·00 | . | 1·19 |
| 0·025 M-H$_2$PO$_4^{2-}$ | . | . | 1·00 | 1·43* | 0·08 |
| Substrate: *p*-nitrophenyl sulphate | | | | | |
| optimum pH | 6·2 | 7·1 | 8·8 | 8·3 | 6·2 |
| optimum substrate conc. (mM) | 15 | 10 | 2·5 | 45 | 100 |
| $K_m$ (mM) | 0·17 | 3·3 | 0·48 | . | 55 |
| Relative activity in: | | | | | |
| 0·025 M-SO$_4^{2-}$ | 1·00 | 1·00* | . | 1·00* | 0·52 |
| 0·025 M-Cl$^-$ | 1·00 | . | . | . | 2·28 |
| 0·025 M-H$_2$PO$_4^{2-}$ | 0·96 | 0·45* | . | 0·94* | 0·07 |

* 0·01 M

The arylsulphatase of *A. metalcaligenes* is the only arylsulphatase from any source for which detailed information on the specificity is available. Dodgson, Spencer & Williams (1956 *a*) investigated the hydrolysis of a wide range of phenyl sulphates by this enzyme and showed that there was a direct relationship between the Hammet constant of the substituent and both the apparent affinity of the enzyme for the substrate (measured as $1/K_m$) and the relative rate of hydrolysis of the substrate. In other words, the more electrophilic the substituent, the more rapid the hydrolysis, a situation exactly analogous to that pertaining in the acid-catalysed hydrolysis of aryl sulphates (Burkhardt, Ford & Singleton, 1936), which has already been considered in § 2.3.1. From the variation of $K_m$ with pH, Dodgson, Spencer & Williams (1955) concluded that groupings having pK values of 8·2 and 9·4 were involved in the binding of the substrate to the enzyme. There was another group with a pK of 7·5–8·0 present in the enzyme-substrate complex. These groups were not identified but they suggested that one at least might be an amino group. This arylsulphatase is, like sulphatase A, inhibited by carbonyl reagents such as the cyanide ion or hydrazine and here the inhibition is of the rather rare uncompetitive type (Dodgson, Spencer & Williams, 1956 $b$).

The arylsulphatase of *A. oryzae* has been studied for a longer period than any of the other microbial arylsulphatases but it unfortunately has been least purified. Since the now classical work of Neuberg, the specificity of the enzyme has been briefly considered only by Boyland, Manson, Sims & Williams (1956) who investigated the rates of hydrolysis of a number of *o*-aminoaryl sulphates: they concluded that only those compounds which have amino groups with $pK_a$ values less than 4·3 were hydrolysed and that the rate of hydrolysis decreased as the basicity of the amino group increased. Unfortunately the conditions chosen for the hydrolyses were empirical and no study was made of the corresponding *m*- or *p*-aminoaryl sulphates so that the interpretation of these results is doubtful. It also is inhibited by carbonyl reagents (Rosenfeld & Ruchelman, 1940; Działoszyński, 1951; Robinson *et al.* 1952) but the mechanism of this inhibition remains unknown.

All three of the above arylsulphatases obviously have certain analogies with the sulphatase C of mammalian liver—for instance, they are not inhibited by sulphate ions—and they seem fairly clearly to belong to the type I arylsulphatases although the enzyme from *A. aerogenes* is rather sensitive to phosphate ions. There is only one report of a microbial arylsulphatase related to the type II enzymes of mammalian tissue. This is the arylsulphatase of *Proteus vulgaris* (Dodgson, 1959 *a*) which is inhibited by phosphate and sulphate ions and, most characteristically, whose hydrolysis of *p*-nitrophenyl sulphate is powerfully activated by $Cl^-$ ions. The arylsulphatase of *P. vulgaris* is therefore quite similar in this respect to a sulphatase B and it is the only known example of such an enzyme from a source other than the livers of eutherian mammals. It is perhaps surprising that the closely related organism *P. rettgeri* apparently contains a quite different type of enzyme which is more analogous to that of *A. aerogenes* (table 7.4). The most characteristic feature of this enzyme was its rather powerful activation by phosphate ions (some 65% in 0·1 M-phosphate) when the substrate was nitrocatechol sulphate: when *p*-nitrophenyl sulphate was the substrate phosphate ions inhibited slightly (Milazzo & Fitzgerald, 1966). Unfortunately it is difficult to interpret these results with certainty because intact cells were used as the source of the enzyme so that complications could have been caused by permeability factors.

## 7.1.5 General properties of the arylsulphatases

The properties of many of the known arylsulphatases have been summarized in tables 7.1 to 7.4 and it is obviously impossible to generalize to any extent about such a diverse group of enzymes. Two points should perhaps be stressed. First, the usual assumption that the arylsulphatases are rather unspecific towards the phenolic component of their substrates may well be unjustifiable and it should be remembered that even nitro-catechol sulphate, which is often regarded as an ideal substrate, is not hydrolysed by all arylsulphatases. For example, the arylsulphatases of maize seedlings or of rat liver microsomes will hydrolyse $p$-nitrophenyl sulphate but not nitrocatechol sulphate (or at least only very poorly in the latter case). Secondly, many arylsulphatases are pronouncedly influenced by the composition and concentration of the buffer used in their assay. For example, increasing concentrations of acetate buffer may decrease the apparent $K_m$ of the substrate (sulphatase B of human liver), increase the rate of the enzyme reaction (arylsulphatases of *Proteus vulgaris* and *Helix pomatia*) or decrease the rate of the reaction (arylsulphatase of *Proteus rettgeri*). Again, the replacement of tris ions by $Na^+$ or $K^+$ ions decreases the activity of the sulphatase C of ox liver (Roy, 1956$c$) (see also § 7.3.1).

As already pointed out, the distinction between the type I and type II arylsulphatases probably only has any real meaning when the arylsulphatases of mammalian livers are considered but it is also true that most microbial arylsulphatases show definite affinities with the type I rather than the type II enzymes. The most constant distinguishing feature of the type I enzymes appears to be their insensitivity to inhibition by $SO_4^{2-}$ ions, that is, to product inhibition: this feature would appear to be a rather fundamental one and it may well reflect some important difference in the mechanisms of the reactions catalysed by the two types of enzyme.

Taking the arylsulphatases as a group, the only general inhibitor is the $SO_3^{2-}$ ion which powerfully inhibits all known arylsulphatases, the inhibition being competitive, at least in all those cases which have been examined in detail. Phosphate ions inhibit the majority of arylsulphatases and here again the inhibition appears to be competitive although the situation may in fact be more complex than suspected because Wortman (1962) showed that the inhibition of beef corneal arylsulphatases by phosphate ions was reversed, in some cases to the extent of a net activation, by the addition of corneal mucopolysaccharides. This most interesting effect has not been examined in detail so that no explanation for it is possible unless it simply be that the mucopolysaccharides, which were used in the form of their calcium salts, competed in some way for the phosphate ions.

Most, if not all, arylsulphatases appear to be inhibited by carbonyl reagents such as hydroxylamine or phenylhydrazine, but not all are inhibited by cyanide ions—in general terms the type I enzymes are cyanide sensitive, whereas the type II enzymes are not.

It is interesting that Hsu & Tappel (1965) claim that the arylsulphatase activity shown by extracts of several rat tissues, particularly of rat colon, is increased by retinol in sharp contrast to the activity of $\beta$-glucuronidase which is inhibited by this compound. From the information available it is not possible to decide whether the action is on sulphatase A or B, or both, but the known participation of hydrophobic bonding in maintaining the structure of sulphatase A makes it tempting to suggest that this enzyme may be the one most affected. The effect of retinol seemed to be a direct one on the enzyme protein and not an indirect result of some action on the lysosomal membrane as was suggested by Guha & Roels (1965) to be a possible explanation of their results which clearly showed an inter-relationship between the effects of vitamin A and $\alpha$-tocopherol on the level of sulphatases A and B in rat liver. The action of vitamin A on the sulphatases, as on ATP-sulphurylase, seems to be complex and the results at present available are contradictory. Błeszyński (1967) reported that vitamin A at a concentration of 250 $\mu$g/ml (approximately 0·9 mM) inhibited sulphatase A by 90% and activated sulphatase B by 20% but work in this laboratory has shown that such concentrations of retinol are without effect on the purified sulphatase A of ox liver. More recently Błeszyński & Leźnicki (1967) have reported that 0·1 mM vitamin A is without action on sulphatases A or B. Obviously more detailed investigations are required to clarify the situation.

### 7.1.6 Distribution of the arylsulphatases

Detailed information about the distribution of the arylsulphatases in different species or in different tissues is scanty because little attention has in the past been paid to the multiplicity of these enzymes and only the overall arylsulphatase activity has usually been determined. How this is made up is in most cases unknown.

*Micro-organisms.* In micro-organisms the arylsulphatases seem to be of only sporadic occurrence (Barber, Brooksbank & Kluper, 1951; Harada,

1952; Whitehead, Morrison & Young, 1952; Hare, Wildy, Billet & Twort, 1952; Harada, Kono & Yagi, 1954; Köhler, Ghatak, Rische & Ziesche, 1966) although no very serious search has been made by reliable techniques and a potential source of error is the extensive use of phenolphthalein disulphate as a substrate. This has several practical advantages and apparently has given useful information in the *Mycobacteria* (see chapter 12) (Kubica & Vestal, 1961; Tarshis, 1963, 1965) but it is not a general substrate for the arylsulphatases: it is not hydrolysed by, for instance, the arylsulphatase of *A. aerogenes* nor by sulphatase C. In light of the great amount of recent work on the importance of the cultural conditions on the amount of arylsulphatase produced by a micro-organism (Cherayil & Van Kley, 1961, 1962, 1963; Rammler, Grado & Fowler, 1964; Harada & Spencer, 1962, 1964; Harada, 1964; Hussey & Spencer, 1967; Milazzo & Lougheed, 1967) it would seem that the presently available information about the occurrence of the enzyme in micro-organisms must be accepted with some reserve.

There is no information on the localization of the arylsulphatases within the bacterial cell except for *A. aerogenes* where the enzyme is intracellular and not bound to the cell structure in any way.

*Invertebrates.* In these the arylsulphatases appear to be very widespread and they have been found in all species so far examined (Ney & Ammon, 1959; Corner, Leon & Bulbrook, 1960). The phyla from which representatives have been studied are the Porifera, Coelenterata, Annelida, Platyhelminthes, Arthropoda (Crustacea, Insecta and Arachnida), Mollusca, Echinodermata, Tunicata and Cephalocordata so that the only major phylum for which information is lacking is the Protozoa. Of these phyla, however, the only one to have been reasonably exhaustively investigated is the Mollusca and certainly here the arylsulphatases occur widely (Soda, 1936; Dodgson, Lewis & Spencer, 1953; Corner, Leon & Bulbrook, 1960; Orzel, 1966).

There is little knowledge of the distribution of the arylsulphatases in the different tissues of invertebrates apart, once again, from the Mollusca in which the major source is the hepatopancreas (digestive gland) from which the enzyme is secreted into the gut. All the other tissues examined —mucous gland, lung, muscular foot, blood and eggs—contained aryl-sulphatase activity but in much lesser amounts (Soda, 1936; Dodgson, Lewis & Spencer, 1953; Ney & Ammon, 1959). In the few insects which have been studied arylsulphatase activity has been detected in the crop

151

fluid, the gut contents, the haemolymph and the ecdyseal fluid (Robinson, Smith & Williams, 1953; Kikal & Smith, 1959).

Information on the intracellular localization of the arylsulphatases in invertebrates is also lacking apart from the claim by Jackson & Black (1967) that these enzymes occur in both the granular and soluble fractions of homogenates of the unfertilized eggs of the sea urchin *Arbacia punctata*. The arylsulphatases in this tissue do not appear to be associated with the acid phosphatases as they are in mammals.

*Vertebrates.* Here the situation is somewhat clearer from a qualitative point of view, but quantitative data are still scanty especially in view of the recent demonstrations that there are at least four arylsulphatases present in many mammalian tissues. In general it may be said that aryl-sulphatases, of unidentified type, are present in most tissues of most species but that the richest source—as yet unexamined in detail—is probably the kidney of the golden hamster (Rutenburg & Seligman, 1956). The available data on the distribution of sulphatases A, B and C in the livers of different vertebrate species is too scanty to allow any very valid conclusions to be drawn but it does appear that there has been an increase in the complexity of the arylsulphatases during the evolution of the vertebrates (Roy, 1958, 1963 b).

Quantitative figures for the distribution of the individual enzymes in the various tissues are few. Values for sulphatase C are available for the rat (Dodgson, Spencer & Thomas, 1953), the human (Dodgson, Spencer & Wynn, 1956) and the human foetus (Pulkkinen, 1961) in which all tissues appear to contain sulphatase C, but in very different amounts. In general, liver is the richest source followed by the adrenals, kidney, spleen and pancreas which have about 30% of the activity of the liver. Other tissues are much less active. No figures are available for the separate distribution of sulphatases A and B in various tissues of a single species but again the values of Dodgson, Spencer and Wynn give approximate figures for the sum of sulphatase A and B activities in the human. In this case kidney is apparently the richest source with about twice the activity of liver. Again, other tissues are very much less active. Quantitatively, these figures must be treated with caution because they were obtained from autopsy material and there is no certainty that post-mortem changes would be similar in all the tissues and this may have falsely distorted the values for the arylsulphatase activity.

As might be expected, the body fluids contain sulphatases A and B but not the insoluble sulphatase C. In the human, sulphatases A and B have been shown to occur in urine (Dodgson & Spencer, 1956 *a*) and in serum (Dodgson & Spencer, 1957 *b*). Arylsulphatase activity, of unidentified type, occurs in saliva (Chauncey, Lionetti, Winer & Lisanti, 1954) and in human milk (Poczekaj & Wenclewski, 1963).

### 7.1.7 Histochemistry and cytochemistry of the arylsulphatases

The intracellular distribution of the arylsulphatases has been extensively studied by biochemical methods only in the livers of higher mammals. There is now no doubt that in this tissue sulphatase C is localized exclusively in the microsomes (Dodgson, Spencer & Thomas, 1954; Gianetto & Viala, 1955; Roy, 1958) and sulphatases A and B in the lysosomes (Roy, 1954 *b*, 1958; Viala & Gianetto, 1955). The latter two enzymes require slightly different conditions for their solubilization (Roy, 1960 *b*; Ugazio & Pani, 1963) which suggests that they may be differently bound to the lysosomes or perhaps even present in slightly different types of particle (Ugazio, 1960 *a*; Franklin, 1962). It seems likely that this distribution of the arylsulphatases will be common to all tissues and certainly it has been clearly shown by Shibko & Tappel (1965) that sulphatases A and B occur in the lysosomes of rat kidney. There is also good evidence that arylsulphatases occur in the lysosomes of rat prostate gland (Lasnitzki, Dingle & Adams, 1966) and of human epidermis (Dicken & Decker, 1966).

The earliest study of the histochemistry of the arylsulphatases was that of Seligman, Nachlas, Manheimer, Friedman & Wolf (1949) who used 6-bromo-2-naphthyl sulphate as substrate and detected the liberated phenol by post-coupling with tetraazotized di-*o*-anisidine: incubation times of from 12 to 24 hr were required and the method was not particularly satisfactory. Almost simultaneously Ohara & Kurata (1950) used phenyl sulphate as substrate and, after incubation for 24 hr, detected enzymatic activity by precipitation of the $SO_4^{2-}$ ions either as $PbSO_4$ which was subsequently localized as PbS, or as benzidine sulphate which was detected by its reaction with naphthoquinone sulphonic acid. Rutenburg, Cohen & Seligman in 1952 introduced 6-benzoyl-2-naphthyl sulphate as a substrate for the histochemical detection of arylsulphatases and pointed out that it was much more satisfactory than the bromo derivative because shorter times of incubation—although still amounting to several hours—were required. This technique has been much used by, for instance, Austin & Bischel (1961) and by Austin, Balasubramanian,

153

Pattabiraman, Saraswathi, Basu & Bachhawat (1963), but the substrate is still by no means an ideal one because its low solubility necessitates its use at concentrations much below the optimum. Moreover, its structure is such that it is only rapidly hydrolysed by sulphatase C or, if $Cl^-$ ions be present, by sulphatase B: it is hydrolysed only relatively slowly by sulphatase A under any conditions (Roy, 1962 b).

Thomson & O'Connor (1966) have carried out the only detailed study of the histochemistry of the arylsuphatases and have obtained some most important results. The tissues used were the kidneys of the rat and of the golden hamster fixed by the method of Holt (1959) which is known to preserve the structure of the lysosomes. Initially they used 6-benzoyl-2-naphthyl sulphate as substrate and localized the phenol liberated from this by coupling it with tetraazotized di-o-anisidine: by this method aryl-sulphatase activity could be localized in the lysosomes of hamster kidney with incubation times as short as 2 min. In the rat kidney, on the other hand, arylsulphatase activity could not be satisfactorily localized in the lysosomes by this method, presumably because the long incubation time of 2–3 hr which was necessary allowed diffusion of the liberated phenol to occur. More recent experiments by these authors have been carried out with 1-naphthyl sulphate as the substrate and with hexaazotized pararosaniline as the coupling agent: with this technique arylsulphatase activity could clearly be localized in the lysosomes of both rat and hamster kidney, the former tissue requiring incubation times of only 20 min. It is not clear which arylsulphatases give these reactions but the media contain $Cl^-$ ions so that sulphatases B could be important.

Several other techniques were also investigated by Thomson & O'Connor (1966) and shown to be unsatisfactory. In particular they could detect no arylsulphatase activity in rat or hamster kidney using either naphthol AS–BI sulphate or naphthol AS–MX sulphate as substrate, despite the use of several compounds of this type by Woohsmann & Hartrodt (1964 a, b) and by Woohsmann & Brosowski (1964). It is also surprising that although Woohsmann detected arylsulphatase activity in several rat tissues by this technique none was found in rat liver. Another method which proved unsatisfactory was that of Goldfischer (1965): this uses nitrocatechol sulphate as substrate and localizes sulphatase activity by precipitating the liberated $SO_4^{2-}$ ions as $PbSO_4$. It might have been predicted that this would have been a most satisfactory technique, using as it did the classical substrate for the arylsulphatases, nitrocatechol sulphate, and being readily adapted for electron microscopy but Thomson

& O'Connor (1966) showed that deposition of lead could occur in the absence of any added substrate, probably by the formation of lead phosphate arising from the action of phophatases on endogenous substrates. Results obtained with this type of method (e.g. Ericsson & Helminen, 1967) must therefore be interpreted with care.

Very recently an indigogenic method has been described by Wolf, Horwitz, Vazquez, Chua, Pak & Von der Muehll (1967): the substrate was 5-bromo-4-chloro-3-indolyl sulphate and with incubation times of 2 hr arylsulphatase was localized in the lysosomes of rat kidney. The enzymatic activity was inhibited by $SO_3^{2-}$ ions and it was detected in many tissues from several species. In view of the value of such indigogenic methods for the histochemical detection of other hydrolytic enzymes it seems possible that this technique will prove of considerable value. The synthesis of suitable substrates has been described by Horwitz, Chua, Noel, Donatti & Freisler (1966).

A number of techniques have been developed in which the $SO_4^{2-}$ ions liberated through sulphatase action are precipitated as $BaSO_4$ but no detailed study of these has yet appeared. Such an investigation is certainly required in view of the possibility of barium phosphate being precipitated in a way similar to lead phosphate. Hopsu, Arstila & Glenner (1965) used 8-hydroxyquinoline sulphate as substrate and localized the $BaSO_4$ by electron microscopy. By this technique they were able to show the occurrence of arylsulphatase in some, but not all, the lysosomes of rat kidney. More recently this technique has been modified for use with p-nitrophenyl sulphate and with nitrocatechol sulphate, with which substrates arylsulphatase activity has been localized in the lysosomes of the proximal convoluted tubules of rat kidney (Hopsu-Havu, Arstila, Helminen, Kalimo & Glenner, 1967) and of mouse kidney (Rowden, 1967). Kawiak, Sawicki & Miks (1964) used p-nitrophenyl [$^{35}$S]sulphate as substrate and localized the resulting precipitate of $Ba^{35}SO_4$ by autoradiography. This technique was applied to a study of cartilage and it was shown that the arylsulphatase was apparently randomly distributed through the chondrocytes (Kawiak, 1964). Should the validity of this technique be established then it will undoubtedly become of great value for histochemical studies, especially as nitrocatechol [$^{35}$S]sulphate is now available (Flynn, Dodgson, Powell & Rose, 1967).

The above discussion has been directed only to the histochemistry of animal tissues. Studies of the histochemistry of the arylsulphatases in plants are few but of peculiar importance because virtually the only evidence for the occurrence of this enzyme in the vegetative tissues of higher plants is histochemical in nature. Avers (1961), Olszewska & Gabara (1964), Walek-Czerneka (1965) and Gorska-Brylass (1965) have all used 6-bromo-2-naphthyl sulphate as substrate and found considerable arylsulphatase activity in a number of plant tissues although only Walek-Czerneka showed its inhibition by $SO_3^{2-}$ ions. According to the Polish workers, the arylsulphatase of plant tissues is localized in the spherosomes which are granules containing many hydrolytic enzymes and which have obvious analogies with the lysosomes of animal tissues. Poux (1966), on the other hand, used the method of Goldfischer (1965) with nitrocatechol sulphate as substrate and claimed that arylsulphatase activity could be found not only in the phragmosomes (which are probably identical with the spherosomes) but also in the perinuclear envelope and the endoplasmic reticulum. In view of Thomson & O'Connor's finding of false-positive results with Goldfischer's technique as applied to kidney it is obvious that this finding must be accepted only with some caution, the more so as it is not in agreement with one of the few studies of plant arylsulphatases which has made use of purely biochemical techniques. Semadeni (1967) claimed that arylsulphatases comparable to sulphatases A and B were not detectable in extracts of maize seedlings, nor in the spherosomes isolated therefrom, but that an arylsulphatase did occur bound to fragments of the endoplastic reticulum. Such an enzyme would have obvious affinities with the sulphatase C of mammalian tissues but it differed from the latter in having a pH optimum of 5. It is also a somewhat atypical enzyme in that it will not hydrolyse nitrocatechol sulphate nor 1-naphthyl sulphate.

### 7.1.8 Physiology and pathology of the arylsulphatases in mammals

Considerable interest has been shown in the arylsulphatases in various physiological and pathological conditions but, once again, most of the work is rendered almost valueless because only the over-all arylsulphatase activity has been determined rather than the activity of individual enzymes. Obviously even quite a large change in one enzyme might be masked almost completely if there were no comparable changes in the others. It may therefore not be coincidence that the only significant results in this field have come from work in which arylsulphatases A and B were separately determined. The assay of the individual arylsulphatases in a mixture is not easy, even if attention is restricted only to sulphatases A, B and C without the added complications of the various forms of sulphatase B, and at present no completely satisfactory method exists. Even assuming it were possible to obtain a quantitative solubilization of the lysosomal arylsulphatases the problem of endogenous inhibitors, especially phosphate ions, is troublesome. There is no substrate known which can be considered as 'specific' for any one arylsulphatase, although there are certainly differences in the relative rates of hydrolysis of the various substrates by different enzymes, so that there seem to be only two possible approaches to the problem of a specific assay: either a quantitative separation of the various enzymes must be achieved or their activities must be modified in some way. The first approach has not yet been utilized but it seems possible that chromatographic methods based on the preparative techniques of Nicholl & Roy (1964) and of Allen & Roy (1968) could be used: for example, at pH 7 sulphatase A is strongly held by DEAE-Sephadex while sulphatase B is not, or at the same pH these two enzymes can be clearly separated by chromatography on Sephadex G-200. It appears that a method based on those observations might be more satisfactory than those based on the use of modifiers because it could easily be used to separate endogenous inhibitors from the enzyme. Of the methods using modifiers the first is that of Baum, Dodgson & Spencer (1959) which is based on the following two observations: one, that sulphatase A shows approximately 'normal' kinetics (see § 7.1.1) in the presence of 0·25 mM-sodium pyrophosphate, a concentration which strongly inhibits sulphatase B, and second, that $Cl^-$ ions strongly inhibit the action of sulphatase B on nitrocatechol sulphate without having a significant effect on sulphatase A. Therefore, by choosing strictly standardized conditions it is possible to separately determine sulphatases A and B. It must be stressed that this method was developed for use with human urine, which does not contain the insoluble sulphatase C, and that its validity when applied to tissues containing sulphatases A, B and C has not been investigated. Nevertheless, the method has provided valuable information when applied to human tissues by Austin, McAfee, Armstrong, O'Rourke, Shearer & Bachhawat (1964), as is discussed below. The second method using modifiers is that of Działoszyński, Błeszyński & Lewosz (1966) which depends upon the different effects of 3 mM-$K_2SO_4$ and 10 mM-basic aluminium acetate on sulphatases A and B. By carrying out the assays in the presence of these modifiers the separate activities of the two enzymes were calculated by means of empirical formulae. This method was devised for use with ox brain and again it has not been shown to be of general validity but the control determinations carried

157

out by Działoszyński *et al.* certainly suggest the method might be of considerable value.

Even admitting that such methods are tedious and of uncertain significance it cannot be too highly stressed that they should be tested and used rather than the simple assay of total arylsulphatase activity, a procedure which is still too often the method adopted. However, even when these more complex methods are used it must still be remembered that at the best they assay sulphatases A and B: they do not determine the separate components of sulphatase B, nor do they determine the insoluble sulphatase C. Further studies of this problem of determining the individual arylsulphatases in tissues are most certainly urgently required.

Considering first physiological changes, the most detailed study is that of Pulkkinen (1961) who measured the sulphatase C activity of several tissues in the rat before and after birth and showed that during this time there was a general and continuous increase in the activity. This increase was maintained at least during the early stages of extra-uterine life. In adult male rats the sulphatase C activity of the liver is nearly double that of the female but this sex difference is not apparent in other tissues (Dodgson, Spencer & Thomas, 1953). Pulkkinen also investigated the sulphatase C activity of the human foetus and suggested that there a similar increase in activity with age also occurred. With sulphatase A and B the information is scanty in the extreme although Roy (1958) and Ugazio (1960 *a*) have both suggested that the sulphatase A and B activity of rat liver decreases with age. Kuczynski, Pydzik & Wenclewski (1964) have, on the other hand, claimed that the type II arylsuphatase activity of human foetal liver and brain increases some four-fold between the seventh and twenty-fourth weeks of pregnancy.

There have been many studies on the influence of endocrine glands and of steroid hormones on arylsulphatase activity (Bianchi, 1955; Shimodo, 1957; Roy, 1958; Satake, 1960; Kodama, 1961) but the only detailed study in which full account was taken of the presence of a number of arylsulphatases is that of Pulkkinen (1961) and Pulkkinen & Hakkarainen (1965). The results were complex, but they clearly showed that the arylsulphatase activity of liver and kidney was influenced by the sex hormones but not by any of the tropic hormones investigated. In general, oestrogens had a greater action on the liver than on the kidney, while androgens primarily influenced the kidney. There is no evidence to show that these hormonal effects are direct ones on the arylsulphatases and it is much more likely that they are simply reflections of more general metabolic changes.

When the pathology of the arylsulphatases is considered the results have been, on the whole, uninteresting (Działoszyński & Gniot-Szulżycka, 1967). In the past a great deal of attention has been devoted to the effect of compounds such as carbon tetrachloride on the arylsulphatases of liver (Shimodo, 1957; Roy, 1958; Ugazio, 1960 b; Ugazio, Artizzu, Pani & Dianzani, 1964) but little of value has emerged. Studies have been made of the arylsulphatase levels in various tumours (Huggins & Smith, 1947; Działoszyński & Zawielak, 1955; Rutenburg & Seligman, 1956; Shimodo, 1957; Sugimoto & Aoshima, 1964; Działoszyński, Kroll & Fröhlich, 1966), in normal and pathological urine (Boyland, Wallace & Williams, 1955; Działoszyński, 1957; Ammon & Ney, 1957; Ammon & Keutel, 1960; Poczekaj, Hejduk & Wenclewski, 1963; Clausen & Asboe-Hansen, 1967), in skin (Pepler, Loubser & Kooij, 1958; Sachs & Braun-Falco, 1960), in leucocytes (Tanaka, Valentine & Fredricks, 1962) and in arterial tissue (Utermann, Lorenzen & Hilz, 1964) but again little of interest has arisen. More important is the work of Tappel, Zalkin, Caldwell, Desai & Shibko (1962) which has shown that in muscular dystrophies of various types there is a rise in the arylsulphatase activity of muscle; however, as this is associated with a rise in the activity of most lysosomal enzymes, there is no reason to believe that any direct relationship exists between muscular dystrophy and arylsulphatases.

The first significant observation in the pathology of the arylsulphatases has come from the work of Austin on the leucodystrophies and related diseases. Austin et al. (1963) showed that in metachromatic leucodystrophy the total arylsulphatase activity of brain was low and this was followed by the demonstration (Austin et al. 1964) of an almost complete lack of sulphatase A, measured specifically by the method of Baum, Dodgson & Spencer (1959), in the brain, liver and kidney of patients suffering from this disease. This fall in sulphatase A activity was not associated with a general fall in lysosomal enzymes because the acid phosphatase and β-galactosidase activities were normal in these patients. The behaviour of sulphatase B in metachromatic leucodystrophy was less constant but at least in nervous tissue it was always lowered. More detailed studies (Austin, Armstrong & Shearer, 1965) have confirmed the original findings but have somewhat complicated the issue by showing that there may well be several types of metachromatic leucodystrophy which differ in the degree of hyposulphatasia; some cases apparently are

159

deficient not only in sulphatase A and sulphatase B but also in sulphatase C. The deficiency of sulphatase A, however, seems to be a constant factor. This work is the first demonstration of an alteration in the level of a specific sulphatase in a specific disease and it is of course tempting to relate the fall in arylsulphatase to the accumulation of cerebroside sulphate which occurs in the tissues of patients suffering from metachromatic leucodystrophy. It has previously been stated (Austin *et al.* 1963) that sulphatase A cannot hydrolyse cerebroside sulphates but this statement was based on work with relatively crude enzyme preparations. Mehl & Jatzkewitz (1968), using either their own highly purified preparations from pig kidney or homogeneous preparations of sulphatase A from ox liver (prepared by Roy), have claimed that sulphatase A does indeed show cerebroside sulphatase activity although nitrocatechol sulphate is hydrolysed by this enzyme some 30 times faster than cerebroside 3-sulphates. Obviously it is impossible to be certain that a single enzyme is catalysing both reactions but it is perhaps pertinent that sulphatase B (from pig kidney) shows little cerebroside sulphatase activity and that the arylsulphatase of *Patella vulgata* shows none. Further consideration of this most interesting development is given in § 7.3.6.

It is, however, clear that greater attention should be paid to the role of arylsulphatases in the metabolism of the central nervous system. That arylsulphatases are present therein in appreciable amounts has been shown by both classical enzymological (Błeszyński & Działoszyński, 1965; Błeszyński, 1967) and by histochemical (Kozik & Wenclewski, 1965; Mietkiewski & Kozik, 1966) methods, the latter clearly showing a concentration of the arylsulphatases in the myelin sheaths.

Austin *et al.* (1964) have also investigated the arylsulphatase activity in gargoylism (Hurler's syndrome), a disease characterized by a general alteration in the metabolism of mucopolysaccharides. In all the tissues studied there was a rise in the sulphatase B activity, but not in that of sulphatase A nor of other lysosomal enzymes. Once again, it is impossible at this stage to relate the altered metabolism of mucopolysaccharides to the increased activity of sulphatase B and much further investigation will be needed to establish the relationship, if indeed there be a direct one.

### 7.1.9 Functions of the arylsulphatases

In micro-organisms it appears that the arylsulphatases are involved in some way in the general sulphur metabolism of the cell, in particular, in the conversion of the inorganic forms of sulphur to the sulphur-containing amino acids. This seems to have been demonstrated quite clearly for several fungi (Harada & Spencer, 1962) and for *Aerobacter aerogenes* (Rammler, Grado & Fowler, 1964; Harada & Spencer, 1964) in which the synthesis of arylsulphatase is repressed by cysteine and by all compounds generally believed to occur on the metabolic pathway from sulphate to cysteine. Cysteine is almost certainly the true repressor, although full details of the system are still not clear.

The repression of arylsulphatase in *A. aerogenes* is interesting as it can be derepressed not only by compounds such as methionine but also by tyramine. This is the explanation of the work of Harada which showed several years ago (Harada, 1957, 1959, 1963; Harada & Hattori, 1956) that tyramine was an inducer of arylsulphatase in this species.

It is now clear that some bacteria (Dreyfuss & Pardee, 1966) and fungi (Scott & Spencer, 1965; Yamamoto & Segel, 1966) can, like rat kidney cortex (Winters, Delluva, Deyrup & Davies, 1962), actively transport $SO_4^{2-}$ ions and it seemed possible that arylsulphates could play a part in this process. However, Pardee (1966) has purified the sulphate-binding protein from *Salmonella typhimurium* and shown that it cannot hydrolyse *p*-nitrophenyl sulphate so that it is unlikely to be an arylsulphatase.

Cherayil & Van Kley (1961, 1962, 1963) have claimed that in the case of *Aspergillus oryzae* cultured on wheat bran the amount of the arylsulphatase produced can be increased a thousand-fold by the inclusion in the culture medium of sodium phosphate, tungstate or molybdate—all of which are competitive inhibitors of the enzyme. The latter two compounds inhibit the growth of the mould, but this can be reversed by the addition of sodium sulphate or apparently of arylsulphatase itself to the medium.

In the invertebrates nothing whatever is known of the function of the arylsulphatases. At least in the molluscs and insects they appear to be secreted into the intestinal juices and so might be considered to be digestive enzymes but there is no clue to any possible substrate which might occur in the diet. As far as the Mollusca are concerned, it has been pointed out (Corner, Leon & Bulbrook, 1960) that there is a correlation between a high level of arylsulphatase activity and a herbivorous habit but this still provides no indication of possible substrates apart from the carbohydrate sulphates which occur in many marine algae. If the arylsulphatases are indeed digestive enzymes then one is tempted to suggest that their specificity *in vivo* must be very different from that attributed to them *in vitro*. As will be discussed later in connection with the glycosulphatases, Egami has suggested a possible sulphotransferase activity for the arylsulphatase

161

of *Charonia lampas* but much more experimental evidence is required before such a role can be taken as proven.

The physiological function of the arylsulphatases in mammals is also unknown and only recently has direct evidence been produced to show that they can function *in vivo* (Flynn, Dodgson, Powell & Rose, 1967). It has been suggested that they may serve a regulatory function, controlling the equilibrium between a hormone and its storage or transport form—for instance, triiodothyronine (Roche, Michel, Closon & Michel, 1959) or serotonin (Kishimoto, Takahashi & Egami, 1961)—but there is little to support such a view. Should the suggestion that at least some arylsulphatases possess cerebroside sulphatase activity be shown to be correct, then obviously the whole question of the function of the former enzymes must be considered in a new light and a role for them in the metabolism of cerebrosides would become highly probable.

It is obvious that the problem of the physiological role of the arylsulphatases is one requiring much further investigation. The widespread distribution of these enzymes and their multiplicity certainly suggest a rather fundamental role, but what this is remains unknown. It is not even clear whether the arylsulphatases have only one function common to all organisms or whether their functions differ in micro-organisms, in vertebrates and in invertebrates. Not the least of the difficulties associated with the physiological function of these enzymes is the problem of the nature of their physiological substrates: although many arylsulphates occur in urine, these are generally regarded as detoxication products which in some sense are non-physiological and so rather unlikely to be the normal substrates of arylsulphatases. Information on the arylsulphates present in tissues is scanty indeed but it could well be that the role of the arylsulphatases in metabolism will only be clarified when the chemistry and distribution of their potential substrates is understood. It is now clear that metabolic interconversions of steroids can occur at the level of their sulphate esters and it may be that the arylsulphates play a part in the metabolism of phenols—as yet unidentified—in living organisms.

Further speculation is useless at this stage, but there is no doubt that it is in this area that the richest prizes lie in the study of the arylsulphatases.

## 7.2 Steroid sulphatases

These enzymes, which catalyse the hydrolysis of a number of different types of steroid sulphate, have been much studied because of their potential usefulness for the hydrolysis of such compounds prior to the separation of the free steroids for analysis. Obviously for such a use to be justifiable the specificity of the enzymes must be clearly understood, a situation which is certainly not the case. None of the steroid sulphatases has been purified to any extent so that there is considerable doubt as to the number of enzymes concerned and further developments must await the isolation of the individual enzymes.

The distribution of steroid sulphatase is rather strange: it occurs in many molluscs and in mammals but apparently it is absent from other invertebrates and vertebrates (Roy, 1958, 1963 b; Ney & Ammon, 1959; Corner, Leon & Bulbrook, 1960). There is no obvious explanation for this very sporadic occurrence of steroid sulphatase in nature, but it must be kept in mind that the searches for the enzyme have, up to the present, not been exhaustive.

It should be noted that oestrone sulphate (7.3), is an arylsulphate and can be hydrolysed by the type I arylsulphatase of A. oryzae (Butenandt & Hofstetter, 1939; Cohen & Bates, 1949). In animal tissues it might therefore be predicted that oestrone sulphate would be most readily hydrolysed

(7.3)

by sulphatase C and the finding that the microsomal fraction of rat liver and kidney had a greater capacity to hydrolyse oestrone sulphate than did the mitochondrial fraction (Pulkkinen & Paunio, 1963) could be taken as support of this. There is, however, some evidence that animal tissues may contain a specific oestrone sulphatase. The first suggestion of this came from the work of Pulkkinen himself who showed (1961) that although foetal rat organs showed a pronounced arylsulphatase activity (probably of type II) they could not hydrolyse oestrone sulphate. More direct evidence has been provided by French & Warren (1967): this is discussed in more detail below (§ 7.2.1) and although the existence of a separate oestrone sulphatase cannot be taken as proven, it is at least highly probable and in view of the importance of oestrogen sulphates in metabolism it is to be hoped that further investigations will soon be forthcoming.

### 7.2.1 Androstenolone sulphatase

This is the enzyme occurring in molluscan and mammalian tissues which is generally referred to as steroid sulphatase, although a more descriptive trivial name is androstenolone sulphatase because the substrate commonly used in the assay of the enzyme is potassium 17-oxoandrost-5-en-3$\beta$-yl sulphate (*7.4*, dehydro*epi*androsterone sulphate, androstenolone sulphate). Any decision on a systematic name should wait until the specificity of a pure enzyme has been investigated.

(*7.4*)

*Molluscan androstenolone sulphatases.* The molluscan enzymes have been extensively studied and those of the European limpet *Patella vulgata* (Roy, 1956 *a*) and of the Roman snail *Helix pomatia* (Jarrige, 1963) have been partially purified. Other species which have been examined are the African land snail *Otala punctata* (Savard, Bagnoli & Dorfman, 1954) and *Charonia lampas* (Takahashi & Egami, 1961 *a*). All these species are gastropod molluscs and it has been pointed out (Corner, Leon & Bulbrook, 1960) that in marine molluscs androstenolone sulphatase is apparently restricted to the sub-class Prosobranchia of the Gastropoda. Not all representatives of this sub-class contain the enzyme and there is no correlation between its presence or absence and any obvious biological property.

These enzymes all have pH optima in the region of 4·5 and they are inhibited by sulphate, sulphite and phosphate ions. The steroid sulphatase of *P. vulgata* hydrolyses only the 3$\beta$-sulphates of the $\Delta^5$ and 5$\alpha$ series of steroids, as shown in table 7.5. In particular, the enzyme hydrolyses *epi*-androsterone and androstenolone sulphates but not androsterone sulphate nor the aetiocholanolone sulphates, at least not at a rate detectable under the experimental conditions used by Roy (1956 *a*). Androsterone sulphate in fact inhibits the hydrolysis of androstenolone sulphate. The enzyme does not hydrolyse 17$\alpha$-, 17$\beta$-, 20$\alpha$-, nor 20$\beta$-steroid sulphates or the more complex compounds, ranol or scymnol sulphates. Cortisone 21-sulphate is hydrolysed, as is discussed below in § 7.2.2.

TABLE 7.5    *The hydrolysis of sulphate esters of steroids by enzyme preparations from* Patella vulgata *and from* Helix pomatia

|  |  | P. vulgata | H. pomatia | | |
|---|---|---|---|---|---|
|  |  | *Activity | *Activity | †V | $K_m$ |
| Sulphate of: |  |  |  |  |  |
| 3α-hydroxy-5α-androstan-17-one | (androsterone) | 0 | 0 | . | . |
| 3β-hydroxy-5α-androstan-17-one | (*epi*androsterone) | 44 | 71 | 1·28 | 0·032 |
| 3α-hydroxy-5β-androstan-17-one | . | 0 | 9 | 0·8 | 3·07 |
| 3β-hydroxy-5β-androstan-17-one | (aetiocholanolone) | 0 | 6 | 0·77 | 6·35 |
| 3β-hydroxyandrost-5-en-17-one | (androstenolone) | 100 | 100 | 1·00 | 0·028 |
| 17α-hydroxyandrost-4-en-3-one | (*epi*testosterone) | 0 | . | . | . |
| 17β-hydroxyandrost-4-en-3-one | (testosterone) | 0 | 0 | . | . |
| 3α-hydroxy-5α-pregnan-20-one | . | 0 | 0 | . | . |
| 3β-hydroxy-5α-pregnan-20-one | . | 9 | . | . | . |
| 3α-hydroxy-5β-pregnan-20-one | . | 0 | 3 | . | . |
| 3β-hydroxy-5β-pregnan-20-one | . | 0 | 0·1 | . | . |
| 3β-hydroxypregn-5-en-20-one | (pregnenolone) | 95 | 99 | 1·75 | 0·023 |
| 20α-hydroxy-5β-pregnane | . | 0 | . | . | . |
| 20β-hydroxy-5β-pregnane | . | 0 | . | . | . |
| 3β-hydroxycholest-5-ene | (cholesterol) | 5 | . | . | . |
| 21-sulphate of: |  |  |  |  |  |
| cortisone | . | 3 | 295 | 18 | 0·2 |
| desoxycorticosterone | . | . | 295 | . | . |
| 17-hydroxydesoxycorticosterone | . | . | 295 | . | . |

\*   Expressed (relative to androstenolone sulphate) as extent of hydrolysis of the ester in 1 hr under arbitrary conditions.

†   Expressed in arbitrary units relative to that of androstenolone sulphate.

In 1960 Corner, Leon and Bulbrook surveyed British molluscs for sources of steroid sulphatase and they noted that preparations from three species, the terrestrial *H. pomatia* and the marine *Buccinum undatum* and *Nassarius reticulatus* (the latter two species being representatives of the order Stenoglossa) could hydrolyse aetiocholanolone sulphate (potassium 17-oxo-5β-androsten-3α-yl sulphate, *7.5*) although only at a rate much lower than that of the hydrolysis of androstenolone sulphate. They were unable to separate the activities towards the two different types of

(7.5)

steroid sulphate but they implied that two separate enzymes were involved, basing this claim on the different distribution of the two activities in various molluscan species (Leon, Bulbrook & Corner, 1960).

The steroid sulphatase of *H. pomatia* has since been studied in more detail by Jarrige, Yon & Jayle (1963) who have confirmed that preparations from this species can hydrolyse not only the steroid sulphates hydrolysed by the enzyme of *P. vulgata* but also the aetiocholanolone

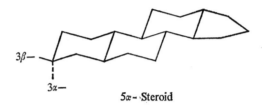

$3\beta-$

$3\alpha-$

5α- Steroid

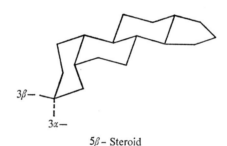

$3\beta-$

$3\alpha-$

5β- Steroid

Fig. 7.2    The conformations of the 5α- (androsterone) and 5β- (aetiocholanolone) series of steriods.

sulphates. The results of this study are given in table 7.5. It should be noted that although the relative values of $k_3$, the rate of breakdown of the enzyme-substrate complex, are similar for aetiocholanolone sulphates and for androstenolone sulphate, the values of the $K_m$ for the former are so large that in practice the rate of hydrolysis of the aetiocholanolone sulphates is very low. Jarrige was also unable to separate the two activities and although they have very different kinetic parameters, and the substrates have very different conformations (fig. 7.2), he believes that only one enzyme is responsible for the two types of reaction. This may be so, but further studies are certainly required in view of the comparative studies of Leon, Corner and Bulbrook and of the kinetic data furnished by Jarrige, Yon and Jayle themselves. For instance the latter workers showed that aetiocholanolone sulphate competitively inhibits the hydrolysis of androstenolone sulphate by the steroid sulphatase of *H. pomatia* with a $K_i$ of 0·09 mm: this value is very different from the $K_m$ of 3·07 mm for the hydrolysis of aetiocholanolone sulphate by the same enzyme preparation and it seems difficult to escape the conclusion that two separate enzymes are in fact involved.

Little can be said of the possible function of steroid sulphatase in the mollusc. The enzyme is apparently secreted by the digestive gland into the intestinal tract which would seem to imply a role in digestion, but there are no known steroid sulphates which might be considered as dietary constituents. The amounts of the enzyme, or enzymes, present undoubtedly vary from species to species, but in no way which can be related to any obvious biological feature of the animals. It is perhaps of interest to recall that some molluscs contain rather unusual sterols (Bergmann, 1962): the biogenesis of these compounds has not been studied but if, as seems likely, they are metabolic products of cholesterol, then it is tempting to suggest that their formation may occur at the level of the sulphate esters in a way analogous to the conversion of cholesteryl sulphate to androstenolone sulphate in the mammal. If this indeed be the case, then an obvious function for the steroid sulphatase would be the production of the free steroid from its metabolic precursor, the sulphate ester. However, to postulate such a function only begs the question as the role of the sterols themselves is quite unknown.

More pertinent, perhaps, are the recent observations by Gottfried & Lusis (1966) and by Gottfried, Dorfman & Wall (1967) that the eggs and spermatheca gland of the slug *Arion ater* can produce *in vitro* a number of $C_{18}$, $C_{19}$, and $C_{21}$ steroids from endogenous precursors. These appear to be the first demonstrations of the occurrence of such steroids in molluscs and their presence therein certainly provides hope for a more rational explanation of the role of steroid sulphatase in molluscan metabolism.

*Mammalian androstenolone sulphatase.* This enzyme was first detected in rat liver by Gibian & Bratfisch (1956) and was subsequently investigated by Roy (1957 *b*) using the partially purified enzyme from ox liver. It is an insoluble enzyme present in the microsomes and is apparently closely associated with sulphatase C from which it has not been separated. Indirect, and therefore suspect, evidence (Roy, 1957 *b*) suggests that sulphatase C and androstenolone sulphatase are separate entities, but the

167

distinction cannot be regarded as proven until the two activities have been at least partially resolved, a procedure made difficult by the insolubility of the enzymes. Burstein (1967) has recently had some success in solubilizing the steroid sulphatase of liver by treating isolated microsomes with heat-treated snake venom: the degree of purification was small and the soluble enzyme, which had a molecular weight of about 600,000, tended to form insoluble aggregates. Its insolubility apart, the androstenolone sulphatase of mammalian liver is similar to that of *P. vulgata* although the pH optimum is rather higher, between pH 7 and 8. In particular, the specificity is exactly similar to that of the *Patella* enzyme in that only the 3-sulphates of $\Delta^5$- or $5\alpha$-steroids are hydrolysed and although no accurate values of $K_m$ are available they are obviously of the same order as those of the molluscan enzyme.

The distribution of the enzyme in mammalian tissues has not been studied in detail but it is known to occur in the liver, adrenals, testes and ovaries of a number of species (Burstein & Dorfman, 1963). The activity of the enzyme is high in rat liver, low in mouse liver and very low in guinea-pig liver but, in contrast, it is high in guinea-pig testes and low in rat testes. At least in the livers of some strains of rats there is a sex difference in the level of the enzyme activity, that in the male being about double that in the female (Roy, 1957 b; Burstein & Westort, 1967). Foetal rat liver (Burstein & Westort, 1967) and several foetal human tissues (Pulkkinen, 1961; French & Warren, 1965) do not contain significant amounts of androstenolone sulphatase but, at least in the former species, such activity develops rapidly after birth.

The richest known source of androstenolone sulphatase is the human placenta which normally has about five times the activity of human liver but the placentae from anencephalic foetuses are even richer sources and have about twenty times the activity of liver (Warren & French, 1965). This enzyme has been investigated in some detail by French & Warren (1966, 1967) who have shown that it, like the enzyme from liver, is bound to the microsomes and is closely associated with arylsulphatase and oestrone sulphatase activity. They have, however, concluded on the following grounds that all three activities are due to separate enzymes:

1. In experiments in which one substrate was used as an inhibitor of the hydrolysis of another only the inhibition of androstenolone sulphatase by oestrone sulphate was competitive. In other cases the inhibition was of no simple type or was non-existent, as in the action of *p*-nitrophenyl sulphate on androstenolone sulphatase.

2. All three activities had different stabilities at 56°.

3. Ribonuclease treatment decreased the activity of oestrone sulphatase and arylsulphatase without affecting androstenolone sulphatase while treatment with butanol markedly decreased the arylsulphatase activity without altering the oestrone and androstenolone sulphatases.

4. The pH-activity curves were very different with the different substrates.

Certainly these results seem to show the presence of three distinct types of sulphatase activity in these preparations.

Although there is no direct information on the role of androstenolone sulphatase in the mammal, it is not difficult to visualize for it a possible, and a very likely, function. The work of Lieberman (Roberts, Bandi, Calvin, Drucker & Lieberman, 1964; Calvin & Lieberman, 1966; Roberts, Bandi & Lieberman, 1967) and of Baulieu (Baulieu, Corpéchot, Dray, Emiliozzi, Lebeau, Mauvais-Jarvis & Robel, 1965) has shown the possibility of the conversion of cholesterol and of pregnenolone to biologically active steroids taking place at the level of the sulphate esters. If this indeed be the case then the function of androstenolone sulphatase could be the hydrolysis of such steroid sulphates with the formation of the free steroids. Such a function is obviously of particular interest and importance in the foetal–placental unit because of the impermeability of the placenta to steroid sulphates and the presence in that organ of large amounts of androstenolone sulphatase which is in striking contrast to the absence of the enzyme from the foetus (Levitz, Condon, Money & Dancis, 1960; Levitz, 1966).

### 7.2.2 Cortisone sulphatase

This enzyme, or enzyme activity, is apparently quite widely distributed in molluscs (Savard, Bagnoli & Dorfman, 1954; Roy, 1956 a; Dodgson, 1961 b; Jarrige, Yon & Jayle, 1963), but it has not been detected in mammalian tissues (Roy, 1957 b; Pasqualini, Cedard, Nguyen & Alsatt, 1967). Although the cortisone sulphatase activity has not been separated from androstenolone sulphatase some of the results of Jarrige (1962) suggest that two distinct enzymes are involved, as might have been predicted from the very different structure of cortisone sulphate (7.6) and say, androstenolone sulphate (7.4). Such evidence is, for instance, the rather differing responses of the two activities to certain inhibitors (e.g. 2·5 mM-phosphate inhibits androstenolone sulphatase by 96% but

169

cortisone sulphatase by only 66%) and the slightly, but apparently significantly, different behaviour of the activities during chromatography or electrophoresis. Certainly this evidence is not conclusive and the possibility that only one enzyme is involved cannot be ignored. Competing-substrate experiments (Dodgson, 1961 *b*), although not conclusive, would certainly be compatible with the view that cortisone sulphate and androstenolone sulphate are hydrolysed by a single enzyme. Further, the hydrolysis of cortisone sulphate is competitively inhibited by androstenonone sulphate, the $K_i$ of the latter being 0·036 mM and quite comparable with the $K_m$ of 0·028 mM for the hydrolysis of androstenolone sulphate by the same enzyme preparation (Jarrige *et al.* 1963).

(7.6)

In view of the great doubt as to the separate nature of the cortisone sulphatase activity, little can be said of the specificity of the presumed enzyme except that preparations which hydrolyse cortisone sulphate will also hydrolyse desoxycorticosterone sulphate and 17$\alpha$-hydroxydesoxycorticosterone sulphate.

It need not be stressed that the function in molluscs of such a cortisone sulphatase, if it has a separate existence, is quite unknown.

As has already been mentioned, mammalian liver preparations apparently cannot hydrolyse cortisone sulphate, yet when this compound, labelled with $^{35}$S, is administered to rats, about 50% of the label appears in the urine as inorganic sulphate within 24 hr (Dodgson, Gatehouse, Lloyd & Powell, 1965). The mechanism of this transformation has not been elucidated but it is unlikely to be due to a direct attack of a steroid sulphatase.

## 7.3   Glycosulphatases

This group of sulphatases can hydrolyse a rather wide range of carbohydrate sulphates and although it is clear that a number of different enzymes are involved, none of them has been purified and the inter-relationships of the various activities must remain in some doubt. Most of these enzymes appear to be of molluscan or microbial origin and there are few reliable reports of their occurrence in the higher animals. There seems to have been no report of the occurrence of glycosulphatases in the higher plants but the possibility of their being present therein seems quite high. Certainly sulphate esters have been described from this source: for example, one of the pigments of beetroot, prebetanin, is the 6'-sulphate of betanin, a $\beta$-D-glucopyranoside (Wyler, Rosler, Mercier & Dreiding, 1967).

### 7.3.1 Glucosulphatase

This enzyme was first detected in extracts of the mollusc *Eulota* by their ability to hydrolyze glucose 6-sulphate (Soda & Hattori, 1931). A quite intensive study of molluscan glucosulphatases was made by the Japanese workers in the 1930s and was summarized by Soda in 1936. The species investigated in greatest detail was the marine gastropod *Charonia lampas* and it was shown that the digestive and hypobranchial (mucous) glands of this organism could yield active preparations of a glucosulphatase capable of hydrolysing glucose 6-sulphate. Other tissues also contained the enzyme but in rather lesser amounts. A wide range of mono- and disaccharide sulphates, as well as adenosine 5'-sulphate (Yamashina & Egami, 1953), were attacked by the enzyme at pH optima in the region of 5. The enzyme was inhibited by phosphate, sulphate, fluoride and borate ions. Unfortunately, the quantitative significance of much of this early work is in some doubt as many of the substrates prepared by the methods then available were quite inhomogeneous (Dodgson & Lloyd, 1961).

In more recent years Dodgson and his group have studied the glucosulphatase activity of some European marine molluscs, in particular of the periwinkle *Littorina littorea* from which the glucosulphatase was concentrated (Dodgson & Spencer, 1954; Dodgson, 1961 *b*): details of the purification (about thirty-fold) and of the methods of assay have recently been described by Lloyd (1966 *a*). The specificity of the enzyme was relatively low and it hydrolysed glucose 6-sulphate ($K_m = 0.017$ M), glucose 3-sulphate ($K_m = 0.03$ M) and galactose 6-sulphate ($K_m = 0.072$ M) at the relative rates shown in table 7.6. The pH optimum varied with the different substrates but was between 5 and 6. This glucosulphatase showed quite anomalous responses to changes in substrate concentration when the reaction mixture contained sodium acetate–acetic acid buffers but behaved normally when tris–acetic acid buffers were used. The reason for this behaviour is not known but it might appear that Na$^+$ ions interact with the enzyme, acting as inhibitors at low substrate concentrations (Dodgson, 1961 b): such behaviour is similar to that found for crude preparations of the sulphatase C of ox liver (Roy, 1956 c). Detailed studies of the inhibition of this enzyme have not been made but fluoride, phosphate and pyrophosphate ions are all quite powerful inhibitors. None of a wide range of polysaccharides of plant or animal origin were attacked by the enzyme, nor were the oligosaccharides produced by the action of hyaluronidase on chondroitin 4-sulphate or chondroitin 6-sulphate (Lloyd, 1966 a).

The only other European species to have been examined for gluco-sulphatase activity is *Patella vulgata* (Lloyd, Lloyd & Owen, 1962; Lloyd & Lloyd, 1963). Crude enzyme preparations from this species will hydrolyse not only glucose and galactose 6-sulphates but also the algal polysaccharide fucoidin and chondroitin 4-sulphate although a number of other poly-saccharide sulphates were not attacked. This ability to hydrolyse fucoidin and chondroitin 4-sulphate clearly distinguishes the preparation from *Patella* and that from *Littorina*, the inability of which to attack poly-saccharide sulphates has been confirmed by Lloyd & Lloyd (1963). The most likely explanation of the difference would be that the former pre-paration contains a separate polysaccharide sulphatase. Stuart (1966) has measured the relative rates of hydrolysis of a number of monosaccharide sulphates by the enzyme preparation from *Patella*: the results are sum-marized in table 7.6, as are those for the corresponding enzyme from *Littorina*, and it is obvious that several puzzling features require explana-tion, not the least being the very slow rate of hydrolysis of authentic glucose 6-sulphate.

These preparations of glucosulphatases from molluscs apparently have a rather low specificity and it is therefore interesting to compare them with a glucosulphatase present in the mould *Trichoderma viride* which can grow on glucose 6-sulphate or galactose 6-sulphate as sole sources of carbon and sulphur (Yamashina, 1951). Under such conditions, but not when grown on the parent sugars, it produces a glucosulphatase which will hydrolyse glucose and galactose 6-sulphates but not glucose 3-sulphate (Lloyd, Large, Davies, Olavesen & Dodgson, 1968). Either this enzyme must have a much higher degree of specificity than the molluscan gluco-sulphatases or the latter are mixtures of a number of glucosulphatases, a not unlikely situation in view of the occurrence of two chondrosulphatases in molluscs (see § 7.3.5).

TABLE 7.6 *The relative rates of hydrolysis of some monosaccharide sulphates by preparations of glycosulphatases from* Patella vulgata *and from* Littorina littorea

|  | Enzyme from | |
|---|---|---|
|  | *Patella* | *Littorina* |
| glucose 6-sulphate* | 1·00 | 1·00 |
| glucose 2-sulphate | 0·90 | . |
| glucose 3-sulphate | 0·33 | 0·10 |
| galactose 6-sulphate | 0·25 | 0·21 |
| glucose 6-sulphate† | 0·06 | . |
| galactose 4-sulphate | 0·04 | . |
| mannose 4-sulphate | 0·00 | . |

     * Prepared by the direct sulphation of glucose.
     † Authentic glucose 6-sulphate.

Another glycosulphatase which shows a remarkable degree of specificity is that present in *Pseudomonas carrageenovora* (Weigl & Yaphe, 1966). This enzyme can, at pH 7·5, desulphate neocarrabiose sulphate [3-*O*-(3,6-anhydro-α-D-galactopyranosyl)-D-galactopyranose 4-*O*-sulphate] with the formation of the corresponding disaccharide: it can also desulphate galactose 6-sulphate but not, most surprisingly, galactose 4-sulphate although this sugar is present in the neocarrabiose sulphate. Glucose sulphates apparently have not been investigated. The properties of this enzyme have not been studied in detail but it is interesting that its activity is completely inhibited in 0·1 M-acetate buffers of pH 4, conditions quite commonly used in searches for glycosulphatase activity. It is perhaps of interest that the glucosulphatase from *T. viride* likewise shows an optimum activity in the region of pH 7·5 with little activity remaining below pH 6 (Lloyd *et al.* 1968).

All the above examples refer to glucosulphatases in molluscs or micro-organisms and references to such enzymes in higher animals are rare. An interesting report is that of Nonami (1959) who stated that glucosulphatase occurred in the allantoic and amniotic fluids of embryonated hen eggs. No chondrosulphatase activity could be detected. This claim would certainly seem worthy of a further investigation, especially as there apparently was a relationship between the sugar and the sulphate contents of the embryo. Other reports of the presence of glucosulphatase activity in higher animals have been documented by Dodgson (1956) and by Dodgson & Spencer (1956 *b*).

## 7.3.2 Cellulose polysulphatase

Extracts of the digestive gland of *C. lampas* contain a second glyco-sulphatase which has been named cellulose polysulphatase (Takahashi & Egami, 1960, 1961 *b*; Takahashi, 1960 *b*). Although complete separation of this enzyme from glucosulphatase has not been achieved, there seems little doubt that the two activities are due to two different enzymes. Cellulose polysulphatase is characterized by its ability to hydrolyse charonin sulphate, a polysaccharide sulphate which is produced by the hypobranchial gland of *C. lampas*. Charonin is a complex carbohydrate containing both cellulose-like and amylose-like components, with $\beta$-1,4 and $\alpha$-1,4 linkages respectively, and probably also containing a number of $\alpha$-1,6 linkages (Egami & Takahashi, 1962). In charonin sulphate positions 2,3 or 6 may be esterified with sulphuric acid and in the fraction with the highest sulphur content (20%) all the hydroxyl groups must in fact be sulphurylated. It is perhaps not surprising that as well as hydro-lysing charonin sulphate, cellulose polysulphatase will also readily attack cellulose sulphate and slowly attack dextran sulphate. Amylose sulphate is not hydrolysed. There is no real information on the specificity of the enzyme, but Takahashi has suggested that the 2- and 3-sulphates in charonin sulphate are preferentially hydrolysed. This is perhaps the reason why the cellulose-like glucan sulphate from the odontophore of the whelk *Busycon caniculatum* (Lash & Whitehouse, 1960) is not attacked by cellulose polysulphatase because in this compound the sulphate groups probably occur on position 6. The enzyme will not hydrolyse the poly-saccharide sulphates from the marine alga *Chondrus* sp.

## 7.3.3 Functions of glucosulphatase and cellulose polysulphatase

The role of these enzymes seems clear in molluscs such as *C. lampas* which form charonin sulphate. This carbohydrate can be degraded by *Charonia* and it is obvious that, unless there exist completely unknown pathways of carbohydrate metabolism, the sulphate groups must be removed before charonin sulphate can be metabolized. It seems highly probable that cellulose polysulphatase must attack charonin sulphate to give a partially desulphated form which can be hydrolysed by the various carbohydrases present in *Charonia* to give glucose and glucose sulphates, the latter of which is then hydrolysed by glucosulphatase to complete the conversion of the polysaccharide to glucose (Takahashi, 1960 *b*). A similar function may well be important in other molluscs because it is becoming clear that many such species produce carbohydrate sulphates: glucan sulphates have been isolated from the marine gastropods *Busycon caniculatum* (Lash & Whitehouse, 1960) and *Buccinum undatum* (Hunt & Jevons, 1966) while the limpet *Patella vulgata* has been shown to contain a polysaccharide sulphate of unknown structure (Lloyd & Lloyd, 1963). Unfortunately the part played by these compounds in the general carbo-hydrate metabolism of the mollusca does not seem clear. Furthermore, at least some marine molluscs must ingest considerable amounts of poly-saccharide sulphates in their diet of marine algae and it could well be that the glycosulphatases are involved in the digestion of these. It must be admitted, however, that attempts to hydrolyse such polysaccharide sulphates with these enzymes have been singularly unsuccessful in the past, apart from the hydrolysis of fucoidin by crude extracts of *P. vulgata* (Lloyd & Lloyd, 1963).

A possible sulphotransferase activity for the glucosulphatase and the associated arylsulphatase of *C. lampas* has been suggested by Suzuki, Takahashi & Egami (1957) who showed that acetone-dried preparations of the mucous gland of this mollusc could not incorporate $^{35}SO_4^{2-}$ ions into charonin sulphate, but could so incorporate the sulphate group of *p*-nitrophenyl [$^{35}$S]sulphate. This transfer of the sulphuryl group was inhibited by phosphate and fluoride ions which suggested the participation of sulphatases. It was then shown (Suzuki, Takahashi & Egami, 1959) that purified preparations of the sulphatases of the mucous gland could not bring about this transfer, but that partly purified preparations from the digestive gland could do so, provided that the preparations of charonin sulphate used as the acceptor were not highly purified. It was concluded that the relatively crude preparations of charonin sulphate contained some dialysable cofactor necessary for the transfer reaction which Egami & Takahashi (1962) suggested could be represented thus:

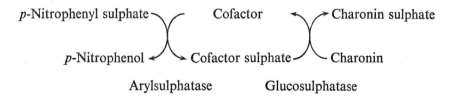

$p$-Nitrophenyl sulphate — Cofactor — Charonin sulphate

$p$-Nitrophenol — Cofactor sulphate — Charonin

Arylsulphatase        Glucosulphatase

Tempting though such a scheme may be it must be realized that there is no real evidence for it: the participation of the sulphatases has not been proven, the only evidence for this being the inhibition of the overall transfer by phosphate ions and, further, there is no indication of the nature of the supposed cofactor. Nevertheless, the work is interesting and important as it is the first real, although slight, indication of a long-suspected transferase activity of the sulphatases.

The function of glucosulphatase in micro-organisms is less obvious but the presence of galactose 3-sulphate residues in the sulphoglycolipid of *Halobacterium cutirubrum* should perhaps be noted (Kates, Palameta, Perry & Adams, 1967).

### 7.3.4 Algal glycosulphatase

There is some evidence (Peat & Rees, 1961) that a carbohydrate sulphatase occurs in marine algae, a not unexpected source of such an enzyme in view of their content of sulphated polysaccharides. Rees (1961 *a, b*) has partially purified from the red alga *Porphyra umbilicis* an enzyme which liberates sulphate from the polysaccharide sulphate porphyran (containing D-galactose, L-galactose, 6-*O*-methyl D-galactose and 3,6-anhydro L-galactose) with the liberation of sulphate. During this reaction 3,6-anhydro L-galactose is formed and Rees has suggested that this may represent the route for the formation of the anhydro-sugar *in vivo*. Perhaps it is also worth recalling the suggestion that the L-galactose of seaweed poly-saccharides is itself formed from D-galactose via D-galactosyl sulphate (Jones & Peat, 1942). However, although there can be no doubt that sulphate ions are liberated from porphyran by enzyme preparations from *Porphyra* there is, as Rees has pointed out, no satisfactory demonstration that a true sulphatase is involved.

### 7.3.5 Chondrosulphatase

This enzyme was first found in putrefactive bacteria by Neuberg & Hofmann (1931) but it received little attention until Dodgson (Dodgson & Lloyd, 1957; Dodgson, Lloyd & Spencer, 1957) began an investigation of the chondrosulphatase of *Proteus vulgaris* and made the important observation that the true substrate of the enzyme is not chondroitin sulphate but the oligosaccharides produced therefrom through the action of either testicular hyaluronidase or the chondroitinase which is present in *Proteus*. The name commonly given to this enzyme, and that recom-mended by the Enzyme Commission, therefore quite wrongly describes its specificity.

A partially purified preparation of chondrosulphatase from *P. vulgaris* will not liberate sulphate from chondroitin 4- or 6-sulphates, dermatan sulphate, keratan sulphate, heparan sulphate, heparin, fucoidin, $\lambda$- or $\kappa$-carrageenin, nor from a number of monosaccharide sulphates. It will liberate sulphate from the oligosaccharides derived from either of the above chondroitin sulphates or from shark cartilage chondroitin sulphate (Lloyd, 1966 b). The pH optimum for chondrosulphatase is around neutrality but there is no reliable information on the kinetic properties of the enzyme because of the difficulty of obtaining an adequately characterized substrate. This difficulty is no longer a significant one since Lloyd, Olavesen, Woolley & Embery (1967) have described the chemical synthesis of N-acetylchondrosine 6-sulphate and shown it to be hydrolysed by chondrosulphatase so that this enzyme can now be studied with a chemically defined substrate.

It should perhaps be mentioned at this stage that Dodgson (Dodgson & Lloyd, 1958; Dodgson, 1959 b) has shown that the action of the chondroitinase present in the strain (NCTC 4636) of P. vulgaris used in the work at Cardiff yields saturated disaccharides from chondroitin sulphate. This has been confirmed by Linker, Hoffman, Meyer, Sampson & Korn (1960) using Dodgson's preparation but these same authors have pointed out that the strain of P. vulgaris used in their own work yields the unsaturated disaccharides which are also produced by the action of bacterial hyaluronidase. The reason for these different findings is not yet clear but they must be kept in mind when chondroitinase-degraded chondroitin sulphate is used as a substrate for chondrosulphatase.

The above results have suggested that the chondrosulphatase of P. vulgaris is rather non-specific and can hydrolyse degradation products of polysaccharides containing either the 4- or 6-sulphates of N-acetylchondrosine. Doubt was cast on this interpretation by Yamagata, Kawamura & Suzuki (1966) who claimed the separation of two chondrosulphatases from P. vulgaris (NCTC 4636), one hydrolysing the 4-sulphate and the other the 6-sulphate of the unsaturated disaccharides produced from chondroitin 4- and 6-sulphates respectively by the action of bacterial hyaluronidase. Yamagata, Saito, Habuchi & Suzuki (1968) have now given details of the separation of these two enzymes, which they have called chondro-4-sulphatase and chondro-6-sulphatase, from extracts of P. vulgaris. Although only a twenty five-fold purification of these enzymes has been achieved it is already apparent that they will be most valuable tools in the study of mucopolysaccharides (Saito, Yamagata & Suzuki, 1968; Suzuki et al. 1968). The two enzymes exhibit pH optima at 7·5 in tris buffers and the 4-sulphatase, but not the 6-sulphatase, is activated by acetate ions. Fairly detailed specificity studies have been carried out and it is clear that chondro-4-sulphatase and chondro-6-sulphatase are specific for the above-mentioned unsaturated disaccharides and for $N$-acetylchondrosine 4-sulphate and $N$-acetylchondrosine 6-sulphate respectively. None of a considerable number of sulphated poly-, oligo- or monosaccharides was hydrolysed except for $N$-acetylgalactosamine 4,6-disulphate which was attacked by chondro-6-sulphatase (to give $N$-acetylgalactosamine 4-sulphate) but not by chondro-4-sulphatase.

The fact that these chondrosulphatases will not hydrolyse intact chondroitin sulphate, but only the disaccharides produced therefrom through the action of a carbohydrase, makes it difficult to place much reliance on reports of the absence of this enzyme from various micro-organisms because it would not be detected unless the appropriate carbohydrase were simultaneously present. Furthermore, it is now clear that the conditions used for the culture of *P. vulgaris* affect its production of chondroitinase and hence its apparent ability to desulphate chondroitin sulphate (Lloyd *et al.* 1967). When the organism was cultured in the presence of glucose or glycerol the production of chondroitinase was greatly depressed although the chondrosulphatase, detected by the hydrolysis of $N$-acetylchondrosine 6-sulphate, was not influenced by such changes in the conditions of growth. On the other hand, Yamagata *et al.* (1968) have shown that the yield of chondrosulphatases from *P. vulgaris* is considerably increased (up to nearly four hundred-fold) when the organism is grown in a medium containing chondroitin sulphates. Bearing these reservations in mind, however, chondrosulphatase does appear to be of very restricted occurrence in bacteria (Dodgson & Spencer, 1956 *b*).

There is some doubt as to the occurrence of chondrosulphatase in molluscs as the reports, even from the same laboratory, are conflicting. Soda & Egami (1938) claimed that *C. lampas* contained a chondrosulphatase and this was apparently confirmed by Takahashi & Egami (1961 *b*) insofar as they showed that relatively crude extracts of the mollusc could liberate sulphate from chondroitin sulphate. On the other hand, Takahashi (1960 *b*) had previously stated that *Charonia* did not contain a chondrosulphatase. Both seasonal variation and nutritional status have been invoked to explain these different results but perhaps the chondrosulphatase of *Charonia* will also only attack degraded chondroitin sulphate and the apparent discrepancies could simply be due to the presence or absence of the appropriate carbohydrases from the various preparations. Nevertheless, it has been stated that extracts of *P. vulgata*

(Lloyd & Lloyd, 1963; Lloyd & Fielder, 1967) can liberate sulphate ions from intact chondroitin sulphate as well as from the oligosaccharides derived therefrom (Lloyd & Fielder, 1968) so that this explanation may well be incorrect.

The situation is also confused in mammals. It has long been known that rats can desulphate chondroitin sulphate *in vivo* (Dziewiatkowski, 1956; Dohlman, 1956) and it has been claimed (Ohmura & Yasoda, 1960) that chondrosulphatase can be detected histochemically in human sweat glands, but until recently all attempts to prepare from mammalian tissues an extract containing a chondrosulphatase have been unsuccessful. In particular, Dodgson & Lloyd (1961) could find no evidence for the hydrolysis of any of a wide range of carbohydrate sulphates by a variety of tissues from a number of mammalian species, which strongly suggested the absence of any glycosulphatase from mammals. Mehl & Jatzkewitz (1963), on the other hand, stated that crude preparations from pig kidney could liberate sulphate from chondroitin sulphate but they made no attempt to characterize the enzyme. Quite recently Held & Buddecke (1967) have purified some eighty five-fold a chondrosulphatase from ox aorta. This enzyme, which had a pH optimum of 4·4, could liberate sulphate ions from chondroitin 4-sulphate but not from chondroitin 6-sulphate. The most interesting feature of this chondrosulphatase was its apparent ability to attack intact chondroitin 4-sulphate: certainly it was clearly shown that the chondrosulphatase activity proceeded independently of hyaluronidase activity, and the latter was in fact completely removed during the purification of the sulphatase, but it was not shown that other carbohydrases were completely absent. Until the second product of the enzymatic reaction has been shown to be intact chondroitin the claim that the chondrosulphatase of the arterial wall attacks intact chondroitin sulphate must be accepted with some reserve. Detailed studies of the specificity of this enzyme have not yet been made but neither chondroitin 6-sulphate, dermatan sulphate nor keratan sulphate were attacked by the enzyme which, incidentally, seems to have been clearly distinguished from the arylsulphatase which is also present in the arterial wall. It would be interesting to know if this enzyme can hydrolyse *N*-acetylglucosamine 6-sulphate and *N*-acetylgalactosamine 6-sulphate, both of which are desulphated *in vivo* by rats (Lloyd, 1961, 1962 *b*) but are not substrates for the chondrosulphatases of *Proteus vulgaris*. In this connection it is interesting that Yamagata (footnote to Yamagata *et al.* 1968) has reported that commercial preparations of testicular hyaluronidase

179

can hydrolyse $N$-acetylgalactosamine 4,6-disulphate to $N$-acetylgalactos-amine 6-sulphate, a reaction which also cannot be brought about by the chondrosulphatases of *P. vulgaris* (cf. p. 178).

### 7.3.6 Cerebroside sulphatase

Fujino & Negishi (1957) first reported the presence of a cerebroside sulphatase in the digestive gland of the abalone, *Haliotis*. No activity was detected in three other molluscan species. The preparation from *Haliotis* could also hydrolyse chondroitin sulphate but not phenyl sulphate nor glucose 6-sulphate. Unfortunately the activity of the preparation was very low and it is difficult to be certain of the significance of the results.

A much more important report is that of Mehl & Jatzkewitz (1963, 1964) who have partially purified a cerebroside sulphatase from the lyso-somes of pig kidney and have shown its presence in other organs, parti-cularly liver and spleen. The enzyme hydrolysed both cerebron and kerasin sulphates with a pH optimum of about 4·5. Sulphite, sulphate and phosphate were rather powerful inhibitors while, surprisingly, 0·025 M-hydroxylamine activated the reaction by a factor of three. No steroid sulphatase, glycosulphatase nor chondroitin sulphatase activity was shown by the preparation but it did have a pronounced arylsulphatase activity, apparently very similar to that of sulphatase A. It was not possible to separate the cerebroside sulphatase and arylsulphatase by high-voltage electrophoresis but this procedure did separate a substance of high molecular weight which activated the cerebroside sulphatase by about thirteen-fold without having any effect on the arylsulphatase. The chemical nature of the 'complementary fraction' has not been elucidated and the only information given was that it was destroyed by treatment with a mixture of ethanol and light petroleum.

Mehl & Jatzkewitz (1968) have now shown that only cerebroside 3-sulphates are hydrolysed by the enzyme with $K_m$ values of 0·1 mM. Synthetic cerebroside 6-sulphates are not hydrolysed. Galactose 3-sulphate was hydrolysed at about 20% the rate of cerebroside 3-sulphate but, as expected, galactose 6-sulphate was not hydrolysed.

As already discussed (see § 7.1.8), Mehl & Jatzkewitz claim that cere-broside sulphatase is identical with sulphatase A, an arylsulphatase. Their evidence for this—the inseparability of the two activities in two quite different preparations, the inhibition of cerebroside sulphatase by nitrocatechol sulphate, and the inhibition of arylsulphatase by cerebroside sulphate—is apparently quite convincing but some points still require clarification. The most important of these is the dependence of cerebroside sulphatase activity on the presence of the complementary fraction which is without action on the arylsulphatase activity. A further anomaly is the activation of cerebroside sulphatase by 0·025 M-hydroxylamine, a concentration which powerfully inhibits sulphatase A (see § 7.1.1). It is to be hoped that explanations of these discrepancies will soon be forthcoming.

Mehl & Jatzkewitz (1963, 1965) also made the important observation that cerebroside sulphatase could not be detected in the kidneys from cases of metachromatic leucodystrophy although it was present in normal human kidney. This is, of course, in keeping with the view (Jatzkewitz, 1960) that metachromatic leucodystrophy arises through the genetically controlled absence of an enzyme capable of destroying cerebroside sulphates. The observation is particularly interesting in view of the deficiency of sulphatase A in metachromatic leucodystrophy (see § 7.1.8) and it certainly gives further evidence for the view that sulphatase A and cerebroside sulphatase are identical. Should this identity be proven then the function of sulphatase A *in vivo* would be much more obvious than it has been hitherto and the unexpected dual specificity would certainly suggest the need for a search for further possible cases of this among the other types of sulphatase.

## 7.4 Myrosulphatase

The activity classically associated with this enzyme has been known for many years (see reviews by Challenger, 1959, and by Kjaer, 1960) and it has long been taken that the over-all myrosinase-catalysed reaction, the fission of a mustard oil glycoside to give an isothiocyanate, glucose and a $SO_4^{2-}$ ion, involved both a carbohydrase and a myrosulphatase (Neuberg & Schonebeck, 1933). Not until recently, however, was definite evidence for this obtained. The study of the enzyme has been complicated by the fact that only in 1956 was the structure of the mustard-oil glycosides elucidated by Ettlinger & Lundeen (1956, 1957): that of a typical mustard-oil glycoside, sinigrin, is shown below (7.7) together with the structure (7.8) which was previously assigned to this compound. It is obvious that the bond split by myrosulphatase, the $=N.OSO_3^-$ bond, is quite different from that attacked by any of the other sulphatases so far studied. As an immediate consequence of the work of Ettlinger & Lundeen it seemed that the so-called myrosulphatase activity could simply be the result of a

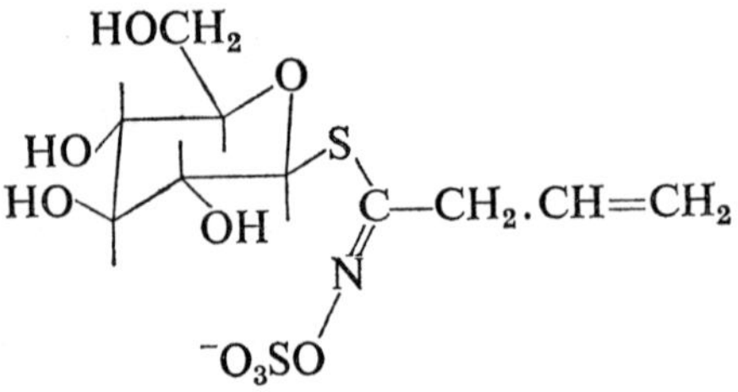

$$\text{(structure 7.7)} \qquad (7.7)$$

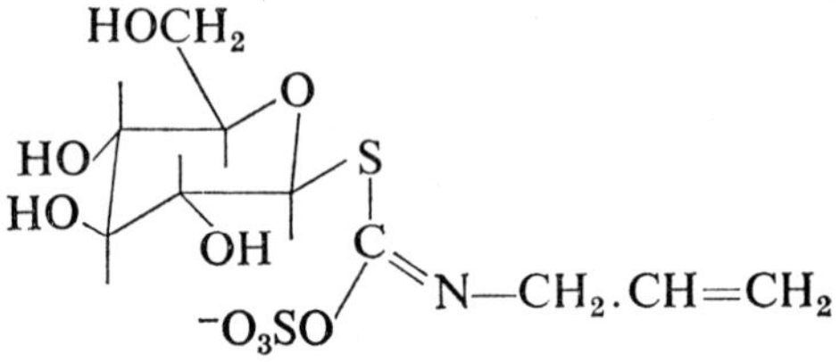

The glucosinolate structure showing HOCH$_2$, HO, HO, OH, O, S, C—CH$_2$.CH=CH$_2$, N, $^-$O$_3$SO

$$\text{(structure 7.8)} \qquad (7.8)$$

The second structure showing HOCH$_2$, HO, HO, OH, O, S, C, N—CH$_2$.CH=CH$_2$, $^-$O$_3$SO

glycosidase- or thioglycosidase-catalysed hydrolysis of the glycoside followed by a spontaneous Lossen rearrangement with liberation of $SO_4^{2-}$ ions (Reese, Clapp & Mandels, 1958) as represented in reaction 7.2.

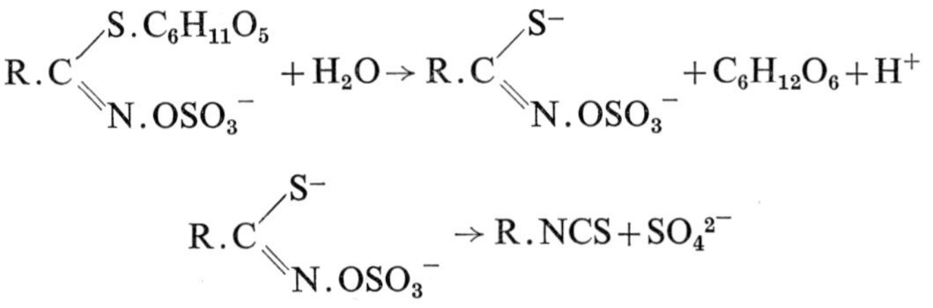

$$\text{R.C}\begin{smallmatrix}\nearrow \text{S.C}_6\text{H}_{11}\text{O}_5 \\ \searrow \text{N.OSO}_3^- \end{smallmatrix} + H_2O \rightarrow \text{R.C}\begin{smallmatrix}\nearrow \text{S}^- \\ \searrow \text{N.OSO}_3^- \end{smallmatrix} + C_6H_{12}O_6 + H^+$$

$$(7.2)$$

$$\text{R.C}\begin{smallmatrix}\nearrow \text{S}^- \\ \searrow \text{N.OSO}_3^- \end{smallmatrix} \rightarrow R.NCS + SO_4^{2-}$$

The classical source of myrosinase is the seeds of the Cruciferae: it also occurs in the vegetative tissues of these plants but it has not been detected in a large number of species representing thirty-seven other families (Nagashima & Uchiyama, 1959 a). Two species have been intensively studied, the Black Mustard, *Brassica niger*, and the Oriental Yellow Mustard, *B. juncea*. Considering that both myrosinase and its substrate occur in the same tissues of these plants it is not surprising that the two are spatially separated, myrosinase occurring in a particular type of cell. the idioblast, as was shown in what must have surely been the first histochemical investigation of any sulphatase (Peche, 1913). The enzyme is therefore inactive until the tissues are disintegrated.

The assay of myrosulphatase (or myrosinase) is inherently more difficult than that of the other sulphatases because of the complex nature of the reaction and the fact that changes in the experimental conditions can qualitatively alter the nature of the products (Schwimmer, 1960). Methods based on the determination of the liberated glucose, isothiocyanate and $SO_4^{2-}$ ions have all been used but perhaps the most convenient type of assay is that which follows, in the pH-stat, the liberation of $H^+$ ions. Schwimmer (1961) has described a spectrophotometric method which utilizes the rather powerful absorption of sinigrin at $227 \cdot 5$ m$\mu$. These techniques all suffer from the disadvantage that they measure an obviously complex reaction and only following the work of Gaines & Goering (1962) have synthetic substrates which allow the specific determination of *myrosulphatase* activity become available (see below): unfortunately these substrates do not yet appear to have been utilized for this purpose.

Gaines & Goering (1960, 1962) have purified the myrosinase from *B. juncea* and shown it to contain a glycosidase and a myrosulphatase which are separable by chromatography on DEAE-cellulose. Only when both enzymes are present is the classical myrosinase reaction (7.2) obtained. The sulphatase will hydrolyse not only mustard-oil glycosides but also simple compounds such as cyclohexanone oxime *O*-sulphate (*7.9*) or acetophenone oxime *O*-sulphate. The ready availability (Smith, P.A.S.

$$(7.9)$$

1948) of such simple substrates containing the $=N.OSO_3^-$ group should greatly simplify the future study of myrosulphatase. Little can be said of the properties of the myrosulphatase but the preparation obtained by Gaines & Goering could not hydrolyse *p*-nitrophenyl sulphate nor glucose 6-sulphate: it hydrolysed sinigrin at an optimum pH of about 6 and it was not inhibited by phosphate, contrary to the findings of Ishimoto & Yamashina (1949) who, using another species of *Brassica*, obtained a myrosulphatase inhibited by both sulphate and phosphate ions.

Despite the apparently unambiguous results of Gaines & Goering there are still claims that myrosinase is a single enzyme. Many of these stem from the work of Nagashima & Uchiyama (1959 *b*) who showed that the myrosinase of White Mustard is inhibited by SH reagents and by hydroxylamine or phenylhydrazine: both the myrosulphatase and glycosidase activities were inhibited equally and this was given as evidence that these were due to a single enzyme. More recently Tsuruo, Yoshida & Hata (1967) stated that they could not separate the sulphatase and glycosidase activities of the myrosinase from *B. juncea*, even using conditions comparable to those of Gaines & Goering (1962). The reason for this discrepancy is not obvious (Schwimmer, 1960) but it is perhaps pertinent that while the Japanese workers have achieved a purification of only about one hundred-fold Gaines & Goering achieved one of about five thousand-fold.

Tsuruo & Hata (1967) have further investigated the activation of myrosinase by ascorbic acid which was first described by Nagashima & Uchiyama (1959 $b$). They showed that this activation was not dependent upon the oxidation-reduction reactions of ascorbic acid but was due to the formation of a complex between the enzyme and the activator with a $K_m$ of 1 mM-ascorbic acid. The ascorbic acid increased not only V for the hydrolysis of sinigrin (by a factor of ten at a concentration of 0·01 M-ascorbic acid) but also the $K_m$ for the substrate, values of 0·93 mM and 0·19 mM-sinigrin being obtained in the presence and absence of ascorbic acid respectively. These same values for $K_m$ were obtained whether the reaction velocity was determined by measuring the production of glucose or of $SO_4^{2-}$ ions and this again was given as evidence for the existence of only a single enzyme in myrosinase. It should be noted that Ettlinger, Dateo, Harrison, Mabry & Thompson (1961) have given a detailed discussion of the action of ascorbic acid on myrosinase activity but the relationship of this to *myrosulphatase* activity is obscure because they themselves state that the ascorbate-activated myrosinase (glucosinolase) is a specific thioglycosidase.

It has been reported that the mollusc *Charonia lampas* contains a myrosulphatase (Ishimoto & Yamashina, 1949; Takahashi, 1960 $a$) although in much lesser amounts than it does the other sulphatases. This molluscan myrosulphatase has very different properties from the plant enzyme, particularly a much lower specificity: it can hydrolyse not only sinigrin but also tetraacetyl sinigrin and tetramethyl sinigrin (Nagashima & Uchiyama, 1959 $c$), neither of which are hydrolysed by classical myrosulphatase. Much further work is needed to establish the validity of the myrosulphatase of *Charonia* and it seems possible that its supposed activity could be due to the action of one of the many glycosidases present in molluscan tissues followed by a spontaneous Lossen rearrangement and liberation of sulphate ions.

There has been no report of the occurrence of myrosulphatase in mammalian tissues since the early claim by Neuberg & Wagner (1927) that it occurred in horse liver, a finding which could not be confirmed, although admittedly under very different experimental conditions, by Baum & Dodgson (1957).

It should be quite clear that the present status of myrosulphatase is a most unsatisfactory one and further studies using oxime *O*-sulphates as substrates are imperative. Only then will the confusion surrounding this enzyme be removed, this confusion almost certainly arising from the fact that the mustard oil glycosides are susceptible to attack by several different enzymes all of which give the same reaction products.

## 7.5  Choline sulphatase

The existence of a choline sulphatase was suggested by the well-known ability of many higher fungi to utilize choline *O*-sulphate (*7.10*) as a source of sulphur (Egami & Itahashi, 1951), but early attempts to detect the enzyme in such fungi were unsuccessful (Spencer & Harada, 1960; Itahashi, 1961). More recently Segel & Johnson (1963 *a*) detected a very weak choline sulphatase activity in *Penicillium chrysogenum* and in *Aspergillus sydowi*, both of which produce and utilize choline sulphate.

$$\begin{array}{c} H_3C \diagdown \quad \diagup CH_2.CH_2.OSO_3^- \\ \overset{+}{N} \\ H_3C \diagup \diagdown CH_3 \end{array} \qquad (7.10)$$

Scott & Spencer (1968) have now shown that appreciable amounts of choline sulphatase occur in the mycelia of *Aspergillus nidulans* grown in a sulphur-deficient medium. The enzyme has an optimum pH of 7·5 and at pH 8·3 (where the activity was only about 60% of the maximum) the optimum substrate concentration is greater than 0·25 M and the $K_m$ is 0·035 M-choline sulphate. This choline sulphatase is completely inhibited by 10 mM sulphite, phosphate, cyanide and cysteine: the immediate product of the enzyme reaction, sulphate, inhibited by only about 25% when present at this concentration. Although choline sulphatase can readily be detected in the mycelia of *A. nidulans* grown on a sulphur-deficient medium, the activity of the enzyme rapidly falls when any of several compounds of sulphur are present. Such compounds were taurine, sulphite, methionine, sulphate or cysteine but Scott & Spencer (1968) concluded that the true co-repressor of the synthesis of choline sulphatase was cysteine. Cysteine, therefore, can control the activity of choline sulphatase in two ways: firstly by repressing the synthesis of the enzyme and secondly by directly inhibiting its activity. The second function is shared by sulphite. It should be noted that in *Neurospora crassa* the formation of choline sulphatase is repressed not by cysteine but by sulphate and Metzenberg & Parson (1966) suggest that $S^{2-}$ ions are the co-repressor in this case.

185

*Pseudomonas nitroreducens*, which can utilize choline sulphate although it cannot synthesize it, forms an intracellular choline sulphatase when grown on a medium containing this ester (Takebe, 1961). This enzyme therefore differs from the fungal enzyme in that it is inducible by its substrate whereas the fungal choline sulphatase is repressible and apparently not subject to induction by its substrate (Scott & Spencer, 1968; Spencer, Hussey, Orsi & Scott, 1968). The choline sulphatase from *P. nitroreducens* has been partially purified and shown to have an optimum pH of 8 and a $K_m$ of 0·04 M-choline sulphate. It is inhibited by sulphate but not, surprisingly, by phosphate which powerfully inhibits most other sulphatases. The enzyme was apparently rather specific because the relatively crude preparations of it which were available could not hydrolyse a number of aryl sulphates, glucose 6-sulphate nor ethyl sulphate. *Pseudomonas aeruginosa* can also form a similar choline sulphatase when it is grown in the presence of choline sulphate but not of any of several other sulphate esters (Harada, 1964).

Choline sulphatase is one of the few sulphatases which have a well authenticated role *in vivo*, in this case of hydrolysing choline sulphate ester to give sulphate ions which can be reduced and so incorporated into the sulphur-containing amino acids (Spencer *et al.* 1968). It should be noted that Spencer & Harada (1960) formerly suggested that the utilization of choline sulphate ester by the higher fungi involved the formation of 3′-phosphoadenylyl sulphate through the reverse action of choline sulphotransferase (§ 6.4). Such a reaction would be very unlikely to occur because of the energy relationships between the compounds and Orsi & Spencer (1964) have since shown that the reaction catalysed by choline sulphotransferase is in fact irreversible so that the participation of a choline sulphatase in the metabolism of the ester can be regarded as essential.

## 7.6 Alkylsulphatase

Although no alkysulphatase has yet been purified there is now little doubt that such enzymes occur. The first indication of their occurrence was the observation by Vlitos (1953) that *Bacillus cereus mycoides* could hydrolyse the herbicide 3′,5′-dichlorophenoxyethyl sulphate to dichlorophenoxyethanol: as the ester is quite stable the action of an alkylsulphatase must be presumed. Some of the higher fungi can utilize dichlorophenoxyethyl sulphate instead of choline *O*-sulphate as a sulphur source (Spencer & Harada, 1960) which might suggest that choline sulphatase is a general alkylsulphatase but this is not borne out by the findings of Takebe (1961).

Hsu (1963, 1965) has isolated from sewage several strains of *Pseudo-monas* which, when cultured in the presence of sodium dodecyl sulphate, readily adapted to yield extracts containing a sulphatase capable of hydrolysing this and similar long-chain alkyl sulphates. A similar enzyme was also found by Payne, Williams & Mayberry (1965) in a *Pseudomonas* sp. grown in a medium containing sodium dodecyl sulphate. The enzyme was concentrated by ammonium sulphate precipitation and shown to have a pH optimum of 7·5 and to be inhibited by phosphate. With a reaction time of ten minutes the temperature optimum for the enzyme was 70°, a remarkably high value which, having in mind that hydrophobic interactions are strengthened by a rise in temperature, perhaps argues for a binding of this type between the enzyme and its substrate. This enzyme preparation could not hydrolyse dichlorophenoxyethyl sulphate nor the aryl sulphates *p*-nitrophenyl sulphate and phenolphthalein sulphate.

These observations therefore suggest the occurrence in certain micro-organisms of at least two different alkylsulphatases which are capable of hydrolysing two rather different types of primary alkyl sulphate.

Payne, Williams & Mayberry (1967) have also produced evidence for the occurrence of a secondary alkylsulphatase. This activity was detected in extracts of *Aerobacter cloacae* which had been cultured in the presence of a mixture of $C_{10}$–$C_{20}$ secondary alkyl sulphates. These extracts could hydrolyse both pentan-3-yl sulphate and sodium dodecyl sulphate: that a separate enzyme was responsible for the two reactions was suggested by the fact that extracts from *Pseudomonas* could hydrolyse sodium dodecyl sulphate but not pentan-3-yl sulphate.

The existence of these alkylsulphatases in micro-organisms has recently become more understandable with the discovery of the natural occurrence of the disulphate ester of docosane-1,14-diol (*7.11*) as the 'sulpholipid'

$$
\begin{array}{c}
\text{O.SO}_3^- \\
| \\
\text{CH}_3.(\text{CH}_2)_7.\text{CH}.(\text{CH}_2)_{12}.\text{CH}_2\text{O.SO}_3^-
\end{array}
\qquad (7.11)
$$

of *Ochromonas danica* (Mayers & Haines, 1967). Presumably this alkyl sulphate would be a substrate both for the primary alkylsulphatase which hydrolyses sodium dodecyl sulphate and for the secondary alkylsulphatase which hydrolyses pentan-3-yl sulphate although this has not been shown. A possible physiological role for these alkylsulphatases can therefore now be more readily visualized.

It is not known whether such alkylsulphatases occur in animals but

187

there is a recent observation which suggests that they might at least be associated with animal tissues. Knaak, Kozbelt & Sullivan (1966) have shown that the rat and the rabbit can, *in vivo*, liberate $^{35}SO_4^{2-}$ ions from 2-ethylhexyl [$^{35}S$]sulphate. However, it must be kept in mind that difficulties may arise in interpreting results obtained with living organisms or with crude extracts derived therefrom because desulphation can certainly occur through reactions not involving sulphatases. This is exemplified by the desulphation of serine $O$-sulphate which can take place either *in vivo*, or *in vitro* when the ester is incubated with extracts of many mammalian tissues (Dodgson & Tudball, 1961; Tudball, 1962*b*; Tudball, Noda & Dodgson, 1964, 1965) or of *Pseudomonas aeruginosa* (Harada, 1964). Thomas & Tudball (1967) have now purified some three-hundred-fold the enzyme responsible for this reaction and have shown it to be distinct from a number of other serine-metabolizing systems. The reaction products have been clearly established to be pyruvate, ammonia and sulphate ions so that a simple sulphatase cannot be involved. The specificity of the system is high: it attacks L-serine $O$-sulphate and a number of peptides thereof but not D-serine sulphate nor several structurally related esters. Threonine sulphate was hardly attacked although this compound quite rapidly yields $SO_4^{2-}$ ions when it is administered to intact rats (Tudball, 1965). Surprisingly, 3-chloro-L-alanine was attacked by the enzyme to give pyruvate and ammonia. The mechanism of this desulphation is obscure but it should be noted that it can be catalysed *in vitro* by pyridoxal 5'-phosphate and metal ions, probably by a reaction quite analogous to the similarly catalysed hydrolysis of other serine $O$-esters (Metzler & Snell, 1952; Longenecker & Snell, 1957). No evidence could be obtained for the participation of pyridoxal phosphate in the enzymatic reaction but the possibility of its occurrence has not been completely excluded. It need not be stressed that the physiological role, if any, of this system is obscure because there is so far no evidence for the occurrence of serine $O$-sulphate *in vivo*.

These studies on the desulphation of serine $O$-sulphate make it essential that both products of a supposed sulphatase-catalysed reaction be isolated before the existence of a sulphatase can be proven, a requirement which seems to have been met in the case of the alkylsulphatases mentioned above. The existence of these enzymes is certain although their relationship to other sulphatases requires investigation. It might be pointed out that these alkylsulphatases, alone among the sulphatases, may have a considerable practical importance because of their ability to destroy the long-chain primary and secondary alkyl sulphates which are used to such a large extent as detergents and which have caused so many problems through their relative resistance to bacterial action.

## 7.7 Sulphamatase

The existence of an enzyme capable of hydrolysing the sulphamate linkage has not been conclusively demonstrated but there are indications that such does exist. The first suggestion came from the work of Korn & Payza (1956) on the degradation of heparin by *Flavobacterium heparinum* to give glucosamine. As the amino groups in heparin are present exclusively as sulphamate groups it appears that the action of a sulphamatase must be invoked to account for the fission of the N—S bond. More convincing evidence for the occurrence of such an enzyme is required however, because other interpretations of the data are possible.

Such evidence has been provided by studies on animal tissues. Despite the fact that 2-[$^{35}$S]sulphoamino-2-deoxy-D-glucose (glucosamine sulphamate) administered to rats is excreted virtually quantitatively (Lloyd, 1964), Lloyd, Embery, Wusteman & Dodgson (1966) showed that the administration to rats of [$^{35}$S]sulphoamino-heparin lead to the excretion in the urine of about 40% of the radioactivity as $^{35}SO_4^{2-}$ ions in the following 48 hr. That a sulphamatase might be involved in this reaction is suggested by observation (Lloyd, Embery, Powell, Curtis & Dodgson, 1966) that such [$^{35}$S]heparin was desulphated when it was incubated at pH 4 or 5 with homogenates of rat spleen but not of other rat tissues. It is to be hoped that further information on this activity will be forthcoming.

# 8

---

## RHODANESE AND 3-MERCAPTO-
## PYRUVATE SULPHURTRANSFERASE

Rhodanese (thiosulphate : cyanide sulphurtransferase, E.C. 2.8.1.1) and 3-mercaptopyruvate sulphurtransferase (3-mercaptopyruvate: cyanide sulphurtransferase E.C. 2.8.1.2) are two enzymes which appear to be involved in the intracellular turnover of reduced sulphur. Although their importance in general sulphur metabolism is by no means certain they are readily purified and assayed, and have proved very amenable to chemical and kinetic studies. As a result fairly extensive investigations into the mechanism of action of these enzymes, particularly rhodanese, have been carried out.

### 8.1 Rhodanese

Lang (1933) discovered the enzyme, rhodanese, which catalyses the formation of thiocyanate from cyanide and thiosulphate (equation 8.1).

$$CN^- + S_2O_3^{2-} \rightarrow SCN^- + SO_3^{2-} \qquad (8.1)$$

The enzyme is particularly active in mammalian liver and kidney but it has been detected in most mammalian tissues (for example: Himwich & Saunders, 1948; Bénard, Gajdos & Gajdos-Török, 1948 a, 1949; Saunders & Himwich, 1950; De Ritis, Coltorti & Giusti, 1954). Rhodanese activity has also been reported in plants (Gemeinhardt, 1938, 1939; Castella Bertran, 1954) and a number of bacteria (see § 8.1.7).

In mammalian tissues rhodanese is mainly associated with the mitochondria (Ludewig & Chanutin, 1950, Sörbo, 1951 a).

### 8.1.1 Assay of rhodanese

Rhodanese is commonly assayed by measuring the rate of formation of thiocyanate from cyanide and thiosulphate. The method described here is that of Sörbo (1953 a).

*Procedure.* The reaction mixture contains 125 $\mu$moles of $Na_2S_2O_3$, 100 $\mu$moles of $KH_2PO_4$, 125 $\mu$moles of KCN and the enzyme in a final volume of 2·5 ml. It is incubated at 20° and the reaction stopped by the addition of 0·5 ml of 38% formaldehyde (which also prevents the appearance of the blue iron-thiosulphate complex): 2·5 ml of ferric nitrate reagent (100 gm of $Fe(NO_3)_3.9H_2O + 200$ ml of 65% (w/v) $HNO_3$ per 1000 ml) and 25 ml of water are added. The optical density at 460 m$\mu$ is measured and compared against a standard curve for thiocyanate. One micro-equivalent of thiocyanate gives an optical density of 0·104 in a cell with a 1 cm pathlength.

To prevent inactivation the enzyme should be diluted in 0·0126 M-$Na_2S_2O_3$ containing 0·025% albumin. Crude extracts or tissue homogenates may be used and any turbidity produced on adding the ferric nitrate reagent should be removed by centrifuging.

## Continuous assay of rhodanese

Smith & Lascelles (1966) have described an assay for rhodanese which allows continuous recording of the enzymatic activity. The method depends on the spontaneous reduction of 2′,6′-dichlorophenolindophenol (DCIP) by sulphite in the presence of $N$-methylphenazonium methosulphate (PMS).

*Procedure.* The reaction mixture contains 300 $\mu$moles of tris buffer, pH 8·7; 150 $\mu$moles of $Na_2S_2O_3$; 150–240 $\mu$moles of half-neutralized NaCN (NaCN/HCl, 2·5/1; mole/mole); 0·5 $\mu$moles DCIP; 0·25 mg PMS and the enzyme in a volume of 3 ml. The reaction is initiated by the addition of cyanide and DCIP reduction followed by measuring the decrease in absorption at 600 m$\mu$; the reference cuvette contains the reaction mixture without thiosulphate.

### 8.1.2 Purification and properties of mammalian rhodanese

Partial purification of mammalian rhodanese was achieved by Lang (1933) and by Cosby & Sumner (1945). The enzyme from beef liver was finally crystallized by Sörbo (1953 a, b). Westley & Green (1959) crystallized rhodanese from beef kidney and reported that this enzyme is indistinguishable from the liver enzyme in respect of crystal form, specific activity and sedimentation rate. Westley (1959) confirmed the identity of the two enzymes by showing that the pH responses, cyanide affinities, heats of activation, heat inactivation kinetics and guanidine inactivation kinetics are similar for both enzymes.

Crystalline mammalian rhodanese, which contains labile sulphur (see below) is a protein showing a typical single absorption peak at 280 m$\mu$. Its amino acid composition is shown in table 8.1 and some of its physical properties in table 8.2. Volini, De Toma & Westley (1966, 1967) showed that the crystalline enzyme with a molecular weight of 37,000 is a dimer which is in a rapid, pH-dependent equilibrium with a monomeric species having a molecular weight of 19,000 (see also Volini & Westley, 1966). Analysis of the protein by peptide mapping indicated that the monomers are identical. Under mildly oxidizing conditions a stable (oxidized?) dimer of rhodanese is formed, possibly by disulphide bridges between the monomers. Mixed disulphide complexes between monomeric rho-danese and dihydrolipoate (Volini, De Toma & Westley 1966, 1967) and mercaptoethanol (Wang & Volini 1967) may be formed.

TABLE 8.1  *Amino acid composition of mammalian rhodanese and 3-mercaptopyruvate sulphurtransferase*

Rhodanese data from Sörbo (1963 b).
3-mercaptopyruvate sulphurtransferase data from Fanshier & Kun (1962).

| | Residues/molecule | |
| Amino acid | Rhodanese (mol. wt. 37,000) | 3-mercaptopyruvate sulphurtransferase (mol. wt. 10,000) |
| --- | --- | --- |
| aspartic acid | 24 | 8 |
| threonine | 14 | 5 |
| serine | 23 | 5 |
| glutamic acid | 30 | 9 |
| proline | 21 | 5 |
| glycine | 28 | 6 |
| alanine | 26 | 6 |
| valine | 26 | 6 |
| methionine | 5 | 1 |
| isoleucine | 7 | 4 |
| leucine | 27 | 7 |
| tyrosine | 12 | 2 |
| phenylalanine | 16 | 4 |
| lysine | 16 | 6 |
| histidine | 8 | 2 |
| arginine | 21 | 4 |
| cysteine | 4 | 1 |
| tryptophan | 11* | . |
| NH$_3$ | 16 | 10 |
| cyanide-labile sulphur | 1† | . |

*   Davidson & Westley (1965) report 8 tryptophan residues/mole.
†   1·3 labile S/mole (Sörbo, 1962 b); 1·5–1·9 (Westley & Nakamoto, 1962; Green & Westley, 1961).

TABLE 8.2  *Physical properties of crystalline mammalian rhodanese*

|  |  |  | References |
|---|---|---|---|
| crystal form | : | plates or needles | 1 |
| pH | : | broad optimum pH 8 to 9. no activity at pH 7 or 10 | 1 |
| temperature | : | optimum 50° | 1 |
| apparent activation energy | : | 7900 calories | 1 |
| molecular wt | : | 37,100 | 1 |
| $S_{20, w}$ | : | 3·2 S | 2 |
|  |  | 3·0 S | 1 |
| absorption spectrum | : | single peak at 280 m$\mu$ | 1 |
| turnover no. | : | 20,000 molecules SCN⁻ per min per molecule enzyme at 20°, pH 8·6 | 1 |
| diffusion coefficient | : | $7·5 \times 10^{-7}$ cm² sec⁻¹ | 1 |
| partial specific volume | : | 0·74 | 1 |

1, Sörbo (1953 *a, b, c*).     2, Westley & Green (1959).

Crystalline rhodanese, prepared according to Westley & Green (1959), contains one equivalent of zinc ion per 19,000 molecular weight (Volini, De Toma & Westley, 1966, 1967). The significance of this zinc content in relation to rhodanese activity is doubtful since in a footnote to their paper Volini *et al.* (1967) reported the preparation of zinc-free rhodanese which had full activity in the thiosulphate-cyanide assay system.

## Miscellaneous properties of rhodanese

Mammalian rhodanese is inhibited by cyanide and sulphite if they are added to the enzyme prior to thiosulphate (Lang, 1949; Saunders & Himwich, 1950; Sörbo, 1951 *a*); inhibition by these compounds is prevented by cysteine (Sörbo, 1951 *a*). Cysteine also prevents inhibition of rhodanese by hydroxylamine, phenylhydrazine and semicarbazide; little or no inhibition is caused by sulphide, cysteine, diethyldithiocarbamate, azide, acetonitrile, cyanate, fluoride, pyrophosphate or thiosemicarbazide (Sörbo, 1951 *a*). Arsenate and bile salts (Bénard, Gajdos-Török & Gajdos, 1947) and certain monovalent anions (Sakai, 1960; Mintel & Westley, 1966 *b*) have been reported to inhibit rhodanese. According to Sakai the order of inhibition by anions is thiocyanate > iodide > nitrite > bromide > chloride > fluoride. Inhibition by salts may be due, in part, to an increase in the ionic strength of the medium (Mintel & Westley, 1966 *b*). Rhodanese activity has been reported to be reduced by high thiosulphate concentrations (Funaki, Shibata, Yamoaka & Watanabe, 1958) and enhanced by surface active agents (Oike, 1958).

### 8.1.3 Mechanism of the rhodanese-catalysed thiosulphate-cyanide reaction

There have been many reports that mammalian rhodanese is inhibited by thiol-reacting compounds (Sato & Hayashi, 1952; De Ritis, Coltorti & Giusti, 1954; Coltorti & Giusti, 1956; Saunders & Himwich, 1950; Sörbo, 1951 $a$; Sörbo, 1957 $a, b$, 1963 $a$) which suggested that thiol groups are essential for rhodanese activity. In one instance (Coltorti & Giusti, 1956) it was reported that preincubation of the enzyme with thiosulphate prevented inhibition by $N$-ethylmaleimide (see also p. 199). Sörbo (1951 $b$) proposed a possible mechanism for rhodanese action to account for inhibition by thiol-binding reagents, which involved a reaction between thiosulphate and the enzyme to form a sulphenyl thiosulphate intermediate (equations 8.2 and 8.3).

$$E\diagdown\!\!\!\!\!\begin{array}{c}S\\\\S\end{array} + SSO_3^{2-} \rightarrow E\diagdown\!\!\!\!\!\begin{array}{c}S-S-SO_3^-\\\\S^-\end{array} \tag{8.2}$$

$$E\diagdown\!\!\!\!\!\begin{array}{c}S-S-SO_3^-\\\\S^-\end{array} + CN^- \rightarrow E\diagdown\!\!\!\!\!\begin{array}{c}S\\\\S\end{array} + SCN^- + SO_3^{2-} \tag{8.3}$$

This mechanism also accounted for the fact that cyanide and sulphide inhibit rhodanese when added before thiosulphate (see p. 193): these reagents were assumed to react with the enzyme to form inactive complexes (equations 8.4 and 8.5).

$$E\diagdown\!\!\!\!\!\begin{array}{c}S\\\\S\end{array} + CN^- \rightarrow E\diagdown\!\!\!\!\!\begin{array}{c}S-CN^-\\\\S^-\end{array} \tag{8.4}$$

$$E\diagdown\!\!\!\!\!\begin{array}{c}S\\\\S\end{array} + SO_3^{2-} \rightarrow E\diagdown\!\!\!\!\!\begin{array}{c}S-SO_3^-\\\\S^-\end{array} \tag{8.5}$$

Szczepkowski (1961 $a$) criticized Sörbo's hypothesis on the grounds that compounds containing sulphenyl thiosulphate groups such as alanine sulphodisulphane and its glutathione analogue react with cyanide to liberate thiosulphate (equation 8.6).

$$R-S-S-SO_3^- + CN^- \rightarrow RSCN + S_2O_3^{2-} \tag{8.6}$$

Inorganic thiocyanate is not formed. This result is consistent with known $S$-nucleophilicities of cyanide, sulphite, and thiosulphate (Foss, 1947;

see chapter 2). Szczepkowski suggested that a double displacement mechanism could account for the various reactions catalysed by rhodanese (equations 8.7 and 8.8 where $XS^-$ is thiosulphate, organic thiosulphonate, inorganic disulphide or organic disulphide, see p. 196).

$$E\underset{S}{\overset{S}{\diagdown\mkern-6mu|\mkern-6mu\diagup}} + XS^- \rightarrow E\underset{S}{\overset{S}{\diagdown\mkern-6mu\diagup}} S + X^- \tag{8.7}$$

$$E\underset{S}{\overset{S}{\diagdown\mkern-6mu\diagup}} S + CN^- \text{ (or } SO_3{}^{2-}) \rightarrow E\underset{S}{\overset{S}{\diagdown\mkern-6mu|\mkern-6mu\diagup}} + SCN^- \text{ (or } S_2O_3{}^{2-}) \tag{8.8}$$

The double displacement mechanism was confirmed by Green & Westley (1961). They showed polarographically that one mole of rhodanese (mol. wt. 37,000) reacts with sulphite or cyanide to form two moles of thiosulphate or thiocyanate. They concluded that the crystalline enzyme is a complex (rhodanese-$S_2$) which contains two atoms of labile sulphur per mole and which is formed from thiosulphate and rhodanese. The overall reaction was described by equations 8.9 and 8.10.

$$\text{Rhodanese} + 2SSO_3{}^{2-} \rightleftharpoons \text{rhodanese-}S_2 + 2SO_3{}^{2-} \tag{8.9}$$

$$\text{Rhodanese-}S_2 + 2CN^- \rightleftharpoons \text{rhodanese} + 2SCN^- \tag{8.10}$$

It may be noted in parenthesis that while Green and Westley write equation (8.10) as a reversible reaction no evidence has yet been obtained for the formation of thiosulphate and cyanide from thiocyanate and sulphite.

Westley & Nakamoto (1962) confirmed the existence of labile sulphur in rhodanese by demonstrating the transfer to the enzyme of $^{35}S$ from the sulphane sulphur of radioactive thiosulphate; no transfer occurred from the sulphonate sulphur of thiosulphate. Treatment of the labelled enzyme with cyanide, sulphite, trichloracetic acid or heat released the $^{35}S$; in the case of cyanide and sulphite, $^{35}S$-labelled thiocyanate and thiosulphate respectively were formed. Inhibition of rhodanese by sulphite and cyanide was accounted for by the fact that the labile-sulphur free enzyme is unstable.

From a kinetic analysis of thiocyanate formation by rhodanese with thiosulphate and methanethiosulphonate (see § 8.1.4) as substrates, Mintel & Westley (1966 a) deduced that the scission of the sulphur–sulphur bond is the rate limiting step in the overall enzymic cyanolytic reaction. They also obtained evidence that the catalytic activity of rhodanese involves, in part, an electronic shift away from the sulphur–sulphur bond of thiosulphate.

195

### 8.1.4 Other reactions catalysed by mammalian rhodanese

The cyanide-thiosulphate reaction catalysed by rhodanese appears to be irreversible (Goldstein & Rieders, 1953). An apparent equilibrium constant

$$K = \frac{[SCN^-][SO_3^{2-}]}{[CN^-][S_2O_3^{2-}]}$$

of $1 \times 10^{10}$ was obtained by Sörbo (1953 $c$), a result that confirmed the essential irreversibility of the reaction. A number of other sulphur compounds may substitute for thiosulphate in the cyanolytic reaction notably 'disulphide' (Sörbo, 1960), thiocystine (Szczepkowski & Wood, 1967) and organic thiosulphonates, $R.SO_2S^-$, (Sörbo, 1953 $c,d$; 1962 $a$; Mintel & Westley, 1966 $a$). The latter are, in fact, considerably more active than thiosulphate; Mintel & Westley (1966 $a$) report the following rates relative to thiosulphate $= 1$: $p$-toluenethiosulphonate $= 37$, ethanethiosulphonate $= 56$, naphthalene-2-thiosulphonate $= 67$, methanethiosulphonate $= 81$, $p$-bromobenzenethiosulphonate $= 85$. The following compounds do not replace thiosulphate: sulphide, thiourea, methionine, cystine, cysteine, ethanedithiol, diphenylsulphide, diphenyldisulphide, thiouracil, dithiobiuret, and naphthylurea (Himwich & Saunders, 1948); $S$-ethyl thiosulphate ($C_2H_5S.SO_3^-$), ethyl xanthate, diethyldithiocarbamate (Sörbo, 1953 $d$); and $S$-sulphocysteine (Villarejo & Westley, 1963 $b$). Hydrosulphite shows a slight reactivity (Sörbo, 1953 $d$) as does $S$-sulphoglutathione (Eriksson & Sörbo, 1967).

Lang (1933) reported that elemental sulphur could replace thiosulphate in the rhodanese reaction in liver homogenates. Purified rhodanese, however, does not use elemental sulphur (Sörbo, 1953 $d$) and the formation of thiocyanate from sulphur and cyanide has since been shown to be due to a separate enzyme, rhodanese-S (De Ritis, Coltorti & Giusti, 1956; Sörbo, 1955).

*Formation of thiosulphate by rhodanese*

Sörbo (1957 $c$) demonstrated that rhodanese catalyses the formation of thiosulphate from sulphite in the presence of ethanethiosulphonate or toluenethiosulphonate (equation 8.11).

$$SO_3^{2-} + R.SO_2S^- \rightarrow S.SO_3^{2-} + R.SO_2^- \tag{8.11}$$

The formation of thiosulphate from disulphide and sulphite in the presence of rhodanese had also been demonstrated (Szczepkowski, 1961 $b$). Szczepkowski also showed that rhodanese catalyses the formation of

196

hiosulphate from sulphide and sulphite in the presence of cystine or oxidized glutathione. The reaction mechanism was assumed to be that described by equations 8.12 and 8.13.

$$R.S.S.R + S^{2-} \xrightarrow{\text{chemical}} R.S.S^- + RS^- \qquad (8.12)$$

$$R.S.S^- + SO_3^{2-} \xrightarrow{\text{rhodanese}} R.S^- + S.SO_3^{2-} \qquad (8.13)$$

Isotope exchange between sulphite and the sulphonate sulphur of thiosulphate in the presence of rhodanese was demonstrated by Sörbo (1962 a).

## 8.1.5 Thiosulphate reductase activity of rhodanese

Crystalline beef liver rhodanese catalyses the reduction of thiosulphate to sulphide and sulphite (Villarejo & Westley, 1963 a, b). Several reducing agents such as borohydride, dithionite, 2,3-mercaptopropanol, dihydrolipoate and dihydrolipoamide are effective electron donors for the reduction. According to Koj (1968) cysteine and glutathione are also active donors although, in an earlier report, Villarejo & Westley (1963 b) claimed that these compounds are inactive. Mercaptoethanol, NADH and NADPH are inactive (Villarejo & Westley, 1963 b).

The overall reaction with dihydrolipoate is described by equation 8.14.

$$\text{dihydrolipoate} + S_2O_3^{2-} \rightarrow H_2S + \text{lipoate} + SO_3^{2-} \qquad (8.14)$$

The reaction is apparently specific for one optical isomer of lipoate since the utilization of DL-dihydrolipoate never exceeds fifty per cent (Villarejo & Westley, 1963 a). The reverse reaction, namely the formation of thiosulphate from sulphide, sulphite and oxidized lipoate in the presence of rhodanese was demonstrated by Villarejo & Westley (1963 b) as was also the enzyme-catalysed synthesis of thiocyanate from sulphide, cyanide and lipoate (equation 8.15).

$$\text{lipoate} + H_2S + CN^- \rightarrow SCN^- + \text{dihydrolipoate} \qquad (8.15)$$

Villarejo & Westley (1963 b) proposed a mechanism for the rhodanese-catalysed thiosulphate-lipoate reaction which involves the formation of rhodanese-$S_2$ from rhodanese and thiosulphate (equation 8.9), the transfer of sulphur from rhodanese-$S_2$ to dihydrolipoate to form lipoate persulphide and the decomposition of the latter to lipoate and sulphide (equations 8.16 and 8.17).

$$\text{rhodanese-S}_2 + 2\text{ lip}\overset{\text{SH}}{\underset{\text{SH}}{\diagdown}} \rightarrow \text{rhodanese} + 2\text{ lip}\overset{\text{S—SH}}{\underset{\text{SH}}{\diagdown}} \tag{8.16}$$

$$2\text{ lip}\overset{\text{S—SH}}{\underset{\text{SH}}{\diagdown}} \rightarrow 2\text{ lip}\overset{\text{S}}{\underset{\text{S}}{\diagdown}}\big| + 2\text{H}_2\text{S} \tag{8.17}$$

This proposal was based on the following evidence:

(1) Substrate concentrations of crystalline beef liver rhodanese (rhodanese-$S_2$, see § 8.1.3) react with dihydrolipoate liberating stoichiometric amounts of sulphide and oxidized lipoate.

(2) Lipoate reacts with sulphide in alkaline solution to form lipoate persulphide with an absorption maximum in the region of 335–40 m$\mu$ similar to that found with other organic persulphides (cf. Rao & Gorin, 1959).

(3) Spectrophotometric and polarographic evidence indicated that persulphides are intermediates in the spontaneous chemical reduction of benzenethiosulphonate under alkaline conditions by dihydrolipoate and other thiols (equation 8.18).

$$\phi.\text{SO}_2\text{S}^- + 2\text{R}.\text{SH} \rightarrow \phi.\text{SO}_2^- + \text{H}_2\text{S} + \text{R}.\text{S}.\text{S}.\text{R}. \tag{8.18}$$

Although Villarejo and Westley were unable to detect the formation of lipoate persulphide during rhodanese-catalysed thiosulphate reduction this may have been due to its instability at the pH (8·6) employed in the enzymic experiments.

A kinetic analysis of the rhodanese-catalysed thiosulphate-lipoate reaction was reported by Volini & Westley (1965, 1966). The enzyme was shown to form kinetically significant binary complexes with both thiosulphate and dihydrolipoate. The first product (sulphite) is discharged before the second substrate (dihydrolipoate) is attached to the enzyme. Variations in kinetic coefficients with pH suggested that an enzymic thiol group may be associated with the binding of lipoate.

Villarejo & Westley (1963 a) were unable to detect bound lipoate in cystalline rhodanese. It would seem unlikely therefore that lipoate mediates sulphur transfer in the rhodanese-catalysed cyanide-thiosulphate or cyanide-thiosulphonate reactions.

### 8.1.6  The active site of rhodanese

The mechanism of thiosulphate binding and the catalytic site of rhodanese have been the subjects of a number of investigations. As the result of these studies, thiol, aromatic, cationic and electrophilic groups have been implicated in the enzymic reactions of rhodanese.

## Thiol groups

Szczepkowski (1961 a) assumed that the enzyme-bound sulphur in the rhodanese reaction was present in a trisulphide group (see equations 8.7 and 8.8). Sörbo (1962 b), while conceding the double displacement hypothesis, continued to implicate thiol groups in the reaction mechanism. He reported that the sulphite inhibition of rhodanese was accompanied by binding of sulphite to the enzyme. The sulphite was released from the enzyme by thiols and the sulphite-inhibited enzyme contained approximately three thiol groups compared with four in the untreated enzyme. Sörbo proposed that rhodanese contains an active thiol group and that the labile sulphur of rhodanese is a persulphide (E.S.SH) stabilized by secondary bonding to the enzyme. Villarejo & Westley (1963 b) have criticized this hypothesis since rhodanese-$S_2$ lacks the absorption bands at 335–40 m$\mu$ indicative of persulphide linkages. They also pointed out that persulphides decompose to sulphide in acid solutions whereas rhodanese-$S_2$ liberates elemental sulphur under these conditions (Westley & Nakamoto, 1962). Moreover, Davidson & Westley (1965) found that when conditions known to favour the stability of the tertiary structure of rhodanese are used, little inactivation of the enzyme by organic mercurials occurs. They implied that earlier results of thiol-binding reagent inhibition of rhodanese (see above) may have been due at least partly to the inherent instability of the enzyme under the conditions used in those experiments.

Wang & Volini (1967), however, have recently reported evidence which indicates an important role for thiol groups in rhodanese activity. They found that alkylating agents, aromatic nitro compounds and aliphatic mercaptans completely inactivate rhodanese and that this inactivation is accompanied by the loss of one of the two thiol groups in the rhodanese monomer. Moreover the inactivation by all three groups of inhibitors is prevented by the presence of the substrate, thiosulphate.

## Aromatic groups

Davidson & Westley (1965) obtained the following evidence that tryptophan is involved in the active site in rhodanese.

(1) Rhodanese is inactivated by N-bromosuccinimide, a mild oxidizing agent which has a high affinity for the indole ring of tryptophanyl residues (Patchornik, Lawson & Witkop, 1958; Fuller, 1963).

(2) N-1-(4-pyridyl)-pyridinium chloride, a reagent forming reversible charge transfer complexes with indole ring systems (Kosower, 1956;

Cilento & Tedeschi, 1961; Kanner & Kozloff, 1964), is a competitive inhibitor of rhodanese with respect to thiosulphate.

(3) The fluorescence spectrum of rhodanese-$S_2$ relative to the labile-sulphur free enzyme indicates quenching by sulphur in a manner similar to the quenching of indole fluorescence by sulphur.

Wang & Volini (1967) reported that aromatic ions, but not the corresponding aliphatic ions, inhibit rhodanese activity competitively which, they suggested, confirms the role of an aromatic group in the catalytic activity. Davidson & Westley (1965) proposed that the mechanism of sulphur binding involves the formation of a charge-transfer complex in which electronic charge from the $\pi$ system of the indole ring of tryptophan is transferred to an overlapping non-bonding orbital of the sulphur atom.

## Cationic and electrophilic groups

Davidson & Westley (1965) reported that high ionic strengths reduce the velocity of the rhodanese-catalysed thiosulphate-cyanide reaction in the presence of non-saturating concentrations of thiosulphate. Since the effect of ionic strength persisted at pH values of about 8·8 a basic residue was implicated. Later Mintel & Westley (1966 b) showed that the affinity of rhodanese for thiosulphate is increased by high dielectric constants and decreased by high ionic strengths of the reaction medium. The maximum velocity of the rhodanese reaction is unaffected by changes in the dielectric constant or ionic strength of the medium.

These results indicated that charge neutralization is involved in the thiosulphate-binding process and it was inferred that a cationic binding group for thiosulphate is present at the active site of the enzyme. The evidence for an electrophilic group at the active site is based on kinetic analysis of the thiosulphate-cyanide reaction (Mintel & Westley, 1966 a; see § 8.1.3).

The nature of the cationic and electrophilic groups is unknown. Two possibilities for the cationic group, the $\epsilon$-amino group of lysine or a guanidino group of arginine, appear to be discounted since the ionization of these groups would probably not be sufficiently altered by the ionic strength of the medium at the pH of the enzyme assay (pH 8·6) to account for the effects of ionic strength on enzyme activity (Mintel & Westley, 1966 b). Another possible candidate, the phenolic group of tyrosine, appears to be eliminated on chemical grounds (Davidson & Westley, 1965).

It was suggested that zinc ions associated with crystalline rhodanese may fulfil both the cationic and electrophilic functions of the active site of rhodanese (Volini, DeToma & Westley, 1966, 1967; Leininger & Westley, 1967, 1968). The preparation of zinc-free rhodanese with full activity however (see § 8.1.2) casts considerable doubt on this hypothesis.

### 8.1.7 Bacterial rhodanese

Rhodanese activity (thiosulphate-cyanide) has been detected in a number of bacteria (Bénard, Gajdos & Gajdos-Török, 1948 b, 1949) including *Bacillus* spp. (Villarejo & Westley, 1963 b, 1966), *Thiobacillus* spp. (McChesney, 1958; Bowen, Butler & Happold, 1965 a, b; Sargeant, Buck, Ford & Yeo, 1966; Lé John, Van Caeseele & Lees, 1967; Kelly, 1968), Thiorhodaceae and Athiorhodaceae (Smith & Lascelles, 1966) and *Pseudomonas aeruginosa* and *Alcaligenes* (Hall & Berk, 1968). Stearns (1953) reported that rhodanese is present in *Escherichia coli* but Villarejo & Westley (1963 b) were unable to confirm this. Rhodanese from *Bacillus subtilis* was partly purified by Villarejo & Westley (1966) and was shown to catalyse both the thiosulphate-cyanide reaction and thiosulphate reduction (see § 8.1.5).

Smith & Lascelles (1966) partly purified rhodanese from *Chromatium*; two active fractions were obtained by chromatography of extracts of *Chromatium* on DEAE-cellulose or CM-cellulose. Rhodanese from *Chromatium* has an optimum pH at 8·7 and Michaelis constants for thiosulphate and cyanide of 0·6 and 20 mM respectively; it is not affected by EDTA, fluoride, azide, thiocyanate, cadmium ions, arsenite or iodo-acetate (Smith & Lascelles, 1966). Tetrathionate, however, inhibits the thiosulphate-cyanide reaction, possibly by reacting chemically with cyanide (see chapter 4).

Rhodanese from *Thiobacillus denitrificans* has been studied by Bowen, Butler & Happold (1965 a,b). They purified a 'monomeric' form of the enzyme with a molecular weight of about 38,000 similar to that of the dimeric mammalian rhodanese. In addition an active 'subunit' with a molecular weight of about 9,000 and a 'tetramer' (molecular weight of about 150,000) were detected by gel filtration of *T. denitrificans* rhodanese. Treatment of the tetramer with mercaptoethanol produced enzymically active fragments with molecular weights in the order of 2,000 and 7,000. No subunit with a molecular weight of 19,000, which would correspond to the monomeric form of mammalian rhodanese, was detected. Bowen *et al.* suggested that the low molecular weight fragments of *T. denitrificans* rhodanese are linked by disulphide bridges to form the 'monomer' which can then aggregate by forming further disulphide bridges.

The enzyme from *T. denitrificans* is strongly inhibited by a number of thiol-binding reagents and is stimulated by reduced glutathione or dihydrolipoate. Its maximum activity occurs between pH 8 and 9; copper is apparently not present. Bowen *et al.* (1965 *b*) proposed a mechanism for the action of *T. denitrificans* rhodanese similar to that advanced by Szczepkowski (1961 *a*) (see § 8.1.3), for mammalian rhodanese, involving the formation of a trisulphide group on the enzyme molecule as in the following scheme.

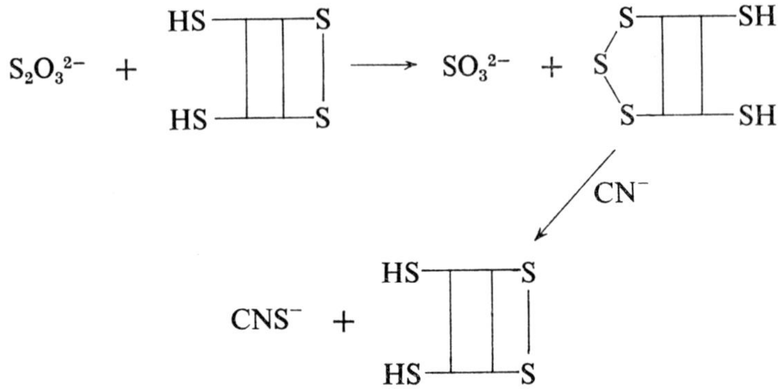

## 8.1.8 Physiological role of rhodanese

Lang (1933) suggested that rhodanese may function in mammalian tissues in cyanide detoxication and this notion has been expressed by a number of workers from time to time. Bénard, Gajdos & Gajdos-Török (1948 *b*) demonstrated that the toxic action of cyanide on yeast respiration is reversed by liver extracts in the presence of thiosulphate and a similar reactivation of cyanide-inhibited mammalian cytochrome oxidase by rhodanese plus thiosulphate was reported by Sörbo (1957 *c*). In *T. denitrificans*, rhodanese activity is increased several fold by the inclusion of cyanide in the growth medium, indicating a possible role for this enzyme in cyanide detoxication (Bowen, Butler & Happold, 1965 *b*). There is no direct evidence, however, that the primary function of rhodanese is in cyanide detoxication and, moreover, the high activities in some mammalian tissues appear to be inconsistent with the extremely low levels of cyanide to which the organisms are normally exposed.

The demonstration of the thiosulphate reductase activity of rhodanese and the fact that 'more physiological' compounds such as lipoate, disulphides and thiocystine are substrates for the enzyme suggest that rhodanese may function in the intracellular transfer of 'reduced sulphur' in the normal synthetic and degradative metabolism of the organism. This view is supported by a certain degree of parallelism between the distribution of cysteine desulphydrase and rhodanese activities in animal tissues (Koj & Frendo, 1962; Frendo, Koj & Górniak, 1963; Koj, Frendo & Borysiewicz, 1964). Schneider & Westley (1963) have suggested that rhodanese may provide a mechanism whereby inorganic sulphur is made available for the synthesis of cysteine in mammals; they demonstrated the incorporation of the sulphane sulphur of thiosulphate into cysteine by lysed rat liver mitochondria without, apparently, the intermediate formation of free sulphide ions. The possibility that rhodanese may function in the terminal stages of cysteine biosynthesis from sulphate in microorganisms and plants has been advanced by Torii & Bandurski (1967) (see chapter 10). Szczepkowski & Wood (1967) demonstrated that rhodanese and cystathionase form a coupled system which is able to use cystine sulphur for transulphuration. The reaction was thought to involve the intermediate formation of thiocystine (R—S—S—S—R in equations 8.19–8.21) although a direct transfer of sulphur from thiocysteine, the product of cystathionase action, was not eliminated.

$$R\text{—}S\text{—}S\text{—}R + H_2O \xrightarrow{\text{cystathionase}} R\text{—}S\text{—}SH + \text{pyruvate} + NH_3 \quad (8.19)$$

$$R\text{—}S\text{—}SH + R\text{—}S\text{—}S\text{—}R \rightarrow R\text{—}S\text{—}S\text{—}S\text{—}R + RSH \quad (8.20)$$

$$R\text{—}S\text{—}S\text{—}S\text{—}R + CN^- \xrightarrow{\text{rhodanese}} R\text{—}S\text{—}S\text{—}R + SCN^- \quad (8.21)$$

or $\quad R\text{—}S\text{—}S\text{—}S\text{—}R + SO_3^{2-} \rightarrow R\text{—}S\text{—}S\text{—}R + S_2O_3^{2-}$

A possible function for rhodanese in the photosynthetic and chemosynthetic sulphur bacteria, *Chromatium* and *Thiobacillus*, may be the cleavage of thiosulphate as an early reaction in the oxidation of this compound to sulphate (Smith & Lascelles, 1966; Charles & Suzuki, 1966 b; see also chapter 9). A role for rhodanese in the sulphur metabolism of *Thiobacilli* is indicated by the findings of Lé John, Van Caeseele & Lees (1967) that the enzyme is apparently induced by thiosulphate in the facultative sulphur autotroph, *Thiobacillus novellus*, during the change from heterotrophic to autotrophic growth. On the other hand, the rhodanese activity of *Chromatium* is similar in bacteria grown on thiosulphate or on organic substrates and also is in the same order as the rhodanese activities of the non-sulphur photosynthetic bacteria, *Rhodo-*

*spirillum rubrum* and *Rhodopseudomonas spheroides* (Smith & Lascelles, 1966). The precise role of rhodanese in the sulphur bacteria, therefore, awaits further clarification.

## 8.2 3-Mercaptopyruvate sulphurtransferase

Wood & Fiedler (1953) reported that 3-mercaptopyruvate reacts with cyanide to form thiocyanate in the presence of crude extracts of acetone powders of pig liver. They attributed this activity to rhodanese but later work showed that it is due to a separate and specific enzyme (Sörbo, 1954; Fiedler & Wood, 1956). Meister and his colleagues (Meister, 1953; Meister, Fraser & Tice, 1954) showed that rat liver and other tissues convert 3-mercaptopyruvate to pyruvate and elemental sulphur. A similar reaction has been reported in *E. coli* (Kondo, Kameyama & Tamiya, 1956). Sörbo (1957 *b*) demonstrated the transfer of sulphur from 3-mercaptopyruvate to sulphite or sulphinates by rat tissues to form thiosulphate and thiosulphonates respectively. These desulphuration and transsulphuration activities have since been shown to be catalysed by a single enzyme, 3-mercaptopyruvate sulphurtransferase (Kun & Fanshier, 1958, 1959 *a, b*). These reactions may be described as follows:

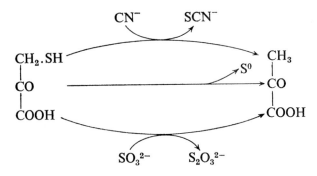

The enzyme is present in many animal tissues and in bacteria (Sörbo, 1957 *b*; Hylin, Fiedler & Wood, 1959; Meister, Fraser & Tice, 1954; Kun & Fanshier, 1959 *b*). Its metabolic function in animals is discussed in chapter 11.

### 8.2.1 Specificity of 3-mercaptopyruvate sulphurtransferase

3-Mercaptopyruvate sulphurtransferase appears to be specific for 3-mercaptopyruvate. Slight activity occurs with 3,3′-dimercaptodipyruvate [*bis*(2-carboxy-2-oxoethyl) disulphide] and ethyl mercaptopyruvate but none with 3-mercaptolactate, 3,3'-dimercaptodilactate [*bis*(2-carboxy-2-hydroxyethyl) disulphide], 2-mercaptoethanol, 2-mercaptoethylamine, L-cysteine, L-alanine 3-sulphinate, L-cystine disulphoxide, 2-iminothiazolidine carboxylic acid, DL-homocysteine, 2-amino-5,6-dihydro-1,3-thiazine-4-carboxylic acid, DL-cystathionine, DL-$\alpha,\alpha'$-dihydroxy-$\gamma,\gamma'$-dimercaptodibutyric acid [*bis*(3-carboxy-3-hydroxypropyl) disulphide] or sulphide ions (Fiedler & Wood, 1956).

### 8.2.2 Properties of purified 3-mercaptopyruvate sulphurtransferase

Kun & Fanshier (1958, 1959 *b*) purified 3-mercaptopyruvate sulphurtransferase from rat liver. It was found to be a copper-containing enzyme and some of its properties are listed in table 8.3. A characteristic feature of the enzyme was a marked stimulation by 2-mercaptoethanol (Meister, Fraser & Tice, 1954; Kun & Fanshier, 1959 *a*, *b*); high concentrations of mercaptoethanol, however, inhibited the enzyme (Kun & Fanshier, 1959 *b*).

TABLE 8.3    *Properties of rat-liver 3-mercaptopyruvate sulphurtransferase*
*(Kun & Fanshier, 1958, 1959 a, b)*

| | |
|---|---|
| isoelectric point | pH 7·4 |
| molecular weight | 35–40,000† |
| absorption maxima (m$\mu$) | 280, 415 |
| sulphur content | 4 atoms/mole |
| thiol (groups/mole) | 2† |
| diffusion coefficient ($D$) | $5\cdot2 \times 10^{-6}$ cm$^2$ sec$^{-1}$ |
| sedimentation coefficient ($S_{20}$) | 2·7 S |
| $K_m$ for CN$^-$ | 1·8 mM |
| $K_m$ for SO$_3{}^{2-}$ | 3 mM |
| $K_m$ for mercaptopyruvate* | 0·5–4 mM |

* Influenced by sulphite concentration.
† Monomeric enzyme has a molecular weight of 10,000 and 1 thiol group per mole.

Fanshier & Kun (1962) later modified their purification procedure for 3-mercaptopyruvate sulphurtransferase to include chromatography on DEAE-cellulose which had been pretreated with 2-mercaptoethanol. The resulting protein had about one fourth the molecular weight of that which had been purified by electrophoresis (Kun & Fanshier, 1958, 1959 *b*). The low molecular weight protein contained one mole of thiol and 1 gram atom of copper per mole of enzyme of molecular weight 10,000. It was suggested that the enzyme exists in an unstable monomeric form and a stable tetrameric form. The turnover number of the 'monomer' was

1,600–1,800 moles of substrate per mole of enzyme per min compared with 360–400 for the tetramer. The amino acid composition of the 'monomer' is shown in table 8.1.

Kun & Fanshier (1958) reported that the loss of copper from 3-mercaptopyruvate sulphurtransferase is accompanied by an equivalent loss of enzymic activity, and the enzyme was shown to be inhibited by metal chelating agents such as 8-hydroxyquinoline, o-phenanthroline, o- and m-mercaptobenzoic acids, and bathocuproine (4,7-diphenyl-2,9-dimethyl-1,10-o-phenanthroline) (Kun & Fanshier, 1961). Van den Hamer, Morell & Scheinberg (1967), however, have recently obtained preparations of the enzyme from rat liver and erythrocytes which have higher specific activities than those of Kun & Fanshier and which do not contain significant amounts of copper.

### 8.2.3 Mechanism of action of 3-mercaptopyruvate sulphurtransferase

Kun & Fanshier (1959 b) and Fanshier & Kun (1962) proposed a mechanism for 3-mercaptopyruvate transulphuration involving the formation of an enzyme-substrate complex and the elimination of pyruvate with the formation of an enzyme-trisulphide or persulphide. The latter reacted either with an acceptor ($CN^-$, $SO_3^{2-}$) or decomposed to enzyme and elemental sulphur. Sörbo (1957 b) had earlier suggested the formation of a persulphide intermediate and drew attention to the fact that all acceptors for this reaction have strong S-nucleophilic properties. Hylin & Wood (1959) showed that 3-mercaptopyruvate sulphurtransferase reacts with 3-mercaptopyruvate to form a non-dialysable sulphur-enzyme complex which can subsequently react with cyanide to form thiocyanate: on the basis of these results a double displacement mechanism similar to that proposed for rhodanese (see § 8.1.3) appears to be indicated for 3-mercaptopyruvate sulphurtransferase (equations 8.22 and 8.23).

$$R—SH + E \rightarrow ES + RH \qquad (8.22)$$
mercaptopyruvate      pyruvate

$$ES + acceptor \rightarrow S\text{-}acceptor + E \qquad 8.23)$$

# 9

## OXIDATION OF INORGANIC SULPHUR COMPOUNDS BY MICRO-ORGANISMS AND PLANTS

Oxidation of inorganic sulphur compounds is widespread in nature. Virtually all organically bound sulphur in decomposing plant, animal and microbial residues is eventually oxidized to sulphate, at least partly after the sulphur has been released as inorganic sulphide. For the most part this oxidation is carried out by the bacterial flora of soils and other environments. A number of bacteria, the sulphur lithotrophs, depend partly, and in some cases entirely, on the oxidation of inorganic sulphur compounds for their supplies of energy and reducing power. In other organisms the oxidations are apparently incidental and their roles, if any, in the physiology of the organisms are unknown.

### 9.1 Oxidation of inorganic sulphur compounds by lithotrophic organisms

The lithotrophic organisms utilize for their synthetic activities either radiant energy (photolithotrophs) or the energy released by the oxidation of inorganic compounds (chemolithotrophs). Many organisms of both groups oxidize reduced inorganic sulphur compounds. The main lithotrophic sulphur bacteria are listed in table 9.1, but the majority of these organisms have not yet been studied in detail and we shall consider in this chapter only those organisms on which research on the metabolism of sulphur compounds has been reported. Several excellent reviews on more general aspects of the lithotrophic sulphur bacteria are available (Bisset & Grace, 1954; Lees, 1955; Schwartz, 1958; Vishniac & Santer, 1957; Kelly, 1967 a).

### 9.1.1 The beggiatoa

The beggiatoa are filamentous, multicellular organisms commonly found in fresh water and marine environments containing hydrogen sulphide. They were originally considered to be strict autotrophs, i.e. utilizing only

TABLE 9.1 *The lithotrophic sulphur micro-organisms**

| Family | General characteristics | Habitat | Genera |
|---|---|---|---|
| | **CHEMOLITHOTROPHS** | | |
| THIOBACTERIACEAE | colourless, coccoid, straight or curved rod shaped bacteria; polar flagellate when motile. Oxidize sulphur compounds and usually deposit sulphur granules within or without the cells | a wide variety of fresh-water and marine environments containing $H_2S$. *Thiobacillus* found in soil, mine waste-waters, sewage, industrial effluents | *Thiobacterium* *Macromonas* *Thiovulum* *Thiospira* *Thiobacillus* |
| BEGGIATOACEAE | colourless cells occurring in trichromes within which they are arranged in chains. The trichromes show a gliding motion when in contact with a substrate. When grown in the presence of hydrogen sulphide the trichromes contain sulphur globules | fresh water and marine environments containing $H_2S$ | *Beggiatoa* *Thiospirillopsis* *Thioplaca* *Thiothrix* |
| ACHROMATACEAE | large spherical, ovoid or short cylindrical cells containing sulphur granules and sometimes inclusions of calcium carbonate | fresh water and brackish mud containing $H_2S$ | *Achromatium* |
| | **PHOTOLITHOTROPHS** | | |
| THIORHODACEAE | unicellular organisms, often developing as cell aggregates or families of variable size and shape. Single cells have the form of spheres, ovoids, short rods, vibrios, spirals and sometimes chains; contain bacteriochlorophyll and carotenoids; strictly photosynthetic; anaerobic or microaerophilic. Colours range from bluish violet through purple to deep red. Sulphur accumulates within the cells | most commonly found in mud and stagnant waters containing $H_2S$, and exposed to light. Many organisms found in sulphur springs | *Thiosarcina* *Thiopedia* *Thiocapsa* *Thiodictyon* *Thiothece* *Thiocystis* *Lamprocystis* *Amoebobacter* *Thiopolycoccus* *Thiospirillum* *Rhabdomonas* *Rhodothece* *Chromatium* |
| CHLOROBACTERIACEAE | small bacteria of varying morphology. Colour green due to presence of chlorophyllous pigments. Frequently deposit elemental sulphur outside the cells. Strictly photosynthetic, anaerobic | marine and fresh water muds containing $H_2S$ and exposed to light; sulphur springs | *Chlorobium* *Pelodictyon* *Clathrochloris* *Chlorobacterium* |

inorganic materials for growth (Winogradsky, 1949) but strains isolated in recent years have been shown to be heterotrophic although their growth is stimulated by sulphide (Cataldi, 1940; Faust & Wolfe, 1961; Scotten & Stokes, 1962; Burton & Morita, 1964). Indeed there is some doubt whether sulphide oxidation is an energy-yielding reaction in these bacteria. Instead sulphide may stimulate growth by reacting with inhibitory hydrogen peroxide which is produced during their metabolism (Burton & Morita, 1964).

During sulphide oxidation by *Beggiatoa*, sulphur globules are deposited within the bacterial cell and later sulphate appears in the external medium. The mechanism of the oxidation is unknown. The suggestion that poly-sulphide, rather than elemental sulphur, is an intermediate (von Deines, 1933) has not been generally accepted (Starkey, 1937).

### 9.1.2 The thiobacilli

The thiobacilli were discovered by Nathansohn (1902). They are small, Gram-negative, generally motile rods and are found in a wide variety of marine, freshwater and terrestial environments. Most are capable of strictly autotrophic growth in simple salt media containing carbon dioxide and an oxidizable sulphur compound. In their carbon metabolism they show a marked similarity to photosynthetic organisms (Santer & Vishniac, 1955; Trudinger, 1955, 1956; Milhaud, Aubert & Millet, 1956; Aubert, Milhaud & Millet, 1956, 1957; Suzuki & Werkman, 1958; Iwatsuka, Kuno & Maruyama, 1962; Gale & Beck, 1967). Facultative forms, capable of adapting to either an autotrophic or heterotrophic environment, are known (Tyulpanova-Mosevich, 1930; Starkey, 1935 b; Lipmann & McLees, 1940; Santer, Boyer & Santer, 1959; London, 1963 a). One recently described organism, *T. perometabolis*, is incapable of strictly autotrophic growth although it derives energy from the oxidation of inorganic sulphur compounds (London & Rittenberg, 1967).

The utilization of organic compounds is, in fact, more widespread among the thiobacilli than had been thought. Several of these organisms concentrate amino acids and glycerol from the external medium (van Niel, Postgate, quoted by Vishniac & Trudinger, 1962) while, by the use of tracer methods, the incorporation of organic compounds including amino acids, Krebs' cycle intermediates and fructose, into the cell material of strictly autotrophic thiobacilli, has been demonstrated (Butler & Umbreit, 1965, 1966; Kelly, 1965, 1966, 1967 b, c; Smith, London & Stanier, 1967). It has been reported that *Thiobacillus thio-oxidans* may, under some circumstances, utilize the energy from the oxidation of glucose for growth (Borichewski, 1965; Borichewski & Umbreit, 1966).

The majority of thiobacilli are strict aerobes but some may grow anaerobically with nitrate as an electron acceptor (see below). One organism, *Thiobacillus ferro-oxidans*, can use the oxidation of ferrous iron as an alternative source of energy (Colmer, Temple & Hinkle, 1950; Temple & Colmer, 1951; Razzell & Trussell, 1963).

Representative 'species' of thiobacilli are listed in table 9.2.

Many additional 'species' have been described in the literature but most of these appear to be identical with, or variants of, one or other of the organisms listed in table 9.2 (Vishniac & Santer, 1957). Indeed, there appears to be considerable doubt whether the tabulated organisms should be regarded as representing true species, particularly where the main distinguishing characteristics are metabolic. For example, the names *Thiobacillus denitrificans* and *Thiobacillus thiocyanoxidans* were originally assigned to organisms which metabolize nitrate and thiocyanate respectively (Beijerinck, 1904; Happold, Johnstone, Rogers & Youatt, 1954). Largely on this basis they were differentiated from *Thiobacillus thioparus*. The ability to grow anaerobically on nitrate and to oxidize thiocyanate, however, appears to be more widespread among thiobacilli than had been thought and the distinction between these three so-called species is no longer clear (van der Walt & de Kruyff, 1955; de Kruyff, van der Walt & Schwartz, 1957; Happold, Jones & Pratt, 1958; Woolley, Jones & Happold, 1962). Baalsrud & Baalsrud (1954), moreover, reported that repeated subculturing of *T. denitrificans* in air led to the loss of its denitrifying activity. The organism then became indistinguishable from *T. thioparus* (Vishniac & Santer, 1957). An unconfirmed report by Johnstone, Townshend & White (1961) suggests that facultative anaerobes, heterotrophs and thiocyanate-oxidizing organisms may develop from pure cultures of *T. thioparus*. The classification of thiobacilli is therefore unsatisfactory. Moreover, ferrous iron-oxidizing organisms belonging to the so-called genus *Ferrobacillus* have many features in common with *T. thio-oxidans* and *T. ferro-oxidans* and the taxonomic position of ferrobacilli has been questioned (Unz & Lundgren, 1961; Marchlewitz & Schwartz, 1961; Ivanov & Lyalikova, 1962). For example *Ferrobacillus ferro-oxidans*, which was originally differentiated from *T. ferro-oxidans* by its inability to oxidize sulphur compounds (Leathen, Kinsel & Braley, 1956) has since been shown to carry out such oxidations (Silverman & Lundgren, 1959; Leathen, quoted by Beck, 1960; Unz & Lundgren,

## TABLE 9.2 *The Thiobacilli*

The members of these genus are Gram-negative, non-sporulating rods measuring 0·5 by 1–3μ. Motile forms are polarly flagellated. All but one are capable of strictly autotrophic growth and most oxidize sulphide, elemental sulphur and thiosulphate to sulphate.

| Organism | Habitat | General characteristics | References* |
|---|---|---|---|
| *T. thioparus* | canal water, mud, soil | growth pH range 7·8–4·5 with an optimum near neutrality; generally aerobic and motile. Some strains grow anaerobically in the presence of nitrate | 1,2,3,4 |
| *T. neapolitanus* (*Thiobacillus X*) | sea water, corroding concrete structures | strict aerobe with properties very similar to those of *T. thioparus* | 1,5 |
| *T. denitrificans* | canal and river water, salt water, peat, composts, mud | optimum growth pH near neutrality; oxidizes sulphur compounds anaerobically in the presence of nitrate; denitrifying ability lost on culturing in air; motile | 2,3,4,6,7 |
| *T. thio-oxidans* | soil | optimum growth pH around 2; withstands 5% $H_2SO_4$; strict aerobe; motile | 8 |
| *T. concretivorus* | corroding concrete structures | very similar to *T. thio-oxidans* | 5,9 |
| *T. ferro-oxidans* | acid mine and soil waters containing hydrogen sulphide | strict aerobe; optimum pH growth range 2·5–5·8; also utilizes oxidation of ferrous iron as a source of energy; motile | 10,11,12 |
| *T. novellus* | soils | facultative autotroph; non-motile; optimum growth pH near neutrality | 13,14 |
| *T. intermedius* | fresh water mud | facultative autotroph; growth pH range 2·0–7·0; motile; autotrophic growth stimulated in the presence of organic matter | 15,16 |
| *T. thiocyanoxidans* | gas works liquor; sewage effluent | very similar to *T. thioparus*; thiocyanate oxidation serves as an energy source; oxidizes formate | 3,17 |
| *T. perometabolis* | soil | motile; no growth in mineral salts without yeast extract or casein hydrolysate; reduced sulphur compounds nevertheless are oxidized to sulphate and the oxidation stimulates growth | 18 |

* *References:* 1, Nathansohn, 1902; 2, Beijerinck, 1904; 3, de Kruyff, van der Walt & Schwartz, 1957; 4, Woolley, Jones & Happold, 1962; 5, Parker & Prisk, 1953; 6, Baalsrud & Baalsrud, 1954; 7, Lieske, 1912; 8, Waksman & Joffe, 1922; 9, Parker, 1947; 10, Colmer, Temple & Hinkle, 1950; 11, Temple & Colmer, 1951; 12, Razzell & Trussell, 1963; 13, Starkey, 1935 *b*; 14, Santer, Boyer & Santer, 1959; 15, London, 1963 *a*; 16, London & Rittenberg, 1966; 17, Happold, Johnstone, Rogers & Youatt, 1954; 18, London & Rittenberg, 1967.

1961). Two additional bacteria with very similar properties which have been described in detail are *Ferrobacillus sulfo-oxidans* (Kinsel, 1960) and an unnamed bacterium from the acidic leaching water in Bingham Canyon, Utah (Beck & Elsden, 1958; Beck, 1960).

A more rational approach to the classification of thiobacilli based on multivariate analysis has recently been initiated by Hutchinson, Johnstone & White (1965, 1966, 1967). By this technique *T. thio-oxidans, T. ferro-oxidans, T. thioparus, T. neapolitanus* and *T. denitrificans* have been identified as distinct species. This classification is supported to a certain extent by analyses of the DNA base compositions of a number of thiobacilli (Jackson, Moriarty & Nicholas, 1968).

The thiobacilli oxidize a number of reduced inorganic sulphur compounds to sulphate and many studies have been made to elucidate the mechanism of this oxidation. It may be said at the outset, however, that we have, as yet, no clear picture of this mechanism. Almost certainly many reactions remain to be discovered and there is much controversy over even the essential features of the oxidative pathway. Difficulties have arisen in the past through the lack of adequate microanalytical methods and through the tendency of many sulphur compounds to undergo secondary chemical reactions. With the advent of modern analytical techniques, particularly radiochemical and chromatographic methods, and by the increasing use of bacterial extracts, more precise information is now becoming available and it is possible that within the near future a more exact picture of sulphur transformations in thiobacilli will emerge. In the following sections our present knowledge of various stages in the oxidation of inorganic sulphur compounds by thiobacilli is reviewed.

### Oxidation of sulphite

The oxidation of sulphite to sulphate is catalysed by extracts of thiobacilli and may be the terminal stage in the oxidation of other, more reduced, sulphur compounds. Peck (1960, 1962*a*) proposed the following pathways for sulphite oxidation (equations 9.1 and 9.2).

$$SO_3^{2-} + AMP \xrightarrow{\text{APS-reductase}} APS + 2e \qquad (9.1)$$

$$APS + P_i \xrightarrow{\text{ADP-sulphurylase}} ADP + SO_4^{2-} \qquad (9.2)$$

AMP is regenerated by the action of adenylate kinase (equation 9.3).

$$2ADP \xrightarrow{\text{adenylate kinase}} AMP + ATP \qquad (9.3)$$

This pathway is supported by the following evidence:

1. The oxidation of sulphite by extracts of *T. thioparus* is stimulated by AMP (Peck, 1961 *b*) and is accompanied by the formation of APS (Peck, Deacon & Davidson, 1965) and the esterification of inorganic phosphate (Peck & Fisher, 1962).

2. The enzymes APS-reductase, ADP-sulphurylase (see chapter 5) and adenylate kinase are present in *T. thioparus* (Peck, 1960) and other thiobacilli (Peck, 1961 *a*).

3. The reversibility of the reaction catalysed by APS-reductase was demonstrated (Peck, 1961 *b*; Peck, Deacon & Davidson, 1965); hitherto only the reduction of APS to sulphite had been established (see chapter 10).

4. During sulphite oxidation by extracts of *T. thioparus* labelled oxygen is transferred from $^{18}O$-labelled inorganic phosphate to the bridge oxygen of ADP as predicted from the known mechanism of the ADP-sulphurylase reaction (equation 9.4; Peck & Stulberg, 1962).

$$Ad-O-\overset{\overset{\textstyle O}{\|}}{\underset{\underset{\textstyle O^-}{|}}{P}}-O-SO_3^- + P^{18}O_4^{3-} \rightarrow Ad-O-\overset{\overset{\textstyle O}{\|}}{\underset{\underset{\textstyle O^-}{|}}{P}}-{}^{18}O-P^{18}O_3^{2-} + SO_4^{2-} \tag{9.4}$$

APS-reductase from *T. denitrificans* was purified by Bowen, Happold & Taylor (1966). The purified enzyme contains flavin-adenine dinucleotide and ferric iron (table 9.3) and appears to be very similar to APS-reductase

TABLE 9.3    *Properties of purified APS-reductase from* T. denitrificans
(*Bowen, Happold & Taylor, 1966*)

| | |
|---|---|
| optimum pH | 7·2 |
| $K_m$ for sulphite | 1·5 mM |
| $K_m$ for nucleotides | AMP, 41 $\mu$M |
| | GMP, 630 $\mu$M |
| | (CMP, UMP inactive) |
| iron content | 4·9–7·8 g atoms per $10^5$ g of protein |
| FAD content | 0·69–0·77 mole per $10^5$ g of protein |
| sedimentation coefficient (uncorrected) | 9·5 S |
| inhibitors | *N*-ethylmaleimide (80% at $10^{-2}$ M). Slight or no inhibition by iodoacetamide, PCMB, arsenite, $CN^-$, dipyridyl, *o*-phenanthroline |

from *Desulfovibrio* (see chapter 10). Guanosine 5'-monophosphate but not cytidine 5'-monophosphate nor uridine 5'-monophosphate can replace AMP in the oxidation of sulphite (equation 9.1). Although *N*-ethyl-maleimide inhibits APS-reductase from *T. denitrificans* the results with other thiol-binding reagents indicate that the enzymic activity does not depend upon thiol groups.

*AMP-independent sulphite oxidation.* There is some doubt as to whether the active sulphate pathway (equations 9.1 to 9.3) is the only, or possibly even the main, mechanism of sulphite oxidation in some thiobacilli. For example T. M. Cook (quoted by Adair, 1966) found that *T. thio-oxidans* lacks APS-reductase although the presence of this enzyme in *T. thio-oxidans* had been claimed by Peck (1961a, 1962a). Rapid sulphite oxidation, in the absence of added AMP, is catalysed by extracts of *T. denitrificans* (Milhaud, Aubert & Millet, 1958), by extracts of *T. thioparus* after treatment with activated charcoal (London & Rittenberg, 1964) and also by washed 'cell fragments' of *T. thio-oxidans* (Adair, 1966).

Charles & Suzuki (1965) partly purified from *T. novellus* a sulphite-oxidizing enzyme the activity of which is unaffected by AMP: ferricyanide and cytochrome-*c* (from *T. novellus* or from mammalian sources) but not oxygen act as electron acceptors (Charles & Suzuki, 1966b). (In crude extracts, however, sulphite oxidation is coupled to oxygen uptake through the action of cytochrome oxidase.) The enzyme is strongly inhibited by thiol-binding reagents which indicates a possible role for thiol groups in sulphite oxidation by this enzyme. Organic compounds such as glucose, lactate and glycerol repress the formation of the enzyme when they are added to the growth medium (Lé John, Van Caeseele & Lees, 1967).

A sulphite-oxidizing enzyme from *T. neapolitanus* has also been partly purified (Hempfling, 1964; Hempfling, Trudinger & Vishniac, 1967). In contrast to that from *T. novellus*, the enzyme from *T. neapolitanus* reacts with both ferricyanide and oxygen and the reaction is stimulated $1 \cdot 5$- to 2-fold by AMP. The fact that the relative activities in the presence and absence of AMP remained fairly constant throughout the course of puri-fication indicated that a single enzyme is responsible for both the AMP-dependent and AMP-independent activities. To account for this Hempfling (1964; see also Trudinger, 1967a) suggested that sulphite oxidation by the *T. neapolitanus* enzyme may involve the formation of an enzyme-sulphite intermediate which reacts either with AMP to form APS or, more slowly, with water to form sulphate. Since reduced glutathione and,

TABLE 9.4 *AMP-independent, sulphite-oxidizing enzymes from thiobacilli*

| Source | Preparation | Properties | References[*] |
|---|---|---|---|
| T. denitrificans | lysis with lysozyme; particles after centrifuging at 20,000 × g | (1) electron acceptors; oxygen, cytochrome-c stimulated by hypoxanthine<br>(2) stimulated by hypoxanthine | 1 |
| T. thioparus | Hughes press crude extract | (1) electron acceptor, oxygen<br>(2) no additional cofactors | 2 |
| T. novellus | partly-purified soluble enzyme | (1) electron acceptors; cytochrome-c, ferricyanide; linked to $O_2$ via cytochrome oxidase<br>(2) $K_m$ for sulphide, 2–40 $\mu$M<br>(3) inhibited by thiol-binding reagents<br>(4) repressed in heterotrophically-grown bacteria | 3,4,5 |
| T. neapolitanus | partly-purified soluble enzyme | (1) electron acceptors; oxygen, ferricyanide<br>(2) stimulated about 80% by AMP<br>(3) inhibited by thiols and thiol-binding reagents | 6,7 |
| T. thio-oxidans | passage through French extrusion press; particles after centrifuging at 30,000 × g | (1) electron acceptor, oxygen<br>(2) inhibited by $CN^-$, $N_3^-$ | 8 |

* *References*: 1, Milhaud, Aubert & Millet, 1958; 2, London & Rittenberg, 1964; 3, Charles & Suzuki, 1965; 4, Charles & Suzuki, 1966 *b*; 5, Lé John, Van Caeseele & Lees, 1967; 6, Hempfling, 1964; 7, Hempfling, Trudinger & Vishniac, 1967; 8, Adair, 1966.

under some conditions, thiol-binding reagents inhibit the *T. neapolitanus* enzyme (Hempfling, 1964; Trudinger, 1967*a*; Hempfling, Trudinger & Vishniac, 1967), a disulphide group was suggested as a possible binding site for sulphite on the enzyme molecule. It was not established, however, that APS is formed during sulphite oxidation in the presence of AMP by the *T. neapolitanus* enzyme. Some properties of the sulphite-oxidizing enzymes from the thiobacilli are listed in table 9.4.

## Oxidation of sulphide and elemental sulphur

Sulphide and elemental sulphur are oxidized by most, if not all, thiobacilli. According to Gutiérrez & Ruiz-Herrera (1968) the ability of *T. ferro-oxidans* to oxidize sulphur is induced by growth on sulphur. This result confirms an earlier finding of Margalith, Silver & Lundgren (1966) who reported that the sulphur-oxidizing activity of *T. ferro-oxidans* is doubled in sulphur-grown organisms compared with those grown on ferrous iron: Landesman, Duncan & Walden (1966), however, obtained the opposite result. Sulphur occasionally precipitates during the oxidation of sulphide by growing cultures (Parker & Prisk, 1953) or bacterial extracts (Suzuki & Werkman, 1959; Charles & Suzuki, 1966 *a*) and it has been suggested that elemental sulphur is an intermediate in sulphide oxidation (Parker & Prisk, 1953; Vishniac & Santer, 1957).

Accumulation of elemental sulphur, however, does not invariably accompany sulphide oxidation by thiobacilli. It was not observed by Vishniac & Santer (1957) with washed cell preparations of *T. thioparus* or by London & Rittenberg (1964) who used extracts of *T. thioparus* and *T. thio-oxidans*. In these cases the major products, apart from sulphate, which accumulated were thiosulphate and polythionates: these were also formed in the experiments of Suzuki & Werkman (1959).

Thiosulphate and polythionates have also been assumed to be intermediates in sulphur oxidation by thiobacilli (Vishniac & Santer, 1957). This assumption was based partly on comparisons with sulphur oxidation in other systems such as mixed soil populations (Guittonneau & Keilling, 1932*a*; Gleen & Quastel, 1953) and animal tissues (Fromageot, 1947). Suzuki & Werkman (1959) detected the formation of polythionates during the oxidation of sulphur by extracts of *T. thio-oxidans*; the addition of substrate concentrations of glutathione was necessary before oxidation occurred and the authors suggested that sulphur is reduced to sulphide prior to oxidation (equation 9.5).

$$S^0 + 2GSH \rightarrow H_2S + GSSG \qquad (9.5)$$

A glutathione reductase (equation 9.6) to regenerate the glutathione is present in *T. thio-oxidans* (Suzuki & Werkman, 1960).

$$NADPH + GSSG + H^+ \rightarrow NADP^+ + 2GSH \qquad (9.6)$$

More recently Suzuki (1965 *a*) prepared extracts of *T. thio-oxidans* which catalyse the oxidation of elemental sulphur in the presence of catalytic amounts of glutathione. Other thiols or reducing agents such as mercapto-ethanol, cysteine and ascorbic acid are unable to replace glutathione and the extracts catalyse the quantitative oxidation of sulphide to thiosulphate (equation 9.7).

$$2S + O_2 + H_2O \rightarrow S_2O_3^{2-} + 2H^+ \qquad (9.7)$$

Similar enzymes have been detected in *T. thioparus* (Suzuki & Silver, 1966), *Ferrobacillus ferro-oxidans* (Silver & Lundgren, 1968) and *T. novellus* (Charles & Suzuki, 1966 *a*). In the latter organism the enzyme is repressed by the presence of organic compounds in the growth medium (Lé John, Van Caeseele & Lees, 1967). A soluble sulphur-oxidizing system from *T. thio-oxidans* was also reported by Tano & Imai (1968 *a*): in this case, added glutathione was unnecessary and sulphide, as well as thiosulphate, was formed during sulphur oxidation. The enzyme respons-ible for sulphur oxidation in *T. thio-oxidans* was partly purified (12-fold) by Suzuki (1965 *a*) and the yields of enzyme obtained during fractiona-tion of the extracts showed that it could account for most of the sulphur-oxidizing activity of the intact cell. The partly purified enzyme was free from sulphite oxidase, glutathione reductase and sulphide-oxidizing activities which, together with the fact that oxidized glutathione could not replace GSH, strongly indicates that the mechanism described by equations 9.5 and 9.6 was not operative in this case. The purified sulphur-oxidizing enzyme from *T. thioparus* was found to contain tightly-bound non-haem iron and labile sulphur (Suzuki & Silver, 1966): some evidence was obtained that the iron was necessary for enzymatic activity.

Elemental sulphur reacts non-enzymatically with sulphides such as $Na_2S$, mercaptoethanol and cysteine to form hydropolysulphides (see chapter 2). Suzuki (1965 *a*) showed that, when elemental sulphur was incubated with GSH, soluble compounds were formed with spectroscopic and other properties indicative of hydropolysulphides. He suggested that glutathione hydropolysulphides (G—S—$S_n$H) are the true substrates for the sulphur-oxidizing system (scheme 9.8).

The sulphur-oxidizing system prepared by Suzuki required molecular oxygen: a number of artificial electron acceptors failed to promote the oxidation of sulphur under anaerobic conditions. This result, together with the fact that a small incorporation of $^{18}O$ into thiosulphate occurred during sulphur oxidation under $^{18}O_2$ by extracts of *T. thioparus* (Suzuki, 1965 *b*), led to the suggestion that the sulphur-oxidizing enzyme is an oxygenase. In the intact cell, however, electron acceptors other than oxygen also function. It has been reported (Adair & Umbreit, 1965) that intact cells of *T. thio-oxidans* oxidize sulphur to sulphate and incorporate small amounts of $CO_2$ under anaerobic conditions while *T. denitrificans* oxidizes elemental sulphur to sulphate anaerobically in the presence of nitrate (Beijerinck, 1904; Baalsrud & Baalsrud, 1954).

$$S_8 + GSH \rightarrow GSS_8H \xrightarrow{O_2, H_2O} GSS_6H + S_2O_3^{2-} + 2H^+$$
$$\downarrow {\scriptstyle O_2, H_2O}$$
$$GSS_4H + S_2O_3^{2-} + 2H^+ \qquad (9.8)$$
$$\text{etc.}$$

Suzuki's earlier results suggested that thiosulphate, or a closely related thiosulphate precursor, is a key intermediate in sulphur oxidation. Recent work, however, has led to some modification of this view. Suzuki & Silver (1966) report that sulphite is the initial product of the GSH-catalysed oxidation of sulphur by *T. thio-oxidans* and *T. thioparus* (equation 9.9) and that thiosulphate arises from a non-enzymic condensation of sulphur with sulphite (equation 9.10).

$$S + O_2 + H_2O \xrightarrow[\text{enzyme}]{\text{GSH}} SO_3^{2-} + 2H^+ \qquad (9.9)$$

$$SO_3^{2-} + S \xrightarrow{\text{non-enzymic}} SSO_3^{2-} \qquad (9.10)$$

It may be noted in parenthesis that Imai, Okuzumi & Katagiri (1962) reported the *enzymic* condensation of sulphite and sulphur by *T. thio-oxidans*: this result has not been confirmed.

The results of Adair (1966; see also Shaulis, 1968) also support the notion that thiosulphate is not an obligate intermediate in sulphur oxidation. He prepared an enzyme system from *T. thio-oxidans* which was able to oxidize sulphur and sulphite to sulphate but was unable to attack thiosulphate or polythionates. In contrast to Suzuki's enzyme which was soluble, Adair's sulphur-oxidizing system was particulate and probably arose from the cell wall-membrane fraction of the bacterium. It is tempting to speculate, therefore, the Suzuki's soluble enzyme system and the soluble sulphite oxidase (see p. 214) arose by degradation of a co-ordinated, particulate enzyme complex catalysing the complete oxidation of sulphur to sulphate. A particulate, sulphur-oxidizing fraction from *T. thio-oxidans* has also been briefly reported by Mori, Kodama & Marunouchi (1967): the addition of soluble components was apparently necessary for activity. The properties of the sulphur oxidizing systems are compared in table 9.5.

Since sulphur may under some circumstances arise from sulphide (and *vice versa*) a common early intermediate may be involved in the oxidation of both these compounds to sulphate. The oxidation of sulphide and sulphur therefore may be tentatively described by scheme 9.11 where X may be a derivative of glutathione or perhaps of a membrane-bound thiol (see p. 223).

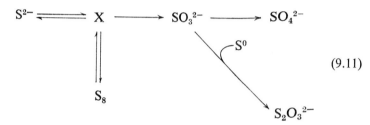

$$(9.11)$$

*Uptake of sulphur by thiobacilli.* A major unsolved problem in sulphur oxidation by thiobacilli is the mechanism cf the initial attack by the organism on the insoluble sulphur particle. It is clear that the physical form of sulphur influences its rate of oxidation although no detailed studies of this effect have been reported. Starkey (1925) showed that sulphur precipitated from a sulphide solution by mineral acids is superior to amorphous or rhombic sulphur as a substrate for *T. thio-oxidans*. Vogler & Umbreit (1941) demonstrated that, within certain limits, the rate of oxidation of powdered sulphur by *T. thio-oxidans* increases with decreasing particle size. Two preparations of sulphur rapidly attacked by thiobacilli are the turbid suspensions produced by acidification of thio-sulphate solutions (with subsequent removal of excess thiosulphate and sulphite by dialysis) and the materials precipitating during the growth of thiobacilli (Vogler, Le Page & Umbreit, 1942). Such preparations are oxidized by washed cell suspensions of *T. thioparus* at rates comparable with those at which soluble compounds such as thiosulphate are oxidized. The superiority of the latter preparations is probably at least partly due to the hydrophilic nature of the sulphur particles. Commercial sulphur preparations (e.g. Flowers of sulphur) are hydrophobic and are oxidized at a slow rate which is increased by the addition of wetting agents such as

TABLE 9.5 *Sulphur-oxidizing systems in thiobacilli*

| Source | Preparation | End-products | Properties | References[*] |
|---|---|---|---|---|
| T. thio-oxidans | partly-purified soluble enzyme | thiosulphate | (1) catalytic amounts of GSH required<br>(2) electron acceptor oxygen<br>(3) may be an oxygenase<br>(4) glutathione polysulphanes may be intermediates | 1,2 |
| T. thioparus | purified soluble enzyme | thiosulphate; sulphite (in presence of HCHO) | (1) probably a similar enzyme to that in *T. thio-oxidans* above<br>(2) contains iron | 3 |
| T. novellus | crude extract | . | (1) similar to *T. thio-oxidans* above<br>(2) repressed in heterotrophic growth | 4,5 |
| T. thio-oxidans | particle preparation from French pressure cell extract | sulphate | (1) membranous<br>(2) inhibited $CN^-$, $N_3^-$, thiol-binding reagents<br>(3) no GSH required | 6 |
| T. thio-oxidans | sonic disruption, supernatant from $130,000 \times g$ for 1 hr | sulphide, thiosulphate | (1) GSH not required<br>(2) inhibited by thiol-binding reagents | 7 |

[*] *References*: 1, Suzuki, 1965 *a*; 2, Suzuki, 1965 *b*; 3, Suzuki & Silver, 1966; 4, Charles & Suzuki, 1966 *a*; 5, Lé John, Van Caeseele & Lees, 1967; 6, Adair, 1966; 7, Tano & Imai, 1968 *a*.

220

Tween 80 and Tergitol 08 (Vogler & Umbreit, 1941; Starkey, Jones & Frederick, 1956; Cook, 1964; Adair, 1966). Jones & Starkey (1961) noted that, during the growth of *T. thio-oxidans*, sulphur originally floating on the surface of the medium becomes wet and settles to the bottom of the culture flask. Surface tension measurements and other tests indicated that surface active agents are released by the bacteria during their most active phase of growth. A major component was later identified as phosphatidyl inositol (Schaeffer & Umbreit, 1963) and a similar compound from brain was shown to wet sulphur and promote its oxidation by *T. thio-oxidans*. Jones & Benson (1965) and Shively & Benson (1967) confirmed the excretion of phospholipids by *T. thio-oxidans* although they were unable to detect phosphatidyl inositol. Instead, the main components were phosphatidyl *N*-methylethanolamine, phosphatidyl glycerol and diphosphatidyl glycerol together with traces of lysophosphatidyl *N*-methylethanolamine and lysophosphatidyl glycerol: the relative proportions of these phospholipids depended upon the ages of the cultures. Whether phospholipids are formed specifically in response to the presence of sulphur remains to be established.

Two main mechanisms may be envisaged for the uptake of sulphur by thiobacilli.

1. Sulphur may react with chemical or enzymic agents excreted by the bacteria with the formation of soluble compounds.

2. A reaction between sulphur and a cellular component may take place at the surface of the bacterial cell.

Most of the evidence to date supports the second mechanism although it cannot be said that the question is definitely resolved. Vogler & Umbreit (1941) showed that the interposition of a semipermeable membrane between the bacteria and substrate prevents sulphur oxidation by *T. thio-oxidans* indicating that solubilization of sulphur by low molecular-weight compounds excreted by the bacteria probably does not occur. The participation of an extra-cellular sulphur-solubilizing enzyme appears to be doubtful since well washed bacteria oxidize sulphur at linear rates without a lag period (Suzuki, 1965 *a*). In support of the 'direct contact' hypothesis Umbreit (1951) claimed that shaking cultures of *T. thio-oxidans* oxidize sulphur less rapidly than still cultures and suggested that shaking dislodges the bacterium from the sulphur particle. Others have failed to confirm this result, but, on the contrary, have found that agitation *increases* the rate of sulphur oxidation by *T. thio-oxidans* (Newburgh,

1954; Starkey, Jones & Frederick, 1956) and *T. neapolitanus* (Trudinger, unpublished results).

Cook (1964) investigated in more detail the effect of shaking on sulphur oxidation (or growth) by cultures of *T. thio-oxidans*. He found that the oxidation was inhibited if the cultures were shaken immediately after inoculation. The inhibition was prevented by the addition of phospholipids or other wetting agents to the medium or if shaking was initiated after a three day stationary incubation during which time phospholipids were assumed to be produced by bacterial metabolism. Cook suggested that wetting agents promote a firm attachment between the sulphur particle and the bacterium and claimed to have demonstrated binding between [14]C-labelled bacteria and sulphur. This claim was criticized by Trudinger (1967 *a*) on the basis that Cook's results did not distinguish between the binding to the sulphur particle of bacteria or of soluble [14]C-labelled compounds excreted from the bacteria.

Nevertheless other evidence suggests that a relatively firm attachment between the bacterium and the sulphur particle does, in fact, occur. Waksman (1932) observed that the surfaces of sulphur granules from cultures of *T. thio-oxidans* are surrounded by bacteria and phase-microscopic studies of mixtures of sulphur with *T. neapolitanus* have revealed bacteria clinging to sulphur particles and an increased concentration of bacteria in the near vicinity of the particles (Trudinger, unpublished results). Schaeffer, Holbert & Umbreit (1963) examined sulphur crystals from cultures of *T. thio-oxidans* by electron microscopy and observed bacteria in contact with the crystal surface. The areas of the crystal occupied by the bacteria exhibited an eroded appearance suggesting that the contact between the bacteria and sulphur had been maintained for a considerable period of time.

The nature of the reaction occurring at the bacterial surface which results in the uptake of sulphur into the organism is unknown. Umbreit, Vogel & Vogler (1942) found that *T. thio-oxidans*, stained by a variety of methods, shows a dipolar appearance (cf. also Waksman & Joffe, 1922) and they isolated from the bacterium a highly unsaturated fat which they concluded to be responsible for the dipolar staining. They suggested that sulphur dissolves in these fat globules and thus becomes available for metabolism by the cell. No mechanism was proposed for the transport of sulphur across the cell membrane which presumably surrounded the fat globules. The dipolar appearance of the bacteria was not evident in electron micrographs of *T. thio-oxidans* (Umbreit & Anderson, 1942).

Knaysi (1943) has made a careful electron microscopic study of *T. thio-oxidans* and has concluded that the 'fat globules' of Umbreit *et al.* are intra-protoplasmic vacuoles containing a mixture of volutin and sulphur (see also Knaysi, 1951). Knaysi demonstrated that, under certain conditions, sulphur exhibits the same staining properties as Umbreit's fat globules, and he concluded 'Regardless of the nature of that reserve material (in the vacuoles) it is difficult to see how such an intra-protoplasmic structure can be placed in contact with sulfur particles in the medium, especially when the cell is imbedded in slime'.

The fine structure of *T. thio-oxidans*, as revealed by electron microscopy, appears to be similar to that of Gram-negative heterotrophic bacteria (Mahoney & Edwards, 1966). Moreover no unusual features have been found in the phospholipid content of the cell envelope of *T. thio-oxidans* (Burke & Jones, 1968) or in the amino acid composition of its cell wall (Crum & Siehr, 1967). The cell envelope of *T. thio-oxidans* does, however, have an unusually high content of acidic amino acids (Burke & Jones, 1968) while the effects of lysozyme and trypsin on the cell wall of *T. thio-oxidans* indicate that the chemical composition of this structure is considerably different from that of the cell wall of the Gram-negative organism, *Escherichia coli* (Marunouchi & Mori, 1968). It remains to be seen whether these unusual features of the structure of *T. thio-oxidans* can be correlated with its ability to oxidize the insoluble sulphur particle.

Sulphur oxidation, by both intact bacteria and cell extracts, is inhibited by thiol-binding reagents (Vogler, 1942; Vogler, Le Page & Umbreit, 1942; Iwatsuka, Kuno & Maruyama, 1962; Adair, 1966; Tano & Imai, 1968 *a*) which together with the requirement for glutathione by the soluble sulphur-oxidizing systems (see p. 217) suggests that thiol groups are necessary for sulphur oxidation. In view of the preparation of sulphur-oxidizing 'membrane' fragments from thiobacilli (see p. 218) the suggestion by Vishniac & Santer (1957) that thiol groups on the bacterial cell membrane may function in the uptake of sulphur takes on added significance. Kaplan & Rittenberg (1962 *b*) found no significant fractionation of stable sulphur isotopes during sulphur oxidation by *T. concretivorus* and they concluded that the $S_8$ molecule *per se* is transported into the cell with no change in valency. While this conclusion may not be tenable the lack of a kinetic isotope effect does indicate that once the initial attack on the sulphur molecule has occurred all the atoms of sulphur are further metabolized and are no longer in equilibrium with the external medium.

## Oxidation of thiosulphate

Thiosulphate is oxidized to sulphate by all species and strains of thiobacilli which have been studied (cf. Pankhurst, 1964) and the mechanism of oxidation has been the subject of many investigations since the original description of this group of organisms.

The overall reaction is described by equation 9.12.

$$S_2O_3^{2-} + 2O_2 + H_2O \rightarrow 2SO_4^{2-} + 2H^+ + 211 \text{ kcal.} \qquad (9.12)$$

Complete oxidation of thiosulphate to sulphate, however, does not always take place. In weakly buffered cultures of *T. thioparus* and related organisms (Nathansohn, 1902; Starkey, 1935 *a*; Parker & Prisk, 1953) the marked fall in the pH of the medium (to pH 3–4) prevents complete oxidation. Trudinger (1964 *e*) found that washed *T. neapolitanus*, grown with low aeration, oxidized thiosulphate to completion only when the bacterial concentration was relatively high; the extent to which thiosulphate was oxidized by low concentrations of organisms was reduced by high substrate concentrations and by high partial pressures of oxygen in the gas phase. The oxidation of thiosulphate by dialyzed norite-treated extracts of *T. thioparus* was also reported to depend upon protein concentration (London & Rittenberg, 1964); concentrated extracts oxidized thiosulphate to completion at a linear rate but dilution of the extracts led to incomplete oxidation. It has been reported that thiosulphate is partially oxidized by *T. thioparus* in the absence of phosphate (Vishniac & Santer, 1957; Santer, Margulies, Klinman & Kaback, 1960; Okuzumi & Kita, 1965) and in the presence of high phosphate concentrations (Jones & Happold, 1961).

*Production of polythionates during thiosulphate oxidation.* The oxidation of thiosulphate by thiobacilli is characterized by the formation of polythionates (table 9.6). The extent of polythionate accumulation, however, varies widely and depends upon a number of experimental factors, including the strain or species of organism employed (Parker & Prisk, 1953; Pratt, 1958; Woolley, Jones & Happold, 1962; Pankhurst, 1964). Santer, Margulies, Klinman & Kaback (1960) reported that polythionates accumulate only in the absence of inorganic phosphate and suppression of polythionate formation from thiosulphate by high phosphate concentrations was found by Jones & Happold (1961): nevertheless phosphate-containing media have been used by most workers to demonstrate

TABLE 9.6  *Production of polythionates from thiosulphate by thiobacilli*

| Organism | Preparation | Polythionates detected* | References† |
|---|---|---|---|
| T. thioparus | cultures | trithionate, tetrathionate, pentathionate | 1–6 |
| | washed bacteria | trithionate, tetrathionate, pentathionate | 7,8 |
| | extracts | trithionate, tetrathionate, | 6,9 |
| T. neapolitanus | cultures | tetrathionate | 10 |
| | washed bacteria | trithionate, tetrathionate, pentathionate | 11–13 |
| | extracts | tetrathionate | 14 |
| T. thio-oxidans | cultures | tetrathionate | 6,10,15 |
| | extracts | trithionate, tetrathionate | 6,15,16 |
| T. denitrificans | cultures | trithionate, tetrathionate, pentathionate | 4 |
| | aged washed bacteria | tetrathionate | 17 |
| T. novellus | extracts | tetrathionate | 18 |
| T. thiocyanoxidans | cultures | trithionate, tetrathionate, pentathionate | 2,5 |
| T. concretivorus | cultures | tetrathionate | 10 |
| | extracts | tetrathionate | 19 |
| T. ferro-oxidans | washed bacteria | trithionate, tetrathionate, pentathionate, hexathionate | 20,21 |
| Thiobacillus strain C | washed bacteria | tetrathionate, trithionate | 22,23 |

* All polythionates detected by various workers are listed. Not every polythionate was detected in every case and amounts accumulated varied considerably. For details the reader is referred to the original references.

† *References:* 1, Nathansohn, 1902; 2, Pratt, 1958; 3, Jones & Happold, 1961; 4, Woolley, Jones & Happold, 1962; 5, Pankhurst, 1964; 6, London & Rittenberg, 1964; 7, Vishniac, 1952; 8, Santer, Margulies, Klinman & Kaback, 1960; 9, Peck & Fisher, 1962; 10, Parker & Prisk, 1953; 11, Trudinger, 1959; 12, Hempfling, 1964; 13, Trudinger, 1964 *e*; 14, Trudinger, 1961 *b*; 15, Okuzumi & Kita, 1965; 16, Tano, Asano & Imai, 1968; 17, Baalsrud & Baalsrud, 1954; 18, Vishniac & Trudinger, 1962; 19, Moriarty & Nicholas, 1968; 20, Landesman, Duncan & Walden, 1966; 21, Sinha & Walden, 1966; 22, Kelly & Syrett, 1966 *a*; 23, Kelly, 1968.

polythionate accumulation. Trudinger (1964 *e*) found that extensive accumulation of polythionates during thiosulphate oxidation by *T. neapolitanus* occurs only with organisms grown with low aeration and that the amounts accumulated are affected by the concentrations of bacteria, substrate and oxygen used in the experiments. The formation of polythionates by organisms grown at high aeration rates was often undetectable even by sensitive radiochemical methods; nevertheless the same organisms oxidized thiosulphate quantitatively to tetrathionate in the presence of thiol-binding inhibitors (Trudinger, 1965).

The variations in polythionate accumulation induced by changes in experimental conditions are presumably due to modifications in the relative rates of the reactions leading to polythionate formation and its subsequent metabolism. Such considerations may help to explain the occasional failures to detect polythionates during thiosulphate oxidation (Starkey, 1934 b; Skarżyński & Szczepkowski, 1959).

Tetrathionate is generally the main polythionate which accumulates during thiosulphate oxidation. Nathansohn (1902) proposed that it arose by direct oxidation of thiosulphate (equation 9.13).

$$2S_2O_3^{2-} + H_2O + \tfrac{1}{2}O_2 \rightarrow S_4O_6^{2-} + 2OH^- \tag{9.13}$$

Tetrathionate is the first polythionate detected during thiosulphate oxidation by washed cells of *T. thioparus*, *T. thio-oxidans* and *T. neapolitanus* (Vishniac, 1952; Trudinger, 1959; Okuzumi & Kita, 1965) and by specially treated extracts of *T. thioparus* (London & Rittenberg, 1964). The quantitative oxidation of thiosulphate to tetrathionate is catalysed by aged *T. denitrificans* (Baalsrud & Baalsrud, 1954), by *T. neapolitanus* in the presence of thiol-binding reagents (Trudinger, 1965) and by crude extracts of *T. novellus*, *T. thio-oxidans*, *T. thioparus* and *T. neapolitanus* (Peck & Fisher, 1962; London & Rittenberg, 1964; Hempfling, 1964; Okuzumi & Kita, 1965; Lé John, Van Caeseele & Lees, 1967).

Trithionate and pentathionate are also formed during the oxidation of thiosulphate (table 9.6). They generally accumulate in relatively small amounts but, under some conditions, they may be the main polythionates found in the medium during the growth of *T. thioparus* (Woolley, Jones & Happold, 1962; Jones & Happold, 1961). Kelly & Syrett (1966 a) reported that trithionate is the major polythionate formed by oxidation of thiosulphate by washed cells of *Thiobacillus* strain C but later work (Kelly, unpublished results) has shown that the type of polythionate accumulated by washed suspensions of this organism depends upon some unknown factor(s) in the medium in which the bacterium is grown: although both trithionate and tetrathionate accumulated during growth, only trithionate persisted long after cessation of growth.

Tamiya, Haga & Huzisige (1941) and Vishniac (1952) assumed that both pentathionate and trithionate arise by the chemical disproportionation of tetrathionate which is enhanced in the presence of thiosulphate.

226

This disproportionation is now thought to result from two nucleophilic displacement reactions (equations 9.14 and 9.15; see chapter 2).

$$S_4O_6^{2-} + S_2O_3^{2-} \rightleftharpoons S_5O_6^{2-} + SO_3^{2-} \qquad (9.14)$$

$$S_4O_6^{2-} + SO_3^{2-} \rightleftharpoons S_3O_6^{2-} + S_2O_3^{2-} \qquad (9.15)$$

In particular, the formation of trithionate by such a mechanism must be seriously considered since the equilibrium of reaction 9.15 is strongly in favour of the displacement of thiosulphate by sulphite and small amounts of the latter have been detected during thiosulphate oxidation by washed cells of *T. novellus* (De Ley & van Poucke, 1961) and *T. neapolitanus* (Trudinger, 1964 *e*).

Two alternative hypotheses for trithionate formation have been advanced by Kelly & Syrett (1966 *a*). The first involves an oxidative reaction between sulphite and thiosulphate (equation 9.16).

$$SO_3^{2-} + SSO_3^{2-} \rightarrow {}^-O_3SSSO_3^- + 2e \qquad (9.16)$$

The second involves an enzymic reaction between thiosulphate and tetrathionate to give trithionate and a sulphodisulphane ($^-S$—$S$—$SO_3^-$) (equation 9.17) with subsequent oxidation of the latter to a second molecule of trithionate (equation 9.18).

$$SSO_3^{2-} + {}^-O_3SSSSO_3^- \rightarrow {}^-O_3SSSO_3^- + {}^-SSSO_3^- \qquad (9.17)$$

$${}^-SSSO_3^- + 1\tfrac{1}{2}O_2 \rightarrow {}^-O_3SSSO_3^- \qquad (9.18)$$

These hypotheses were proposed by Kelly & Syrett to account for the distribution of [35]S within the trithionate formed by oxidation of singly labelled [35]S-thiosulphate in their experiments. No direct evidence for reactions 9.16 to 9.18 was reported. Clearly more work is necessary before the mode of formation of trithionate, and the role of this polythionate in thiosulphate metabolism, is established.

*Formation of elemental sulphur from thiosulphate.* Elemental sulphur is a common product of thiosulphate metabolism by thiobacilli: it accumulates primarily outside the bacterial cell and arises from the sulphane sulphur group of thiosulphate (Skarżyński, Ostrowski & Krawczyk, 1957; Ostrowski & Krawczyk, 1957; Peck & Fisher, 1962; Trudinger, 1964 *e*). The formation of sulphur is governed by experimental conditions. Increases in substrate concentration and oxygen limitation increase the amounts of sulphur formed (Lange-Posdeeva, 1930; Saslawsky, 1927; Vishniac, 1952; Trudinger, 1964 *e*). Under appropriate conditions of pH control, aeration and substrate concentration, thiosulphate may be oxidized to sulphate by *T. neapolitanus* and by some strains of *T. thioparus* without any significant formation of sulphur (Vishniac & Santer, 1957; Hempfling, 1964; London & Rittenberg, 1964; Charles & Suzuki, 1966 *a*). Moreover Kelly & Syrett (1966 *a*) reported that no $^{35}S$ appears in elemental sulphur during the simultaneous oxidation of unlabelled sulphur and sulphane-labelled thiosulphate by *Thiobacillus* strain C. Trudinger (1964 *e*) found that high aeration rates during growth limit the subsequent formation of sulphur by washed-cell suspensions of *T. neapolitanus*.

Tamiya, Haga & Huzisige (1941) and Vishniac (1952) attributed sulphur production to the extracellular decomposition of pentathionate (equation 9.19) formed from tetrathionate (equation 9.14).

$$S_5O_6^{2-} \rightarrow S_4O_6^{2-} + S^0 \tag{9.19}$$

Others (Starkey, 1935 *a*; Skarżyński, Ostrowski & Krawczyk, 1957; Ostrowski & Krawczyk, 1957; Peck, 1960; 1962 *a*) attribute sulphur formation to specific biological reactions and there is a deal of evidence to suggest that mechanisms other than the breakdown of pentathionate operate, at least under some conditions.

1. With washed bacteria, sulphur may precipitate within a few minutes after contact of the bacteria with thiosulphate, and it is doubtful that the chemical disproportionation of tetrathionate and decomposition of pentathionate occur at sufficient rates to account for this (see chapter 2).

2. The formation of sulphur is not correlated with polythionate formation. For example Peck (1960) reported that his strain of *T. thioparus* produced large amounts of sulphur but no polythionates.

3. Precipitation of sulphur does not occur during oxidation of thiosulphate to tetrathionate by heterotrophic organisms (London, 1964; Trudinger, 1967 *b*).

*Formation of sulphate from thiosulphate.* Santer, Margulies, Klinman & Kaback (1960) reported that, at intermediate stages of oxidation, sulphate is formed by *T. thioparus* to an equal extent from both the sulphur atoms of thiosulphate. Other workers have found, however, that there is a preferential conversion of the sulphonate-sulphur to sulphate (Skarżyński, Ostrowski & Krawczyk, 1957; Ostrowski & Krawczyk, 1957; Peck & Fisher, 1962; Trudinger, 1964 *d, e*; Kelly & Syrett, 1966 *a*.)

Labelled sulphate is formed when thiosulphate is oxidized by *T. thioparus* in the presence of $^{18}$O-labelled phosphate (Santer, 1959). Peck & Stulberg (1962) have accounted for this observation by assuming that sulphite is an intermediate in the oxidation. Initially, labelled ADP is formed (p. 213; equation 9.4) from which labelled AMP arises through the action of adenylate kinase (equation 9.20).

$$2Ad{-}O{-}\overset{\overset{\displaystyle O}{\|}}{\underset{\underset{\displaystyle O^-}{|}}{P}}{-}^{18}O{-}P^{18}O_3{}^{2-} \rightleftharpoons Ad{-}O{-}\overset{\overset{\displaystyle O}{\|}}{\underset{\underset{\displaystyle O^-}{|}}{P}}{-}^{18}O^- + [^{18}O]ATP \quad (9.20)$$

Subsequently $^{18}$O is transferred to sulphate by APS-reductase and ADP-sulphurylase (equations 9.21 and 9.22).

$$SO_3{}^{2-} + Ad{-}O{-}\overset{\overset{\displaystyle O}{\|}}{\underset{\underset{\displaystyle O^-}{|}}{P}}{-}^{18}O^- \rightleftharpoons Ad{-}O{-}\overset{\overset{\displaystyle O}{\|}}{\underset{\underset{\displaystyle O^-}{|}}{P}}{-}^{18}O{-}SO_3^- + 2e \quad (9.21)$$

$$Ad{-}O{-}\overset{\overset{\displaystyle O}{\|}}{\underset{\underset{\displaystyle O^-}{|}}{P}}{-}^{18}O{-}SO_3^- + PO_4{}^{3-} \rightleftharpoons ADP + SO_3{}^{18}O^{2-} \quad (9.22)$$

With infinite recycling the specific activity of sulphate (atoms % excess $^{18}$O) would be 25% of that of the phosphate added. Santer (1959) obtained a value of 22% for the sulphate formed by oxidation of thiosulphate which indicates that the major part of the sulphate arose from an intermediate containing a P—O—S linkage.

*Oxidation of thiosulphate by bacterial extracts.* Crude, cell-free preparations of *T. thio-oxidans* (London & Rittenberg, 1964) and *T. novellus* (van Poucke, 1962; Aleem, 1965; Charles & Suzuki, 1966 a) have been described which catalyse the oxidation of thiosulphate to sulphate. In these cases oxidation occurred without added cofactors or special treatments. Others, however, have found that untreated crude extracts of *T. thio-oxidans* (Okuzumi & Kita, 1965), *T. neapolitanus* (Hempfling, 1964) and *T. thioparus* (Peck & Fisher, 1962; London & Rittenberg, 1964) oxidize thiosulphate quantitatively to tetrathionate. Some properties of cell-free thiosulphate-oxidizing systems are compared in table 9.7. London & Rittenberg (1964) reported that extracts of *T. thioparus* oxidize thio-

TABLE 9.7 *Thiosulphate oxidizing systems from extracts of thiobacilli*

| Source | Preparation | End-products | Principal characteristics | References* |
|---|---|---|---|---|
| *T. neapolitanus* | passage through French press followed by $(NH_4)_2SO_4$ fractionation and chromatography on Amberlite IRC50 | tetrathionate | (1) ferricyanide or cytochrome 553.5 act as electron acceptors. Slow reduction of mammalian cytochrome-$c$ <br> (2) coupled to oxygen in the presence of a particle fraction from the bacterium <br> (3) not inhibited by thiol-binding reagents | 1–4 |
| *T. thioparus* | passage through French press | sulphur and sulphate | (1) requires substrate amounts of GSH and AMP <br> (2) electron acceptor oxygen <br> (3) intermediate formation of sulphide and sulphite <br> (4) phosphorylation accompanies oxidation <br> (5) tetrathionate also oxidized probably after chemical reduction to thiosulphate by GSH | 5,6 |
| *T. thioparus* | passage through Hughes press or sonic disruption followed by dialysis against polyethylene glycol and treatment with activated charcoal | sulphate | (1) electron acceptor oxygen <br> (2) no cofactors required <br> (3) omission of dialysis and charcoal treatments results in incomplete oxidation with the accumulation of polythionates <br> (4) sulphide, polythionates and sulphite also oxidized to sulphate <br> (5) tetrathionate and trithionate formed transiently during oxidation of thiosulphate to sulphate | 7 |
| *T. thio-oxidans* | sonic disruption | sulphate | (1) electron acceptor oxygen <br> (2) no treatment of extracts required for complete oxidation <br> (3) sulphide, tetrathionate and trithionate also oxidized <br> (4) traces of trithionate formed during thiosulphate oxidation | 7 |

230

| Organism | Method of preparation | Product | Properties | Reference |
|---|---|---|---|---|
| *T. thio-oxidans* | ultrasonic disruption; particulate preparation after centrifuging at 90,000 × g for 1 hr | tetrathionate; trithionate? | (1) inhibited by $N_3^-$ not by thiol-binding reagents <br> (2) membranous | 8 |
| *T. thio-oxidans* | sonic disruption followed by precipitation with 70% $(NH_4)_2SO_4$ and dialysis against distilled water | tetrathionate (in citrate buffer) | (1) electron acceptor, oxygen <br> (2) optimum pH, 4·5 <br> (3) addition of phosphate promotes oxidation beyond the stage of tetrathionate: oxidation, however, still incomplete | 9,10 |
| *T. concretivorus* | extrusion through French press | tetrathionate | (1) particulate enzyme | 11 |
| *T. novellus* | sonic disruption followed by $(NH_4)_2SO_4$ fractionation and chromatography on calcium phosphate gel. (Thiosulphate-cytochrome-*c* reductase) | not specified | (1) electron acceptor mammalian (and bacterial?) cytochrome-*c* <br> (2) inhibited by metal-binding reagents but not by thiol-binding reagents <br> (3) no additional cofactors required <br> (4) formed adaptively during autotrophic growth on thiosulphate <br> (5) crude extracts oxidize thiosulphate with consumption of oxygen | 12 |
| *T. novellus* | sonic disruption under $N_2$ | sulphate | (1) polythionates not accumulated <br> (2) GSH inhibitory <br> (3) repressed in heterotrophically grown cells | 13,14 |
| *T. novellus* | sonic disruption | sulphate | (1) sulphide and polythionates oxidized <br> (2) endogenous cytochrome-*c* reduced by sulphide, thiosulphate and polythionates | 15 |

* *References*: 1, Trudinger, 1958; 2, Trudinger, 1961 *a, b*; 3, Hempfling, 1964; 4, Trudinger, 1965; 5, Peck, 1960; 6, Peck & Fisher, 1962; 7, London & Rittenberg, 1964; 8, Tano, Asano & Imai, 1968; 9, Imai, Okuzumi & Katagiri, 1962; 10, Okuzumi & Kita, 1965; 11, Moriarty & Nicholas, 1968; 12, Aleem, 1965; 13, Charles & Suzuki, 1966 *a*; 14, Lé John, Van Caesele & Lees, 1967; 15, van Poucke, 1962.

231

sulphate to completion only after extensive treatment with activated charcoal and dialysis against polyethylene glycol which they suggested removed an inhibitor present in the crude extracts: despite this treatment no added cofactors were required for the oxidation of thiosulphate. Tetrathionate and trithionate were formed transiently during the oxidation of thiosulphate to sulphate.

Trudinger (1961 b) separated, and partly purified, a soluble enzyme from *T. neapolitanus* which catalyses the oxidation of thiosulphate to tetrathionate. Ferricyanide or a *c*-type cytochrome isolated from the bacterium act as electron acceptors: mammalian cytochrome-*c* also reacts but at a slower rate. This enzyme has been detected in *T. thioparus* (Santer, unpublished results) and *T. novellus* (Lé John, Van Caeseele & Lees, 1967) and may be similar to the thiosulphate-cytochrome-*c* reductase from *T. novellus* reported by Aleem (1965). The latter enzyme was purified 215-fold and catalyses the transfer of electrons to mammalian cytochrome-*c* and, presumably, to the endogenous cytochromes of the bacterium. The *T. novellus* enzyme is inhibited by metal-binding reagents such as cyanide, *o*-phenanthroline and diethyldithiocarbamate but not by thiol-binding reagents. The enzyme from *T. neapolitanus* is also insensitive to thiol-binding reagents (Trudinger, 1965) but cyanide and diethyldithiocarbamate do not inhibit when ferricyanide is used as an electron acceptor (Trudinger, 1961 b). The end-product of the thiosulphate–cytochrome-*c* reaction catalysed by the *T. novellus* enzyme was not reported. Tano, Asano & Imai (1968) obtained a particulate preparation from *T. thio-oxidans* which oxidizes thiosulphate to tetrathionate and an unidentified compound, possibly trithionate. This enzyme system was inhibited by azide but not by 2,2'-dipyridyl, carbon monoxide or thiol-binding reagents.

Vishniac & Trudinger (1962) suggested that the true function of the thiosulphate-oxidizing enzyme in *T. neapolitanus* is to transfer thiosulphate to a thiol group within the cell to form a sulphenyl thiosulphate which is then further metabolized to sulphate: tetrathionate formation was thought to be due to a non-specific secondary reaction between thiosulphate and an enzyme-bound intermediate (scheme 9.23).

This proposal was based on the fact that intact *T. neapolitanus* oxidizes thiosulphate quantitatively to tetrathionate in the presence of thiol-binding reagents under conditions where little or no polythionate accumulates in the absence of inhibitor (Trudinger, 1965; Kelly, 1968). The accumulation of tetrathionate is also caused by the use of 100% oxygen as the gas phase (Trudinger, 1964 e): this, it was suggested (Trudinger,

1965), might be due to the oxidation of the active thiol to a disulphide. The accumulation of polythionates during thiosulphate oxidation by *T. ferro-oxidans* in the presence of thiol-binding reagents was reported by Sinha & Walden (1966): in this case, however, both tetrathionate and trithionate were formed.

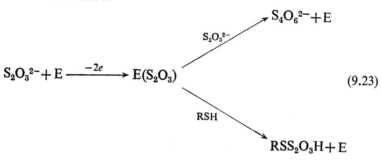

$$S_2O_3^{2-} + E \xrightarrow{-2e} E(S_2O_3)$$

$$S_2O_3^{2-} \nearrow S_4O_6^{2-} + E$$

$$\searrow RSH \; RSS_2O_3H + E$$

(9.23)

Peck and his colleagues (Peck, 1960; Peck & Fisher, 1962) reported that AMP and substrate concentrations of glutathione, homocysteine or cysteine stimulate thiosulphate oxidation by extracts of *T. thioparus* and that sulphur and sulphate, instead of polythionates, are formed. The presence of a glutathione-dependent thiosulphate reductase in the extracts led Peck to propose that thiosulphate is initially reduced to sulphide and sulphite (equation 9.24) which are subsequently oxidized to sulphur and sulphate respectively.

$$S_2O_3^{2-} + 2e + 2H^+ \rightarrow SO_3^{2-} + H_2S \qquad (9.24)$$

The oxidation of thiosulphate in the presence of glutathione and AMP was accompanied by a dinitrophenol-insensitive esterification of inorganic phosphate (Peck & Fisher, 1962) and during the oxidation $^{18}O$ was transferred from $[^{18}O]AMP$ to sulphate (Peck & Stulberg, 1962). These results indicate the formation of a P—O—S linkage by the mechanism described on page 229. At the concentrations of glutathione and thiosulphate (20 mM-GSH; 5 mM-$S_2O_3^{2-}$) used by Peck, however, the oxidation of sulphite is almost completely inhibited (Hempfling, 1964; Hempfling, Trudinger & Vishniac, 1967). Thus it is possible that the P—O—S bond formed during thiosulphate oxidation is not that of APS or that APS is formed from bound rather than free sulphite.

Hempfling (1964) also studied the oxidation of thiosulphate in the presence of glutathione by extracts of *T. neapolitanus*. Although glutathione alone was not oxidized, the oxygen consumption with glutathione plus thiosulphate was greater than with thiosulphate alone: the reaction was

AMP-independent and no sulphate was formed. The stoichiometric relationship between oxygen consumption and the amounts of glutathione and thiosulphate added were consistent with the following mechanism (equations 9.25 and 9.26) by which thiosulphate is first oxidized to tetrathionate and then regenerated by the chemical reduction of tetrathionate by glutathione.

$$2SSO_3^{2-} \rightarrow {}^-O_3S-S-S-SO_3^- \qquad (9.25)$$

$${}^-O_3S-S-S-SO_3^- + 2GSH \xrightarrow{\text{chemical}} 2SSO_3^{2-} + GSSG + 2H^+ \quad (9.26)$$

The reason for the conflicting results of Hempfling and of Peck and his colleagues is not immediately obvious. The situation is further complicated by the fact that glutathione strongly inhibits thiosulphate oxidation by extracts of *T. novellus* (Charles & Suzuki, 1966 *a*).

*The role of rhodanese in thiosulphate metabolism.* The enzyme rhodanese which catalyses the cyanolysis of thiosulphate to thiocyanate and sulphite (equation 9.27) is present in some thiobacilli (see chapter 8).

$$S_2O_3^{2-} + CN^- \rightarrow SCN^- + SO_3^{2-} \qquad (9.27)$$

Charles & Suzuki (1966 *a*) suggest that the function of rhodanese in these organisms is to cleave thiosulphate to 'sulphur' and sulphite which are then oxidized by the sulphur-oxidizing enzyme (p. 217) and sulphite oxidase (p. 214) respectively. Rhodanese may also be responsible for the reduction of thiosulphate by extracts of *T. thioparus* (see equation 9.24) since rhodanese from mammalian tissues and from *B. subtilis* has been shown to catalyse both the cyanolysis and reduction of thiosulphate (see chapter 8). The thiosulphate-oxidizing enzyme from *T. neapolitanus* and other organisms has been suggested to play a role in the cleavage of thiosulphate (Trudinger, 1967 *a*) but its relationship, if any, to rhodanese remains to be determined. The fact that the rhodanese-catalysed cyanolysis of thiosulphate by *T. denitrificans* is inhibited by thiol-binding reagents (Bowen, Butler & Happold, 1965 *b*) whereas the oxidation of thiosulphate to tetrathionate by extracts of *T. neapolitanus* is not (Trudinger, 1965) suggests that the two enzymes are different. In this respect it may be significant that two distinct enzymes appear to be responsible for the cyanolysis of thiosulphate and the oxidation of thiosulphate to tetrathionate by the photosynthetic sulphur-bacterium, *Chromatium* (Smith, 1966).

### 9.1.3 Metabolism of polythionates

Many thiobacilli oxidize tetrathionate and trithionate to sulphate (Parker & Prisk, 1953; Baalsrud & Baalsrud, 1954; Jones & Happold, 1961; Kelly & Syrett, 1964 a; London & Rittenberg, 1964; Trudinger, 1964 a, c; Okuzumi, 1965; Okuzumi & Imai, 1965; Landesman, Duncan & Walden, 1966). Pentathionate and hexathionate are oxidized by *T. thiooxidans* (London, 1964). Dithionate is either not attacked (London & Rittenberg, 1964) or is oxidized at very slow rates (Vishniac, 1952): Cook (1967) briefly reported that dithionate oxidation by *T. denitrificans* required the presence of thiosulphate.

The rate of tetrathionate oxidation by *T. thioparus* has been reported to depend upon the sodium:potassium ratio in the medium (Woolley, 1962). Rapid oxidation of tetrathionate and trithionate by washed *T. neapolitanus* is catalysed only by dense bacterial populations, is inhibited by 100% oxygen and usually requires the addition of a 'sparking' amount of thiosulphate (Trudinger, 1964 a, c). The requirement for thiosulphate is reduced by increasing the bacterial concentration or reducing the oxygen concentration. These effects appear to be due, at least in part, to a requirement for a low concentration of dissolved oxygen for rapid oxidation of the polythionates.

Little is known of the mechanism of oxidation of polythionates. Under some conditions there is a sequential formation of tetrathionate and trithionate from thiosulphate and it has been suggested that trithionate is an intermediate in tetrathionate oxidation (Vishniac, 1952; Vishniac & Santer, 1957; London & Rittenberg, 1964). According to Sinha & Walden (1966) tetrathionate is oxidized by *T. ferro-oxidans* to trithionate and sulphate, the latter being derived entirely from the sulphane groups. They suggested the following mechanism for the breakdown of tetrathionate (scheme 9.28) but in view of the widespread evidence that sulphate is preferentially formed from the sulphonate groups (see p. 228) this suggestion must be accepted with some caution.

$$^{35}SSO_3{}^{2-} \rightarrow {}^-O_3S-{}^{35}S-{}^{35}S-SO_3{}^- \rightarrow {}^-O_3S-{}^{35}S-SO_3{}^- + {}^{35}SO_4{}^{2-} \quad (9.28)$$

Trithionate, tetrathionate and pentathionate are metabolized anaerobically by *T. novellus* (van Poucke, 1962) and *T. neapolitanus* (Trudinger, 1964 b, c; Kelly, 1968). The reaction involving trithionate obeys equation 9.29 and has also been reported to be catalysed by *T. thioparus* (Huzisige & Haga, 1944).

$$^-O_3S-S-SO_3{}^- + OH^- \rightarrow HSO_4{}^- + SSO_3{}^{2-} \quad (9.29)$$

The products of anaerobic tetrathionate and pentathionate metabolism are more complex: in addition to sulphate and thiosulphate, sulphur and other polythionates are formed and the reaction is influenced by the previous growth history of the cells (Trudinger, 1964 *b*). With *T. neapolitanus* the quantitative relationships between tetrathionate metabolized and thiosulphate and sulphate formed indicate that the initial reaction is an asymmetric hydrolysis according to either equation 9.30 or 9.31.

$$^-O_3S—S—S—SO_3^- + OH^- \rightarrow SSO_3^{2-} + S^0 + HSO_4^- \qquad (9.30)$$

$$^-O_3S—S—S—SO_3^- + 2OH^- \rightarrow SO_3^{2-} + S(OH)_2 + SSO_3^{2-} \qquad (9.31)$$

Since a low dissolved oxygen concentration in the medium is apparently necessary for the rapid oxidation of trithionate and tetrathionate by *T. neapolitanus* (Trudinger, 1964 *a, c*) the anaerobic reactions (equations 9.29 to 9.31) may also be the first steps in the oxidation of these compounds.

An enzymic disproportionation of tetrathionate to trithionate and pentathionate (see equations 9.14 and 9.15) by extracts of *T. thio-oxidans* was reported by Okuzumi (1965, 1966 *a*): ferrous, nickel or cobalt ions were required for full activity. According to Imai, Okuzumi & Katagiri (1962), trithionate is reduced by *T. thio-oxidans* to sulphite and thiosulphate (equation 9.32).

$$^-O_3S—S—SO_3^- + 2e \rightarrow SSO_3^{2-} + SO_3^{2-} \qquad (9.32)$$

Since thiosulphate is produced in excess of sulphite during the anaerobic metabolism by *T. thio-oxidans*, Okuzumi (1966 *b*) suggested that both reduction (equation 9.32) and hydrolysis (equation 9.29) of trithionate are catalysed by this organism.

The metabolism of tetrathionate and trithionate, both aerobic and anaerobic, by *T. neapolitanus* is very sensitive to thiol-binding reagents (Trudinger, 1965) while the oxidation of these compounds is inhibited by 100% oxygen (Trudinger, 1964 *a, c*). It was suggested that thiol groups are necessary for polythionate oxidation and that the reduced oxygen tension required for polythionate oxidation might be associated with the generation of thiol groups.

### 9.1.4 The path of sulphur during thiosulphate oxidation by thiobacilli

From the foregoing account it is clear that there is a deal of apparently conflicting evidence on the mechanism of the oxidation of thiosulphate and polythionates to sulphate. The sequences outlined in scheme 9.33 to

9.35 represent alternative pathways of thiosulphate oxidation which are currently in vogue.

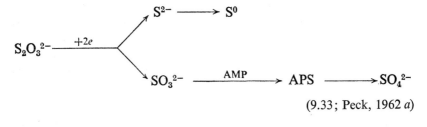

$$S_2O_3^{2-} \xrightarrow{+2e} \begin{cases} S^{2-} \longrightarrow S^0 \\ SO_3^{2-} \xrightarrow{AMP} APS \longrightarrow SO_4^{2-} \end{cases}$$

(9.33; Peck, 1962 *a*)

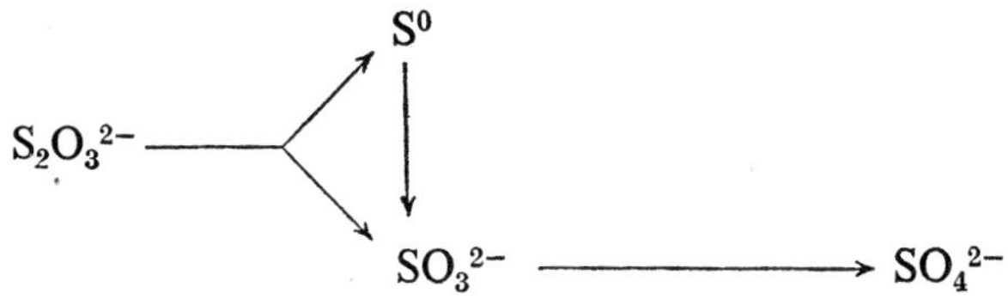

(9.34; Charles & Suzuki, 1966 *a*)

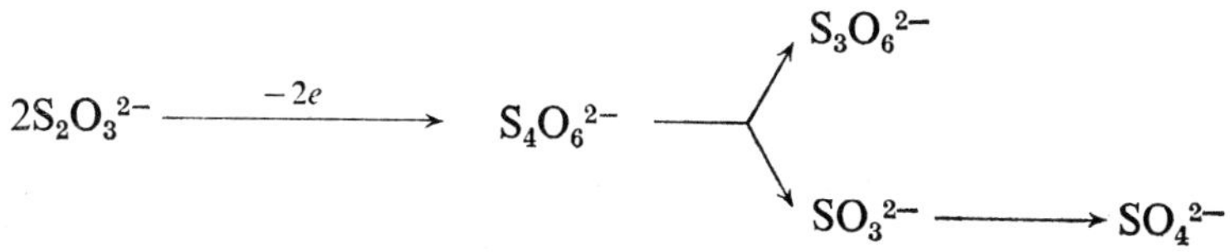

(9.35; Vishniac & Santer, 1957; London & Rittenberg, 1964; and others)

Perhaps the most cogent evidence that the polythionate pathway (scheme 9.35) is not the 'basic' mechanism for thiosulphate oxidation, at least in *T. novellus*, has been provided by Charles & Suzuki (1966 *a*) who showed that extracts of this organism oxidize thiosulphate to sulphate but are unable to metabolize tetrathionate. Moreover since mechanisms exist for the oxidation of sulphide or sulphur and sulphite (see pp. 212 and 216) the cleavage of thiosulphate (schemes 9.33 and 9.34) would appear to be the simplest and most direct route for its oxidation. Nevertheless, any proposed mechanism for thiosulphate oxidation must be able to account for the formation of polythionates. It is, of course, possible that two or more 'basic' mechanisms of thiosulphate oxidation operate but it may be pointed out that the schemes 9.33 to 9.35 are not necessarily incompatible. They are based largely on analyses of the products of thio-sulphate oxidation in experiments involving a variety of organisms, preparations and experimental conditions and, in view of the chemical reactivity of most of the compounds one can imagine to be intermediates in thiosulphate metabolism, it might be surprising if identical or even similar results were obtained under all conditions. Indeed, it has been shown (Trudinger, 1964 e) that the intermediate products of thiosulphate oxidation by washed cells of T. neapolitanus are altered quite extensively by changes in substrate, cell and oxygen concentrations during the experiments and by changes in the growth conditions of the organism.

The possibilities of 'alternative' reactions catalysed by enzymes of the sulphur pathway have already been illustrated with reference to rhodanese and the thiosulphate-oxidizing enzyme (p. 233).

A few speculative attempts to 'unify' the various hypotheses for the mechanism of thiosulphate oxidation have been made (e.g. Lees, 1960; Trudinger, 1967 a). A solution to the problem, however, must obviously await further evidence.

### 9.1.5  Oxidation of thiocyanate

Thiocyanate is oxidized by T. thiocyanoxidans and some other thiobacilli. The mechanism of this oxidation has been studied by Youatt (1954) and appears to involve two hydrolytic reactions with the formation of carbon dioxide, ammonia and sulphide (equations 9.36 and 9.37).

$$CNS^- + H_2O \rightarrow HCNO + HS^- \tag{9.36}$$

$$HCNO + H_2O \rightarrow CO_2 + NH_3 \tag{9.37}$$

The sulphide is then oxidized to sulphate. Youatt reported that T. thiocyanoxidans is unable to oxidize tetrathionate and suggested a pathway for sulphide oxidation involving thiosulphite, thiosulphate, metabisulphite and pyrosulphate as intermediates (scheme 9.38).

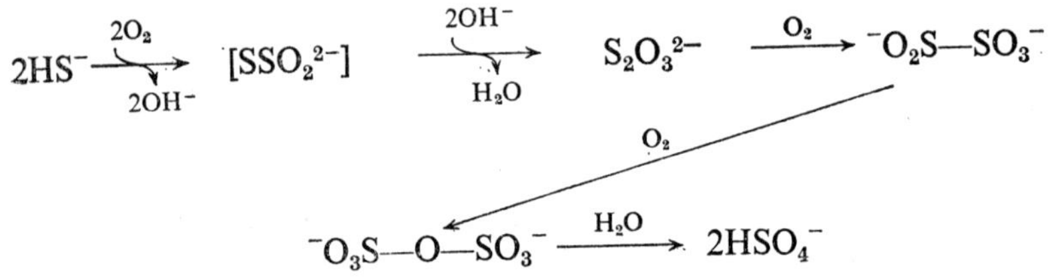

$$\tag{9.38}$$

No evidence of such a pathway has been forthcoming and T. thiocyanoxidans has since been shown to oxidize tetrathionate and to accumulate polythionates during thiosulphate oxidation (Pratt, 1958; Jones & Happold, 1961) indicating an essential similarity between T. thiocyanoxidans and other thiobacilli.

## 9.1.6 Electron transport of thiobacilli

Electron transport appears to be coupled via a cytochrome system to the oxidation of sulphur compounds by thiobacilli. Cytochrome pigments have been detected in a number of species of thiobacilli and several cytochromes of the $c$-type have been partially purified and characterized (table 9.8). Szczepkowski & Skarżyński (1952) reported that *T. thiooxidans* was devoid of cytochromes but later work (Cook & Umbreit, 1963; London, 1963 $b$; Tano, Kagawa & Imai, 1968) has demonstrated the presence of cytochrome pigments in this species confirming an earlier report by Emoto (1933). Cytochromes of the $b$-type have been detected in *T. neapolitanus* (Trudinger, 1961 $a$; Hempfling, 1964) and *T. denitrificans* (Trudinger, unpublished results).

On the addition of reduced inorganic sulphur compounds to intact bacteria or crude extracts the cytochromes become reduced (Baalsrud & Baalsrud, 1954; Skarżyński, Klimek & Szczepkowski, 1956; Trudinger, 1961 $a$; van Poucke, 1962; London, 1963 $b$; Hempfling, 1964; Aleem, 1965; Aleem, Ross & Schoenhoff, 1968): sulphide, thiosulphate, tetrathionate and trithionate catalyse such reductions. As mentioned earlier (pp. 214, 232) cytochromes of the $c$-type act as electron acceptors for partially purified thiosulphate-oxidizing enzyme systems from *T. neapolitanus* (Trudinger, 1961 $b$) and from *T. novellus* (Aleem, 1965) and for the sulphite-oxidizing system of *T. novellus* (Charles & Suzuki, 1966 $b$). Milhaud, Aubert & Millet (1958) demonstrated that the $c$-type cytochrome from *T. denitrificans* mediates electron flow between thiosulphate or sulphite and nitrate. According to Aleem (1965) neither flavins nor cytochrome-$b$ mediate electron flow between thiosulphate and oxygen in *T. novellus*.

Cook & Umbreit (1963) reported that *T. thio-oxidans* contains benzoquinone and the importance of quinones in sulphur oxidation by this organism is indicated by the fact that both sulphur oxidation and cellular ubiquinone are diminished at almost identical rates by ultraviolet irradiation (Adair, 1968).

The terminal oxidase in thiobacilli is for the most part unknown and may differ in different organisms. The oxidation of sulphur by *T. thiooxidans* is inhibited by cyanide and azide (Vogler, Le Page & Umbreit, 1942; Iwatsuka & Mori, 1960; Tano & Imai, 1968 $a$; Adair, 1966) and the reoxidation of reduced cytochromes by a particle preparation of *T. neapolitanus* is inhibited by KCN (Trudinger, 1961 $b$). Sulphite oxidation

TABLE 9.8 *Cytochromes of the c-type in thiobacilli*

| Source | Absorption maxima in reduced state | | | $E'_0$ (V) | Absorption to Amberlite IRC50, XE64 at pH 7 | Auto-oxidation at pH 7 | References[†] |
|---|---|---|---|---|---|---|---|
| | α | β | γ | | | | |
| T. denitrificans | 552 | 522 | 416 | +0·270 | + | − | 1 |
| T. thioparus (cytochrome-s) | 551 | 526 | 422 | +0·140–0·150 | − | + | 2 |
| T. neapolitanus | 550 | 521 | 416 | +0·200 | + | + | 3 |
| | 553·5 | 524 | 418 | +0·210 | −* | + | 3 |
| | 557 | 525 | 419 | +0·155 | + | v. slow | 3 |
| T. neapolitanus (acid fraction) | 550/557 | . | . | +0·2 | − | + | 3 |
| T. thio-oxidans | 550 | 521 | 415 | +0·253 | + | . | 4 |

\* Absorbed to Amberlite at pH 5.

† *References*: 1, Aubert, Milhaud, Moncel & Millet, 1958; 2, Klimek, Skarżyński & Szczepkowski, 1956; Skarżyński, Klimek & Szczepkowski, 1956; 3, Trudinger, 1961 *a*; 4, Tano, Kagawa & Imai, 1968.

by crude extracts of *T. novellus* is inhibited by azide and cyanide (Charles & Suzuki, 1966 *a*). Iwatsuka & Mori (1960) also reported inhibition by CO of the sulphur oxidation by *T. thio-oxidans*: the inhibition was not reversed by light which suggests that the terminal oxidase in this organism differs from classical cytochrome oxidase. On the other hand, the inhibition by CO of thiosulphate oxidation by extracts of *T. novellus* is reversed by light (Aleem, 1965): this, together with the appearance of an absorption band in the region of 600–10 m$\mu$ in reduced extracts, strongly suggests the presence of cytochromes of the *a*- or $a_3$-type in *T. novellus*. Milhaud, Aubert & Millet (1958) reported that difference spectra of reduced *T. denitrificans* extracts with and without CO showed an absorption band with a maximum at 418 m$\mu$. In similar experiments with *T. neapolitanus*, Ikuma & Hempfling (unpublished results) found absorption bands for the reduced CO-binding pigment at 570, 540 and 415–17 m$\mu$. Milhaud *et al.* (1958) interpreted their spectra as indicating the presence of cytochrome-$a_3$. Ikuma and Hempfling have pointed out, however, that the spectra of the CO-treated extracts are typical of cytochrome-*o* (Castor & Chance, 1959; the soret band of the reduced CO-cytochrome-$a_3$ is at 432 m$\mu$; Morton, 1958) and showed, further, that the oxygen affinity of the respiration system of intact *T. neapolitanus* is considerably lower than that of cytochrome-$a_3$. Trudinger (1961 *a*) was unable to detect *a*-cytochromes in *T. neapolitanus*.

For the biosynthesis of organic matter a continuous supply of reduced pyridine nucleotides is necessary which, in the thiobacilli, results from the oxidation of sulphur compounds. Extracts of *T. novellus* reduce $NADP^+$ during thiosulphate oxidation but not anaerobically in the presence of thiosulphate (Aleem, 1965). Aleem (1966 *a*, *b*) demonstrated energy-linked reduction of pyridine nucleotides by *T. novellus* extracts in the presence of thiosulphate, formate, succinate or mammalian cytochrome-*c*. The reduction of $NAD^+$ by succinate was completely inhibited by inhibitors of flavoprotein systems. In the case of the obligate autotroph *T. neapolitanus*, thiosulphate, but not succinate, supported the energy-linked reduction of $NAD^+$. It has yet to be shown whether or not pyridine nucleotide reduction is coupled directly to the oxidation of other compounds in the sulphur pathway.

According to Smith, London & Stanier (1967) *T. thio-oxidans* and *T. thioparus* lack NADH oxidase which, they suggested, may account for obligate autotrophy in these organisms. NADH oxidation linked to cytochrome-*c* or oxygen has, however, been demonstrated in *T. nea-*

*politanus* (Hempfling, 1964; Hempfling & Vishniac, 1965; Trudinger & Kelly, 1968). Tano & Imai (1968 *b, c*) demonstrated NADPH oxidation coupled to cytochrome-*c* reduction in *T. thio-oxidans*. A tentative generalized description of electron flow in thiobacilli is given by scheme 9.39.

$$S_8 \searrow$$

Ubiquinone?

$$
\begin{array}{l}
S_2O_3^{2-} \searrow \\
\qquad \qquad \rangle \text{------ Cytochrome-}c \longrightarrow \left\{ \begin{array}{l} \text{Oxidase} \\ \text{Cytochrome } a, a_3 \\ \text{Cytochrome } o \end{array} \right\} \longrightarrow O_2 \\
SO_3^{2-} \nearrow
\end{array}
$$

Flavin?

$$(\text{NADP})\text{NAD} \xrightarrow{\quad CO_2 \quad} [CH_2O] \qquad (9.39)$$

### 9.1.7 Phosphorylation by thiobacilli

The oxidation of inorganic sulphur compounds by thiobacilli is accompanied by uptake of inorganic phosphate into the cell (Vogler & Umbreit, 1942; Baalsrud & Baalsrud, 1952; Baalsrud, 1954; Umbreit, 1954; Newburgh, 1954), and occurrence of adenosine 5'-triphosphate in thiobacilli was shown by Barker & Kornberg (1954). An earlier report that the ATP of these organisms was adenosine 3'-triphosphate (Le Page & Umbreit, 1943) appears to be erroneous. The incorporation of [32P]-phosphate into ATP during thiosulphate oxidation was demonstrated by Milhaud, Aubert & Millet (1957) for *T. denitrificans* and by Vishniac & Santer (1957) for *T. thioparus*. The fixation of carbon dioxide into 3-phosphoglyceric acid, the major pathway of $CO_2$ assimilation (Aubert, Milhaud & Millet, 1956, 1957; Milhaud, Aubert & Millet, 1956), is coupled to the oxidation of sulphur compounds: in extracts it requires the addition of ATP (Trudinger, 1955, 1956; Suzuki & Werkman, 1958; Iwatsuka, Kuno & Maruyama, 1962; Johnson & Peck, 1965; Aleem & Huang, 1965; Gale & Beck, 1966, 1967; Mayeux & Johnson, 1966; Bruff & Johnson, 1966). Transfer of energy from the oxidative systems to $CO_2$ assimilation appears therefore to take place by the classical mechanism involving phosphorylated nucleotides.

The sequence of reactions postulated by Peck for sulphite oxidation (see p. 212) results in the generation of one high-energy phosphate bond for each sulphite molecule oxidized. Peck & Fisher (1962) showed that in the presence of NaF and EDTA to inhibit hydrolysis of phosphate esters, the product of phosphate esterification during sulphite oxidation by *T. thioparus* is ADP and that under suitable conditions, the stoichiometry approaches 1 mole of phosphate esterified for each mole of sulphate produced. The esterification is not inhibited by the uncoupling agent 2,4-dinitrophenol, indicating that electron-transport phosphorylation is not involved. Peck (1962 *a*) considers that substrate-level phosphorylation associated with sulphite oxidation may play a major role in the energy metabolism of the thiobacilli and also advanced the interesting idea that direct utilization of the high-energy sulphate bond for synthetic purposes might occur in certain cases. This latter speculation was based, in part, on the fact that APS-reductase from thiobacilli catalyses the formation, not only of APS, but also of sulphate esters of other nucleotides (guanosine 5'-phosphate, inosine 5'-phosphate, deoxyadenosine 5'-phosphate) which apparently cannot be utilized by ADP-sulphurylase. At the present time there is no direct evidence that such a direct energy transfer occurs.

Calculations by Vishniac & Trudinger (1962), based on the experimental yields of bacteria in relation to energy supply determined by Bauchop & Elsden (1960), indicate that substrate-level phosphorylation associated with thiosulphate oxidation would be insufficient to account for the yields of *T. neapolitanus* and *T. thioparus* obtained under the best growth conditions so far devised (Hempfling & Vishniac, 1965). Vogler, Le Page & Umbreit (1942) reported that 2,4-dinitrophenol stimulated the respiration of *T. thio-oxidans* on sulphur when added at concentrations in the order of 3 $\mu$M. This result indicated the possibility of electron-transport phosphorylation but it was not confirmed by Iwatsuka, Kuno & Maruyama (1962) and Beck & Shafia (1964) who reported that sulphur oxidation by *T. thio-oxidans* and *T. ferro-oxidans* is strongly inhibited by dinitrophenol above 1 $\mu$M. Nevertheless evidence for electron-transport phosphorylation in thiobacilli has been obtained by Kelly & Syrett (1963; 1964 *a*) who compared the effect of uncoupling agents (dinitrophenol, arsenate) on the oxidation of sulphur compounds and on $CO_2$ fixation by intact *T. thioparus*: $CO_2$ fixation was taken as a measure of the formation of high-energy phosphate. They found that, at concentrations which have little effect on oxygen uptake, both uncoupling agents inhibit $CO_2$ fixation with all substrates tested but that fixation coupled to sulphide oxidation is more sensitive to the inhibitors than that coupled to the oxidation of thiosulphate, tetrathionate or trithionate. These results indicated that both electron-transport phosphorylation and substrate-

243

level phosphorylation occur and that relatively more of the latter is linked to thiosulphate oxidation than to sulphide oxidation. Further support for these conclusions was obtained by Kelly & Syrett (1964*b*, 1966*b*) who showed that the rapid rise in cellular ATP during the first 1–2 min of sulphide oxidation is severely depressed by dinitrophenol at concentrations which have a relatively slight effect on ATP formation linked to thiosulphate oxidation. From these results one might conclude that electron-transport phosphorylation is associated mainly with the oxidation of highly reduced sulphur such as sulphide or the sulphane group of thiosulphate.

Electron-transport phosphorylation linked to the oxidation of mercapto-ethanol has been demonstrated in extracts of *T. neapolitanus* (Hempfling, 1964; Hempfling & Vishniac, 1965); the phosphorylation was sensitive to uncoupling agents and P/O ratios in the order of 0·4 were obtained. Charles & Suzuki (1966 *b*) also reported electron-transport phosphoryl-ation (P/O in the order of 0·1) coupled to sulphite oxidation by extracts of *T. novellus*: similar results were reported by Davis & Johnson (1967) who used extracts of *T. thioparus*. Ross, Schoenhoff & Aleem (1968) have recently reported that DNP-sensitive phosphorylation is coupled to the oxidation of thiosulphate by extracts of *T. neapolitanus*; P/O ratios approaching 1·00 were obtained.

## 9.2 Photolithotrophic bacteria

Van Niel (1931, 1935) developed a general concept of photosynthesis which is expressed by equation 9.40 in which $H_2A$ represents an electron donor the nature of which depends upon the particular photosynthetic system.

$$CO_2 + 2H_2A \xrightarrow{\text{light}} (CH_2O) + 2A + H_2O \tag{9.40}$$

In a number of photosynthetic bacteria the electrons may be donated by an oxidizable inorganic sulphur compound (table 9.1). The two main groups of photolithotrophic sulphur bacteria which have been studied in any detail are the green bacteria, *Chlorobium* spp., and the purple sulphur bacteria, *Chromatium* spp. Some of the so-called purple non-sulphur bacteria, for example, *Rhodopseudomonas palustris* are known to utilize sulphur compounds (van Niel, 1944) but, although various aspects of the metabolism of these bacteria have been very intensively investigated, few studies of their sulphur metabolism have been reported.

The genus *Chlorobium* comprises a group of strictly anaerobic non-motile organisms which are found in marine and freshwater muds and in stagnant waters containing H₂S. They are obligatory photosynthetic organisms and are generally intensely green due to the presence of chlorophyllous pigments. The two best known species, *Chlorobium limicola* (van Niel, 1931) and *Chlorobium thiosulphatophilum* (Larsen, 1952; Shaposhnikov, Kondrat'eva & Fedorov, 1958) depend upon the presence of oxidizable sulphur compounds for growth and are unable to utilize organic compounds as their sole carbon source. Organic compounds may, however, be utilized in the presence of sulphur compounds. Sadler & Stanier (1960) reported that the growth yield of *C. limicola* is increased by the addition of certain organic compounds such as acetate to the medium and Larsen (1953) demonstrated a light-induced carboxylation of propionate to succinate by *C. thiosulphatophilum*. Other green bacteria have been reported in which organic compounds may replace sulphur compounds as electron donors (Mechsner, 1957; Shaposhnikov *et al.* 1958). *C. limicola* and *C. thiosulphatophilum* are differentiated on the basis of their growth characteristics and utilization of sulphur compounds: *C. limicola* uses sulphide, elemental sulphur and hydrogen as electron donors; sulphide is oxidized to sulphate and elemental sulphur which precipitates extracellularly. In addition to the above electron donors *C. thiosulphatophilum* also uses thiosulphate and tetrathionate; all sulphur compounds are oxidized completely to sulphate (Larsen, 1953). The mechanism of oxidation of the sulphur compounds by *Chlorobium* is little known beyond the fact that elemental sulphur is formed during sulphide oxidation (Fedorov & Maksimov, 1965).

The purple sulphur bacteria of the genus *Chromatium* owe their colour to the presence of carotenoid pigments which mask that due to bacteriochlorophyll. They are motile, anaerobic, strictly photosynthetic organisms which exhibit a variety of morphological forms including ovoid, bean and vibrio shaped forms, and short rods. Various organic substances, particularly the lower fatty acids and some hydroxy and dibasic acids may replace sulphur compounds as electron donors in the photosynthetic reaction (van Niel, 1936, 1941). Sulphide, thiosulphate, endogenous and exogenous elemental sulphur and sulphite are oxidized by washed cell suspensions and growing cultures of *Chromatium*. During sulphide and thiosulphate oxidation, elemental sulphur generally accumulates within the bacterial cell (van Niel, 1931, 1936; Eymers & Wassink, 1937; Bregoff & Kamen, 1952; Petrova, 1959; Losada, Nozaki & Arnon, 1961). The form of elemental sulphur was identified as orthorhombic by Trüper & Hathaway (1967); orthorhombic sulphur is also produced from sulphide by *Thiocystis violacea* and *Chlorobium* (Trüper & Hathaway, 1967) and by the colourless sulphur bacterium *Thiovulum majus* (La Riviére, 1963). Light-dependent anaerobic oxidation of sulphide by *Chromatium* depends upon the presence of carbon dioxide, indicating a close coupling between oxidation and photosynthesis (Trüper & Schlegel, 1964).

The formation of sulphur from sulphide and thiosulphate by *Chromatium* suggests that sulphide is oxidized to sulphur and that thiosulphate undergoes cleavage into two one-sulphur moieties.

The latter hypothesis was supported by Smith (1965) and Smith & Lascelles (1966) who showed that the photochemical oxidation of thiosulphate by washed suspensions of *Chromatium* Strain D is biphasic in character. In the initial fast phase about half the sulphur of thiosulphate is converted to sulphate which arises largely from the sulphonate group of thiosulphate. The sulphane-sulphur of thiosulphate accumulates within the cell during this initial phase and is subsequently oxidized to sulphate at a slower rate. Trüper & Pfennig (1966) also found that elemental sulphur arises from the sulphane group of thiosulphate during the photosynthetic metabolism of *Chromatium* and *Thiocapsa floridana*. Smith A. J. (1964) and Smith & Lascelles (1966) isolated rhodanese (see chapter 8) from *Chromatium* strain D which, they suggested, may function as a thiosulphate-cleaving enzyme in the metabolism of thiosulphate (cf. p. 203). The activity of this enzyme in *Chromatium* was, however, about the same in organisms grown on thiosulphate, succinate or pyruvate and was only about twice that of the Athiorhodaceae, *Rhodospirillum rubrum* and *Rhodopseudomonas spheroides* which are unable to utilize thiosulphate.

Tetrathionate has been suggested, largely on indirect evidence, as an intermediate in light-dependent oxidation of thiosulphate by photosynthetic bacteria (van Niel, 1936; Eymers & Wassink, 1937; Wassink, 1941; Larsen, 1952, 1953). Smith (1966) isolated a thiosulphate-oxidizing enzyme from *Chromatium* strain D which catalyses the oxidation of thiosulphate to tetrathionate. A similar enzyme appears to be present in *Chlorobium thiosulphatophilum* (Mathewson, Burger & Millstone, 1968). Smith, however, considers that the enzyme is not involved in the oxidation of thiosulphate to sulphate since its activity occurs within a pH range (below about pH 7, optimum 5·0) which is unfavourable for the growth of *Chromatium*. Moreover, tetrathionate is not metabolized by *Chromatium* strain D and inhibits the formation of sulphur from thiosulphate at pH 7·3. The presence of APS-reductase in *Chromatium* was demonstrated by

246

Peck (1961 a). Thiele (1968) reported high specific activities of APS reductase in *Thiocapsa floridana* and several strains of *Chromatium*. By contrast, little activity of APS-reductase was found in *Chromatium* strain D. ADP-sulphurylase was also present in extracts of *Thiocapsa floridana* strain 6311. None of the strains of photosynthetic bacteria studied by Thiele possessed a sulphite oxidase system similar to that reported in thiobacilli (Thiele, 1968). The path of sulphur in *Chromatium* may be tentatively described by scheme 9.41 which is adapted from Smith (1966).

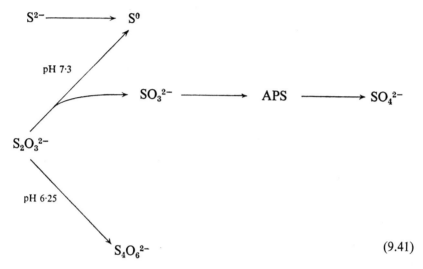

$$(9.41)$$

In *Chromatium* strain 211 (Trüper & Pfennig, 1966) and in *Rhodopseudomonas palustris*, a member of the Athiorhodaceae (Rolls & Lindstrom, 1967 a, b), thiosulphate oxidation is inducible.

Under appropriate conditions thiosulphate causes the dark reduction of cytochromes in *Chromatium* and supports the light-induced reduction of pyridine nucleotides, nitrogen fixation and hydrogen evolution (Arnon, Losada, Nozaki & Tagawa, 1961; Arnon, 1961; Losada, Nozaki & Arnon, 1961).

These results have been interpreted in terms of a generalized mechanism of non-cyclic electron flow in which electrons are transferred from the donor to the acceptor systems via cytochromes and the photochemical apparatus (scheme 9.42, Arnon *et al.* 1961; Ormerod & Gest, 1962).

In the photosynthetic sulphur bacteria the compound $A^-$ in the scheme on page 248 is an oxidizable sulphur compound which is replaced by organic substances in other photosynthetic bacteria and by water ($OH^-$) in green plants. The question of electron flow in photosynthesis is however a complex one and discussion of this aspect is beyond the scope of this book. Some recent publications in this field are those of McElroy & Glass (1961) and of Gest, San Pietro & Vernon (1963).

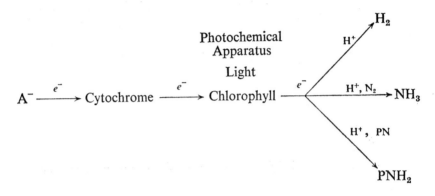

$$A^- = \text{electron donor;} \qquad PN = \text{pyridine nucleotide}$$

$$(9.42)$$

## 9.3 Oxidation of inorganic sulphur compounds by heterotrophic micro-organisms

It has long been recognized that inorganic sulphur compounds are metabolized by a number of heterotrophic organisms from soil (Demolon, 1921; Guittonneau, 1925 a, b; 1926, 1927; Guittonneau & Keilling, 1927). The species involved are mainly *Bacillus* spp., *Pseudomonas* spp., Actinomycetes and moulds, and the oxidations are probably incidental to the general metabolism of these organisms although they may be important in the turnover of sulphur in soils. *Aspergillus niger* (*Sterigmatocystis nigra*) and *Penicillium glaucum* (Guittonneau, 1927) and *Penicillium luteum* (Abbott, 1923) have been claimed to oxidize sulphur to sulphate but, in general, Guittonneau and his colleagues found that mixed cultures were required for the complete oxidation and that soil organisms fell into two main groups; those oxidizing sulphur to thiosulphate (equation 9.43) and those oxidizing thiosulphate to sulphate (equation 9.44).

$$2S^0 + O_2 + OH^- \rightarrow S_2O_3^{2-} + H^+ \qquad (9.43)$$

$$S_2O_3^{2-} + 2O_2 + OH^- \rightarrow 2SO_4^{2-} + H^+ \qquad (9.44)$$

Polythionates are also produced during sulphur oxidation in soils (Guittonneau & Keilling, 1932 *a, b*).

The oxidation of sulphide to elemental sulphur by *Sphaerotilus natans*, *Alternaria* and yeast has been reported by Skerman, Dementjeva & Skyring (1957) and by Skerman, Dementjeva & Carey (1957).

*Pseudomonas fluorescens, Pseudomonas aeruginosa, Achromobacter stutzeri* and the so called *Thiobacillus trautweinii* catalyse the oxidation of thiosulphate to tetrathionate (Trautwein, 1921, 1924; Starkey, 1934 *a, b*; 1935 *a*). Organisms catalysing a similar reaction were also reported by Sijderius (1946) and Baalsrud & Baalsrud (quoted by van Niel, 1953), by London (1964), and by Hall & Berk (1968).

Almost nothing is known of the mechanism of these oxidations. Trudinger (1967 *b*) studied the oxidation of thiosulphate to tetrathionate by two organisms isolated from soil. In one organism the enzyme responsible was constitutive and all attempts to obtain activity in cell-free extracts were unsuccessful. In the second organism the enzyme was induced by thiosulphate and could be extracted from the intact cell as a particle-bound enzyme. The enzyme was characterized by a low affinity for oxygen (apparent $K_m$, 223 $\mu$M). Ferricyanide and mammalian cyto-chrome-*c* replaced oxygen as electron acceptors: the cytochrome-*c* of the bacteria was also reduced by thiosulphate. The enzyme was inhibited by azide, cyanide and cupferron but no direct evidence for the participation of a metal in the reaction was found. The enzyme did not appear to be involved in the incorporation of thiosulphate into cell material or in detoxication and the oxidation of thiosulphate did not support growth.

## 9.4   Oxidation of inorganic sulphur compounds by plants

Very little is known of oxidative reactions on inorganic sulphur compounds in plants. Sulphite is oxidized to sulphate by mitochondria from pea internodes (Arrigoni, 1959), oat seedlings (Tager & Rautanen, 1955, 1956) and pea roots (Stickland, 1961): in the last instance ADP, but not pyridine nucleotides, was reported to stimulate sulphite oxidation. The oxidation of sulphite by pea mitochondria is accompanied by the esteri-fication of inorganic phosphate into ATP (Arrigoni & Rossi, 1961).

Pyridine nucleotides are also not required for sulphite oxidation by oat seedling mitochondria (Tager & Rautanen, 1955): instead cytochrome-*c* and magnesium ions are necessary for maximum activity. Sulphite oxidation by oat seedling mitochrondria is inhibited by cyanide, stimulated by azide, and not affected by dinitrophenol or iodoacetate.

According to Knobloch (1966 *a*, *b*) sulphide is oxidized in the light by a number of algae and green plants and plays a role as a secondary electron donor in the photosynthetic reaction. Several inhibitors of photosynthetic electron transport and of photophosphorylation also inhibit sulphide oxidation which requires ferrous ions for activity. Sulphide inhibits oxygen evolution and its oxidation is increased in manganese-deficient organisms. Since manganese deficiency is thought to impair the mechanism transforming (OH) radicals to molecular oxygen (Kessler, 1957) Knobloch suggested that sulphide competes for these radicals and that its oxidation is expressed by equation 9.45.

$$2(OH^{\cdot}) + H_2S \xrightarrow{\ Fe^{2+}\ } 2H_2O + S^0 \qquad (9.45)$$

# 10

## REDUCTION OF INORGANIC
## SULPHUR COMPOUNDS BY
## MICRO-ORGANISMS AND PLANTS

The ability to reduce inorganic sulphur compounds is widespread amongst micro-organisms and plants. For the most part sulphate and other sulphur compounds are reduced to the level of sulphide, either as free sulphide or combined in the form of cysteine, methionine etc. By analogy with Kluyver's (1953) classification of bacterial nitrate reductions, Postgate (1959 a) suggested the terms *assimilatory sulphate reduction* to describe the small scale reduction of sulphate to sulphur-containing organic cell constituents and *dissimilatory sulphate reduction* to describe the large scale transformation of sulphate to sulphide ions which is linked to energy-yielding processes. Although assimilatory sulphate reduction may, in some circumstances, result in the formation of free sulphide this classification has since been justified on an enzymic basis (see below).

### 10.1 Assimilatory sulphate reduction

Assimilatory sulphate reduction is a property of the majority of bacteria, fungi, yeasts, algae and plants as evidenced by their ability to utilize sulphate as the sole source of cellular sulphur. A number of other sulphur compounds including sulphite, thiosulphate, sulphide, elemental sulphur, alanine 3-sulphinic acid and $S$-sulphocysteine may substitute for sulphate in many instances (eg. Wilson, 1962).

### 10.1.1 Pathway of assimilatory sulphate reduction

Nutritional studies with mutant strains indicated that sulphite and thio-sulphate are intermediates in the reduction of sulphate to cysteine by *Escherichia coli* (Lampen, Roepke & Jones, 1947), *Neurospora* (Horowitz, 1950, 1955; Ragland & Liverman, 1958), *Salmonella* (Clowes, 1958), and *Penicillium notatum* (Hockenhull, 1948). This view was supported by isotope competition experiments with *E. coli* (Cowie, Bolton & Sands, 1950, 1951; Roberts, Abelson, Cowie, Bolton & Britten, 1955) and

251

*Salmonella* (Clowes, 1958) in which sulphite and thiosulphate were shown to suppress the incorporation of $^{35}SO_4^{2-}$ into cell material. The incorporation of labelled sulphate is also suppressed by sulphite in mung bean leaves (Asahi & Minamikawa, 1960) and by thiosulphate in an unnamed soil bacterium (Trudinger, 1967 *b*). Sulphite, thiosulphate and *S*-sulphocysteine were proposed as intermediates in sulphate reduction by *Aspergillus nidulans* (Hockenhull, 1949; Nakamura & Sato, 1960; Nakamura, 1962). Shepherd (1956) concluded that alanine 3-sulphinic acid (cysteine sulphinic acid) rather than *S*-sulphocysteine is the precursor of cysteine in *Aspergillus* but in *E. coli* and *Salmonella typhimurium* the sulphinate appears to be degraded to sulphite before the sulphur is utilized for synthesis (Leinweber & Monty, 1961, 1962).

The proposed role of sulphite (or a compound readily convertible to sulphite) in assimilatory sulphate reduction received further support from direct demonstrations of the formation of sulphite from sulphate in a number of tissues. As early as 1932, Nightingale, Schermerhorn & Robbins demonstrated sulphite formation in the phloem and cambium of sulphur-starved tomato plants on the addition of sulphate to the nutrient medium. Later Fromageot & Perez-Milan (1956, 1959) isolated labelled sulphite from excised tobacco leaves which had been infused with $^{35}SO_4^{2-}$ ions: similar results with mung bean leaves were reported by Kawashima & Asahi (1961).

The formation of sulphite from sulphate by wine-yeasts was shown by Schanderl (1952). Enzymic studies on sulphate reduction have provided strong confirmatory evidence for the intermediate role of sulphite in assimilatory sulphate reduction (see p. 255 *et seq.*).

The situation regarding thiosulphate is not quite so clear. Nakamura & Sato (1960) reported that an enzyme system from yeast synthesizes *S*-sulphocysteine from serine and thiosulphate in the presence of pyridoxal phosphate (equation 10.1). A similar enzyme is present in *A. nidulans* and the activity of this enzyme is low in a mutant which requires either *S*-sulphocysteine, cysteine or methionine for growth (Nakamura & Sato, 1963 *a,b*) while a cysteine-requiring mutant of *A. nidulans* was found to accumulate *S*-sulphocysteine derived from sulphate (Nakamura

$$\begin{array}{ccc} CH_2OH & & CH_2.S.SO_3^- \\ | & & | \\ H.C.NH_2+S_2O_3^{2-}+H^+ & \rightarrow & H.C.NH_2+H_2O \\ | & & | \\ COOH & & COOH \end{array} \qquad (10.1)$$

& Sato, 1962). These results indicated that thiosulphate is an intermediate in sulphate reduction and that it is the direct precursor of organically-bound sulphur. Woodin & Segel (1968) studied the metabolism of S-sulphocysteine by *Penicillium chrysogenum*. S-sulphocysteine was apparently converted to cysteine by a chemical reaction with reduced glutathione. The reaction was driven in the direction of cysteine synthesis by regeneration of reduced glutathione by glutathione reductase. Levinthal & Schiff (1965) and Levinthal (1967) obtained a partially purified preparation from *Chlorella pyrenoidosa* which converted sulphate to thiosulphate in the presence of ATP, $Mg^{2+}$, lipoic acid and NADPH (see also Schiff & Levinthal, 1968). PAPS was identified as an intermediate in the reaction (Hodson, Schiff, Scarsella & Levinthal, 1968). Only the sulpho group of thiosulphate was derived from sulphate: the sulphane group appeared to arise from bound sulphur in the enzyme preparation (Levinthal & Schiff, 1968). The role of this thiosulphate-synthesizing system in relation to cysteine biosynthesis is obscure since added thiosulphate is apparently cleaved by growing cells of *Chlorella* to sulphite and '—SH', the latter being preferentially incorporated into organic cell constituents (Hodson, Schiff & Scarsella, 1968).

Leinweber & Monty (1963) concluded that thiosulphate is not an intermediate in sulphate reduction in *Salmonella* but that thiosulphate is first reduced to sulphite and sulphide prior to conversion to cysteine. Their hypothesis relied strongly on the fact that mutants of *Salmonella* which are unable to utilize sulphite incorporate only the sulphane sulphur of thiosulphate and accumulate sulphite whereas the wild-type bacteria utilize both sulphur atoms, although not to the same extent. The possibility that reductive scission of the thiosulphate molecule occurred after its conversion to an organic form was not excluded. Nevertheless the presence of an enzyme in *Salmonella* and other organisms which catalyses the direct reduction of thiosulphate to sulphide and sulphite (see § 10.1.4) lends weight to the contention of Leinweber & Monty. Hilz, Kittler & Knape (1959) obtained evidence that thiosulphate is not an intermediate in sulphate reduction by yeast. They showed that although thiosulphate inhibits the activation of sulphate by yeast extracts (see p. 255), it does not compete with [35]S-labelled 'active sulphate' in the formation of sulphide. Furthermore thiosulphate is apparently not formed by highly purified sulphite reductase preparations from yeast (see § 10.1.3).

Nutritional studies with mutants of *E. coli* (Lampen, Roepke & Jones, 1947) and *Neurospora* (Horowitz, 1955) indicated that sulphide may be an intermediate in sulphate reduction by these organisms. Sulphide supresses the incorporation of [35]S-labelled sulphate into cysteine and methionine by yeast (Schlossmann, Brüggemann & Lynen, 1962) while mutants of *E. coli* (Lampen, Roepke & Jones, 1947) and of *Salmonella* (Clowes, 1958; Dreyfuss & Monty, 1963 *a*; Mizobuchi, Demerec & Gillespie, 1962) which are unable to grow on sulphide are simultaneously unable to utilize the more oxidized forms of sulphur. These results indicate a key role for sulphide in cysteine biosynthesis.

Yeast contains an enzyme, serine sulphhydrase (cysteine synthase) which catalyses the formation of cysteine from serine and sulphide (equation 10.2; Schlossmann & Lynen, 1957).

$$\begin{matrix} CH_2OH & & CH_2SH \\ | & & | \\ H.C.NH_2+H_2S & \rightleftharpoons & H.C.NH_2+H_2O \\ | & & | \\ COOH & & COOH \end{matrix} \qquad (10.2)$$

The reaction is reversible and pyridoxal phosphate is required for maximum activity. The reaction has also been shown to occur in *E. coli*, *Bacillus subtilis*, *Aspergillus niger*, *Micrococcus aureus*, *Aerobacter aerogenes* and spinach, as well as in some animal tissues (Brüggemann, Schlossmann, Merkenschlager & Waldschmidt, 1962; Brüggemann & Waldschmidt, 1962).

A second mechanism for cysteine formation from sulphide in *E. coli* and *S. typhimurium* was demonstrated by Kredich & Tomkins (1966, 1967). This involves the intermediate formation of *O*-acetylserine from serine and acetyl-CoA followed by the displacement of acetate by sulphide (equations 10.3, 10.4).

$$\begin{matrix} CH_2OH & & CH_2.O.OC.CH_3 \\ | & \xrightarrow{\text{serine transacetylase}} & | \\ H.C.NH_2+\text{acetyl-CoA} & & H.C.NH_2+CoA \\ | & & | \\ COOH & & COOH \end{matrix} \quad (10.3)$$

$$\begin{matrix} CH_2.O.OC.CH_3 & & CH_2SH \\ | & \xrightarrow[\text{sulphhydrase}]{\text{O-acetylserine}} & | \\ H.C.NH_2+H_2S & & H.C.NH_2+CH_3COOH \\ | & & | \\ COOH & & COOH \end{matrix} \quad (10.4)$$

No requirement for added pyridoxal phosphate was demonstrated but highly purified serine transacetylase (which also showed strong *O*-acetylserine sulphhydrase activity) had a spectrum similar to that of pyridoxal-phosphate containing enzymes. Several mutants of *Salmonella* which are unable to utilize sulphide contained negligible amounts of serine trans-acetylase and low $O$-acetylserine sulphhydrase activities, indicating that these enzymes play an essential role in cysteine biosynthesis in the intact organism (Kredich & Tomkins, 1966).

The foregoing results suggest that in *Salmonella*, and probably in a number of other micro-organisms and plants, the pathway of cysteine biosynthesis from sulphate and other inorganic sulphur compounds may be formulated as in the following scheme:

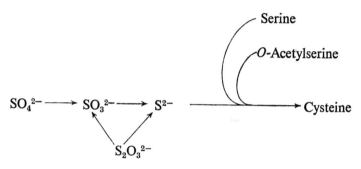

It is possible, however, that alternative pathways involving a more direct role for thiosulphate may operate in certain organisms (Roberts *et al.* 1955; Horowitz, 1950; Ragland & Liverman, 1958; Clowes, 1958).

## 10.1.2   Reduction of sulphate to sulphite

### Activation of sulphate

Cell-free extracts catalysing the reduction of sulphate to sulphate were prepared from yeast by Hilz & Kittler (1958) and by Wilson & Bandurski (1958 *a*). Reduced pyridine nucleotides and ATP were required for full activity. The reduction of sulphate was inhibited by selenate, tungstate and molybdate, the 'classical' inhibitors of ATP-sulphurylase (Hilz, Kittler & Knape, 1959), and subsequent studies with fractionated extracts (Bandurski, Wilson & Asahi, 1960; Wilson, Asahi & Bandurski, 1961) established that ATP is required for 'activation' of sulphate by ATP-sulphurylase and APS-kinase (see chapter 5).

Both Hilz *et al.* and Bandurski *et al.* demonstrated that PAPS is the true substrate for the reductive step.

As a result of these studies the initial steps in the sulphate reduction in yeast were formulated according to equations 10.5 to 10.7.

There is a deal of evidence that this pathway operates in a variety of other organisms. Sulphate activation and PAPS reduction have been demonstrated in *E. coli* (Mager, 1960; Fujimoto & Ishimoto, 1961; Pasternak, 1962; Pasternak, Ellis, Jones-Mortimer & Crichton, 1965), in *Salmonella* (Dreyfuss & Monty, 1963 *a*, *b*), in *B. subtilis* (Pasternak, 1962; Pasternak *et al.* 1965) and in *Chlorella* (Levinthal & Schiff, 1965; Levinthal, 1967; see also Schiff, 1959; Wedding & Black, 1960) while direct or presumptive evidence has been found for the formation of PAPS during sulphate reduction by *Penicillium chrysogenum* (Segel & Johnson, 1963 *b*), by *Rhodospirillum rubrum* chromatophores (Ibanez & Lindstrom, 1962) and by *Euglena* (Abraham & Bachhawat, 1963).

$$SO_4^{2-} + ATP \xrightarrow{\text{ATP-sulphurylase}} APS + PP_i \qquad (10.5)$$

$$APS + ATP \xrightarrow{\text{APS-kinase}} PAPS + ADP \qquad (10.6)$$

$$PAPS + 2e \xrightarrow{\text{PAPS-reductase}} PAP + SO_3^{2-} \qquad (10.7)$$

The situation in plants is not so clear since neither Asahi (1964) nor Balharry & Nicholas (1968) were able to detect PAPS during the reduction of $^{35}$S-labelled sulphate by spinach chloroplasts; the formation of APS was readily demonstrated. It may be relevant, however, that certain organisms such as, for example, *Salmonella pullorum*, possess a 3′-nucleotidase which degrades PAPS: the formation of PAPS during sulphate reduction by extracts of such organisms is readily detected only when a 3′-nucleotide is present to prevent PAPS degradation (Kline & Schoenhard, 1968).

The importance of the sulphate-activation pathway in the intact cell is shown by the existence of mutants of *Aspergillus, Neurospora, Salmonella, E. coli* and yeast which utilize sulphite but not sulphate and which are deficient in either ATP-sulphurylase, APS-kinase or PAPS-reductase (Ragland, 1959; Naiki, 1964; Dreyfuss & Monty, 1963 *a*; Wheldrake & Pasternak, 1965; Hussey, Orsi, Scott & Spencer, 1965; Metzenberg & Parson, 1966; Spencer, Hussey, Orsi & Scott, 1968). Moreover molybdate, a potent inhibitor of ATP-sulphurylase, prevents sulphate incorporation by intact mung bean leaves (Kawashima & Asahi, 1961).

### PAPS-reductase

The reduction of PAPS by yeast appears to specifically require NADPH (Wilson & Bandurski, 1958 *a*) although a slight activity with NADH has been reported (Hilz, Kittler & Knape, 1959). Mager (1960) reported that the PAPS-reductase from *E. coli* differed from the yeast enzyme in being specific for NADH but Fujimoto & Ishimoto (1961) found that the reduction of PAPS by *E. coli* extracts could be coupled to the oxidation of either NADH or NADPH and, according to Pasternak *et al.* (1965), NADPH is about five times as effective as NADH as an electron donor for PAPS reduction in *E. coli*.

Hilz, Kittler & Knape (1959) found that sulphite formation from PAPS in yeast extracts is stimulated by lipoic acid and inhibited by arsenite. They proposed a mechanism for PAPS-reductase involving the formation of a lipoate-sulphenyl sulphite (or thiosulphate ester) as in equations 10.8 to 10.10.

$$PAP.O.SO_3^- + Lip\begin{array}{c} \diagup SH \\ \diagdown SH \end{array} \rightarrow PAP + Lip\begin{array}{c} \diagup S.SO_3^- \\ \diagdown SH \end{array} \qquad (10.8)$$

$$Lip\begin{array}{c} \diagup S.SO_3^- \\ \diagdown SH \end{array} \rightarrow Lip\begin{array}{c} \diagup S \\ | \\ \diagdown S \end{array} + HSO_3^- \qquad (10.9)$$

$$Lip\begin{array}{c} \diagup S \\ | \\ \diagdown S \end{array} + NADPH + H^+ \rightarrow Lip\begin{array}{c} \diagup SH \\ \diagdown SH \end{array} + NADP^+ \qquad (10.10)$$

In support of this mechanism Hilz & Kittler (1960) showed that substrate concentrations of reduced lipoic acid or lipoamide could replace NADPH as an electron donor for PAPS reduction; other thiol compounds such as cysteine, GSH and BAL were not active. Similar results were obtained by Fujimoto & Ishimoto (1961) for the PAPS-reductase from *E. coli* although here reduced methyl viologen and cytochrome-$c_3$ from *Desulfovibrio desulfuricans* were also active as electron donors.

Wilson and his colleagues (Bandurski, Wilson & Asahi, 1960; Wilson, Asahi & Bandurski, 1961; Asahi, Bandurski & Wilson, 1961) isolated from yeast extracts two heat-labile proteins (enzymes A and B) and a heat stable, low molecular weight protein (fraction C) which were required for the reduction of PAPS to sulphite by NADPH. Enzyme A catalysed the reduction of a disulphide group in fraction C in the presence of NADPH; cysteine, lipoate, lipoamide and bovine serum albumin were not reduced. A reduction of GSSG appeared to be due to contaminating glutathione reductase in the partly purified preparations of enzyme A. Lipoate and lipoamide were, however, reduced by enzyme A and NADPH in the presence of fraction C. No lipoate was detected in fraction C and the role of lipoate in PAPS reduction was apparently a secondary one.

Enzyme A exhibited diaphorase activity and catalysed the reduction of methylene blue, ferricyanide and 2,6-dichlorophenolindophenol by NADPH in the absence of fraction C. Both the diaphorase activity and the reduction of fraction C were stimulated by FAD and were inhibited by the thiol-binding inhibitors arsenite, $N$-ethylmaleimide and $p$-chloro-mercuribenzoate. The reduction of fraction C was also inhibited by the flavin inhibitor quinacrine (see also Fujimoto & Ishimoto, 1961).

On the basis of these results the following reaction sequence (equations 10.11 to 10.13) for the reduction of PAPS was proposed.

$$\text{Enzyme A-FAD} + \text{NADPH} + \text{H}^+ \rightarrow \text{NADP}^+ + \text{enzyme A-FADH}_2 \quad (10.11)$$

$$\text{Enzyme A-FADH}_2 + \text{fraction C-SS} \rightarrow \text{enzyme A-FAD} + \text{fraction C-(SH)}_2 \quad (10.12)$$

Fraction C–(SH)$_2$ + PAPS

Enzyme B $\quad\quad\quad\quad\quad\quad\quad$ (10.13)

Fraction C–SS + PAP + SO$_3^{2-}$ + 2H$^+$

Alternative reaction pathways are illustrated by equations 10.14 and 10.15: in the former lipoamide may replace lipoate.

$$\text{Fraction C-(SH)}_2 + \text{Lipoate} \rightarrow \text{fraction C-SS} + \text{Dihydrolipoate} \quad (10.14)$$

$$\text{Enzyme A-FADH}_2 + \text{Dye} \rightarrow \text{enzyme A-FAD} + \text{Dye-H}_2 \quad (10.15)$$

Later Bandurski (1965) suggested that a disulphide group in enzyme A may be reduced and reoxidized during the oxidation of NADPH by dyes or fraction C.

No studies have yet been made on the mechanism of PAPS reduction in other organisms. A protein with properties similar to those of fraction C has, however, been detected in chloroplasts (Asahi, 1963, 1964).

### The end-product of PAPS reduction

Hilz et al. (1959) and Wilson et al. (1961) identified $^{35}$S-labelled sulphite amongst the end-products of $^{35}$S-labelled sulphate reduction by yeast preparations to which unlabelled carrier sulphite had been added. By omitting carrier sulphite Torii & Bandurski (1964) obtained evidence that the true product of PAPS reduction is a non-dialysable sulphur-containing compound, 'bound sulphite', which is exchangeable with free sulphite. They suggested that a thiol or disulphide group may be involved in the formation of 'bound sulphite' since PAPS prevents the appearance of the free thiol groups which are generated in the PAPS-reducing system when NADPH is added in the absence of PAPS. Later Torii & Bandurski (1967) showed that partly purified preparations of 'bound sulphite' are hydrolysed by proteases and react with 5,5'-dithio-*bis*-(2-nitrobenzoic acid) [bis(3-carboxy-4-nitrophenyl) disulphide] to form 3-carboxy-4-nitrophenyl thiosulphate. These results indicate that 'bound sulphite' is a sulpho group attached to a protein molecule, possibly through a thiol group, but a number of properties of 'bound sulphite' (including the fact that only half of the $^{35}$S of the labelled material is released on treatment with unlabelled sulphite) suggest that the derivative is not a simple *S*-sulpho compound.

The nature of the protein moiety of 'bound sulphite' is unknown although fraction C has been suggested as fulfilling this role (Torii & Bandurski, 1967). This claim was based both on the suspected participation of the thiol (or disulphide) groups of fraction C in PAPS reduction and on a calculated molecular weight for 'bound sulphite' of 4,000–8,000 which is in the range previously estimated for fraction C.

### 10.1.3 Sulphite reductase

The reduction of sulphite to sulphide is a complex six-electron reaction which in *Salmonella* is controlled by at least six structural genes (Dreyfuss & Monty, 1963 *a*). The reaction has been shown to be catalysed by extracts of *E. coli* (Mager, 1960; Leinweber & Monty, 1961, 1962; Kemp, Atkinson, Ehret & Lazzarini, 1963; Ellis, 1964 *a*), *S. typhimurium* (Dreyfuss & Monty, 1963 *a*), *A. nidulans* (Yoshimoto, Nakamura & Sato, 1961, 1967), *Neurospora crassa* (Leinweber, Siegel & Monty, 1965; Leinweber & Monty, 1965; Siegel, Leinweber & Monty, 1965), yeast (Wainright, 1961, 1962, 1967; Okuda & Uemura, 1965; Naiki, 1965; Yoshimoto & Sato, 1965, 1968 *a*, *b*) and of a number of plants (Tamura, 1964, 1965; Asada, 1967). In most cases sulphite reductase preparations also possess hydroxylamine, nitrite and sometimes cytochrome-*c* reductase activities which appear to be due to the same enzyme system (table 10.1). The effects of a number of inhibitors on sulphite reductases from different sources are shown in table 10.2. All are particularly sensitive to cyanide but differ in their responses to the thiol-binding reagents *p*-chloromercuribenzoate (PCMB) and arsenite which are discussed in more detail below. Metal complexing reagents appear to have little or no effect. NADPH seems to be the natural electron donor for sulphite reduction although not all

TABLE 10.1   *The properties of sulphite reductases from various sources*

| Source and reference | $K_m$ (SO$_3^{2-}$) | Electron donors for sulphite reduction | | Other reductase activities* | Miscellaneous properties |
|---|---|---|---|---|---|
| | | NADPH | MVH | | |
| E. coli (1)‡ | 8 μM | + | · | NO$_2^-$ (0·4 mM) – MVH<br>SeO$_3^{2-}$ (0·09 M) – MVH<br>cytochrome-c – MVH | $K_m$ (NH$_2$OH)/$K_m$ (SO$_3^{2-}$) = 100 |
| E. coli (2) | · | + | · | NH$_2$OH – NADPH | mol. wt 700,000. Contains 7–8 flavin groups (FMN = FAD), 14–16 iron atoms and 12–16 labile sulphide groups per mole. Contains a haem-like component. Absorption peaks at 385 and 587 mμ |
| E. coli (3) | · | + | · | · | absorption peaks at 380 and 587 mμ |
| yeast (4) | 190 μM | + | + | NH$_2$OH – NADPH<br>NO$_2^-$ – NADPH<br>DCIP – NADPH<br>indophenol – NADPH | |
| yeast (5) | · | + | · | | |
| yeast (6) | 14 μM | + | + | NO$_2^-$ (1 mM) – NADPH<br>NH$_2$OH (4·5 mM) – NADPH<br>cytochrome-c (60 μM) – NADPH<br>ferricyanide – NADPH<br>ubiquinone – NADPH<br>menadione – NADPH<br>DCIP – NADPH<br>indophenol – NADPH | mol. wt 350,000; $S_{20,w}^0$ = 14·8 S. Contains 1 mole FMN, 1 mole FAD and 5 atoms of non-haem iron per mole of enzyme. Absorption peaks at 386, 455 and 587 mμ |
| Aspergillus nidulans (7) | 25 μM | − | + | NH$_2$OH (0·02 M) – MVH<br>menadione – NADPH<br>cytochrome-c – NADPH<br>DCIP – NADPH<br>indophenol – NADPH<br>menadione – NADH | optimum pH 7·2–7·8 in tris, 8 in phosphate. Apparently homogeneous. $S_{20,w}^0$ = 4·23 S. Absorption peaks at 384 and 585 mμ |

| | | | | | |
|---|---|---|---|---|---|
| *Neurospora crassa* (8) | 4·4 μM | + | $NH_2OH$ (0·96 mM) – NADPH | · | $K_m$ (NADPH) 40–50 μM. Stimulated by FAD (see text) |
| *Allium odorum* (9) | 630 μM | – | $NH_2OH$ – MVH | + | optimum pH 7·0–7·5. Apparently homogeneous. $S_{20,w} = 4·5$ S. Absorption peak at 586 mμ |
| *spinach* (10) | 20–670 μM† | – | $NH_2OH$ – MVH $NO_2^-$ – MVH | + | optimum pH varies between 7 and 8 in different buffers. Mol. wt 83,000–85,000. Contains a haem-like component but no flavin. Iron content 0·76 atoms/mole. Absorption peaks at 279, 404 and 589 mμ. |

* Numbers in parentheses are the $K_m$ values of the various substrates.
† This value varies considerably depending upon the pH and the nature of the buffer.
‡ *References*: (1) Kemp, Atkinson, Ehret & Lazzarini (1963). (2) Ellis (1964 *a*); Mager (1960). (3) Kamin, Masters, Siegel, Vorhaben & Gibson (1968). (4) Naiki (1965). (5) Wainright (1961, 1962, 1967). (6) Yoshimoto & Sato (1968 *a*). (7) Yoshimoto, Nakamura & Sato (1961, 1967). (8) Siegel, Leinweber & Monty (1965). (9) Tamura (1965). (10) Asada (1967); Asada, Tamura & Bandurski (1968).

TABLE 10.2 *Inhibitors of sulphite reduction in various organisms. The figures are the percentage inhibitions caused by the various compounds at the concentrations given in the brackets*

| electron donor | Spinach (1)† MVH | Allium odorum (2) MVH | Aspergillus (3) MVH | Yeast (4) NADPH | Yeast (4) MVH |
|---|---|---|---|---|---|
| inhibitor: | | | | | |
| *p*-chloromercuribenzoate | 100% (50 μM)* | 0% (50 μM) | 3% (1 mM) | 100% (50 μM) | 0% (50 μM) |
| KCN | 93% (100 μM)* | 100% (50 μM) | 97% (100 μM) | 100% (100 μM) | 100% (100 μM) |
| $NaAsO_2$ | · | 0% (5 mM) | 78% (100 μM) | 12% (1 mM) | · |
| $Na_2S$ | · | · | 69% (1 mM) | · | · |
| 8-hydroxyquinoline | 34% (500 μM) | 7% (5 mM) | 12% (100 μM) | · | · |
| *o*-phenanthroline | 40% (5 mM) | · | · | · | · |

* This value depends upon the oxidation state of the enzyme.
† *References*: (1) Asada (1967). (2) Tamura (1964). (3) Yoshimoto, Nakamura & Sato (1961, 1967). (4) Naiki (1965).

preparations of sulphite reductase respond to this donor (table 10.1). The artificial electron donors, reduced methyl viologen (MVH) or reduced benzyl viologen (BVH) substitute for NADPH and so provide convenient methods for assaying the enzyme either by direct spectrophotometric measurement of MVH (or BVH) oxidation or by measurement of hydrogen uptake in the presence of MVH (or BVH) and hydrogenase. Yoshimoto, Nakamura & Sato (1967) have drawn attention to the fact that MVH and BVH are 'one electron' carriers of low potential ($E_0' = -0.440$ and $-0.359$ V respectively) and that two electron carriers such as neutral red ($E_0' = -0.325$ V), Nile blue ($E_0' = -0.122$ V), methylene blue ($E_0' = +0.011$ V), toluidine blue ($E_0' = +0.115$ V), thionine ($E_0' = +0.063$ V), and phenazine methosulphate ($E_0' = +0.08$ V) are incapable of replacing the viologen dyes as electron donors for the sulphite reductase from *A. nidulans* (see also Tamura, 1965).

The possibility that an intermediate electron carrier is required to link sulphite reduction with NADPH oxidation was suggested by Siegel & Monty (1964) and Siegel, Click & Monty (1964) on the basis of kinetic evidence and enzyme analyses of cysteine-requiring mutants of *S. typhimurium*. Later, Naiki (1965) demonstrated that NADPH-linked, sulphite reductase activity of partly-purified preparations from *Saccharomyces cerevisiae* is considerably more sensitive to heat, low ionic strengths and PCMB than is MVH-linked sulphite reduction by the same preparations. Naiki also obtained mutants of yeast which were unable to utilise sulphite for growth and which lacked NADPH-linked, sulphite reductase activity.

The yeast system has now been studied in detail by Yoshimoto and Sato (1968 *a,b*). They obtained, from wild-type yeast, a homogeneous protein which catalysed the complete reduction of sulphite to sulphide. The protein contained per mole (based on a minimum molecular weight of 350,000) 1 mole each of FAD and FMN, 5 atoms of non-haem iron and 2 to 3 moles of acid-labile sulphide. In addition a non-flavin chromophore (or chromophores) absorbing at 587 and 386 m$\mu$ was present in the enzyme. By contrast, either FAD or both FAD or FMN were absent from the purified sulphite reductases from the mutants which lacked NADH-linked but not MVH-linked activities. The mutant enzymes were also smaller with sedimentation coefficients in the order of 5·1–6·6 S compared with a value of 14·8 S for the enzyme from wild-type yeast. These results suggest that the sulphite reductase in wild-type yeast is a multicomponent flavoprotein, the flavin being responsible for transferring electrons from NADPH to sulphite.

Henderson & Loughlin (1968) studied strains of *S. typhimurium* with mutations in each of the six cistrons controlling NADPH-linked sulphite reductase activity. They reported that mixtures of extracts of certain pairs of these mutants were able to carry out NADPH-linked sulphite reduction although the extracts themselves were inactive. The components responsible for the complementation were heat-labile and non-dialysable and only a small fraction of the wild-type reductase activity was generated by mixing the extracts. The results suggested that the appearance of activity was due to the interaction of mutant polypeptides or proteins to form a hybrid enzyme molecule containing subunits from each mutant rather than to the cooperative action of two or more enzymes.

Kamin, Masters, Siegel, Vorhaben & Gibson (1968) purified NADPH-linked sulphite reductase from *E. coli* to a homogeneous state. This enzyme had a molecular weight of 700,000 and resembled the sulphite reductase of yeast in containing flavins (7–8 moles per mole of enzyme and equal amounts of FAD and FMN), iron (14–16 atoms per mole), labile sulphide (12–16 moles per mole) and a chromophore(s) absorbing at 385 and 587 m$\mu$. Flavins have also been reported to be present in purified sulphite reductase from *A. nidulans* (Yoshimoto *et al.* 1967) and have been shown to stimulate NADPH-linked sulphite reduction by yeast (Prabhakararao & Nicholas, 1968), *E. coli* (Mager, 1960) and *N. crassa* (Leinweber & Monty, 1965; Siegel, Leinweber & Monty, 1965). On the other hand, sulphite-dependent NADPH oxidation by purified sulphite reductase from *N. crassa* is not affected by FAD (Siegel *et al.* 1965) which might suggest that different stages in the overall reduction require different electron donors.

The inability of purified, MVH-linked sulphite reductase from spinach (Asada, Tamura & Bandurski, 1968), and possibly that of *Allium odorum* (Tamura, 1965), to catalyse NADH-linked sulphite reduction may be attributed to the absence of flavins in these enzymes. The enzyme from *A. nidulans*, however, was also unable to utilize NADH although flavins were, apparently, present (Yoshimoto *et al.* 1967). Nevertheless, in view of the demonstration that FMN-containing proteins from mutants of yeast catalyse only MVH-linked sulphite reduction (see above) it would be necessary to establish the nature of the flavin in the enzyme from *A. nidulans* before concluding that additional factors were required to link sulphite reduction to NADPH in this instance. It may be relevant that the sedimentation coefficient of the *A. nidulans* enzyme (4·2 S) is more in accord with the values for the 'incomplete' sulphite reductases of mutant yeasts than with that of the enzyme from wild-type yeast.

263

The chromophore absorbing in the region of 587–589 m$\mu$ appears to be a general property of sulphite reductases. In addition to the enzymes from yeast and *E. coli* it is present in purified sulphite reductases from spinach (Asada *et al.* 1968), *A. odorum* (Tamura, 1965) and *A. nidulans* (Yoshimoto *et al.* 1967). The nature of the chromophore is uncertain although Asada *et al.* (1968) concluded that it may be an unusual haemo-protein; this conclusion was based on an analysis of the absorption spectrum, metal content and carbon monoxide sensitivity of the enzyme from spinach. Electron spin resonance studies on purified sulphite reductase from *E. coli* have also indicated the presence of a haem-like moiety, containing high-spin, ferric iron, in this enzyme (Kamin *et al.* 1968).

Asada (1967) obtained evidence for two catalytically active sites in purified spinach sulphite reductase. This enzyme is inhibited by cyanide and PCMB only after anaerobic treatment (reduction?) with MVH; no inhibition occurs if the inhibitors are added to the enzyme prior to MVH treatment or if the MVH-treated enzyme is exposed to air before the inhibitors are added. Sulphite protects the enzyme against inhibition by PCMB but not by cyanide. Asada proposed that spinach sulphite reductase contains a PCMB-sensitive, sulphite-protected site (possibly a thiol group) and a cyanide-sensitive site, both of which become exposed to inhibitors on reduction. Earlier (Asada, Tamura & Bandurski, 1966) reported that spinach sulphite reductase can be resolved by fractionation on calcium phosphate gel into two components, one of which is inactivated by PCMB. The nature of the cyanide-sensitive site in spinach sulphite reductase is unknown. Mager (1960) has suggested that a free carbonyl group may occur in the active site of the sulphite reductase from *E. coli* but the spinach enzyme is not inhibited significantly by other carbonyl reagents such as semicarbazide, dimedone, isonicotinylhydrazide or boro-hydride, or by metal chelating agents such as 2,2'-dipyridyl and 8-hydroxyquinoline (Asada, 1967).

Spinach sulphite reductase is stimulated by pyrophosphate, and by nucleoside di- and triphosphates (Asada, Tamura & Bandurski, 1966; Asada, 1967); the mechanism of this stimulation is not known.

Wainright (1961, 1962) claimed to have resolved yeast sulphite reductase into six protein fractions all of which were required for enzymatic activity. It now appears that only one of these fractions corresponds to sulphite reduc-tase (Wainright, 1967) and that the effect of the other five proteins on sulphite reduction may be through their modification of the NADPH/NADP$^+$ ratio.

Although the sulphite reductase preparations described above utilize free sulphite there is some evidence that this may not be a normal intermediate in sulphate reduction. Torii & Bandurski (1967), for example, showed that 'bound sulphite' formed from PAPS by PAPS-reductase (see pp. 258–9) is reduced to sulphide by yeast extracts in the presence of NADPH and Asada & Bandurski (1967) obtained evidence that sulphite must be bound to a thiol group before being reduced. Levinthal (1967) was unable to detect sulphite during the reduction of $^{35}$S-labelled PAPS to thiosulphate by *Chlorella*.

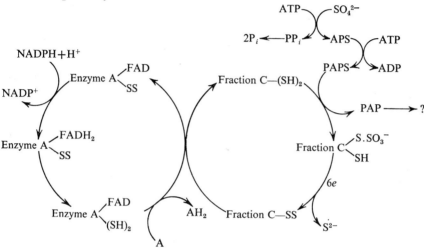

Fig. 10.1 The enzymatic pathway of assimilatory sulphate reduction (adapted from Bandurski, 1965). A and $AH_2$ represent oxidized and reduced dyes or oxidized and reduced lipoic acid.

Results with certain mutants of *Salmonella* (Thompson, 1967) have been interpreted as indicating that enzymes A and B and fraction C of the PAPS-reductase system are also required for sulphite reduction by the intact cell and that they may be concerned with the formation of bound sulphite. In this connection it is interesting that two of the proteins of Wainright's (1962) 'resolved' sulphite reductase from yeast had properties similar to the A and C proteins required for PAPS reduction. Thus it is possible that no free intermediates exist between PAPS and sulphide and, moreover, in view of the existence of sulphurtransferases such as rhodanese (see chapter 8) sulphur may be transferred to cysteine without the formation of free sulphide (Torii & Bandurski, 1967).

The pathway of assimilatory sulphate reduction is summarized in fig. 10.1 which is adapted from Bandurski (1965).

## 10.1.4   Thiosulphate reduction

As has been mentioned earlier (see § 10.1.1) thiosulphate may be reduced by some organisms before being incorporated into cell material. Predictably the sulphane sulphur group is used preferentially during the assimilation of thiosulphate by bacteria (Leinweber & Monty, 1963; Trudinger, 1967 *b*) but such discrimination could arise either from direct scission of the thiosulphate molecule or from cleavage of the sulphur–sulphur bond after conversion of thiosulphate to some organic derivative such as *S*-sulphocysteine. It has also been known for many years that micro-organisms can produce hydrogen sulphide from thiosulphate (Beijerinck, 1900, 1901; Sasaki & Otsuka, 1912): this reaction is widespread amongst yeasts and bacteria (Tanner, 1917, 1918; Clarke, 1953; Olitzki, 1954) and is due to the reduction of thiosulphate according to equation 10.16 (Neuberg & Welde, 1914).

$$S_2O_3^{2-} + 2e \rightarrow S^{2-} + SO_3^{2-} \qquad (10.16)$$

Neither elemental sulphur nor sulphate are formed (Tanner, 1917, 1918). In intact cells the reaction is stimulated by anaerobic conditions and by the addition of appropriate hydrogen donors such as sugars and organic acids (Braun & Silberstein, 1942; Mitsuhashi & Matsuo, 1950; Artman, 1956). Tarr (1933, 1934) and Mitsuhashi & Matsuo (1953) showed that the enzyme responsible for thiosulphate reduction in bacteria is distinct from cysteine desulphurase. Cell free extracts of *E. coli* reduce thiosulphate in presence of pyruvate, $Mg^{2+}$ and co-carboxylase (Artman, 1956) while in *Micrococcus lactilyticus* molecular hydrogen supports thiosulphate reduction (Woolfolk, 1962). Kawakami, Iizuka & Mitsuhashi (1957) reported that pyruvate, formate, lactate and serine support active thiosulphate reduction by extracts of *Proteus vulgaris*: with formate as the hydrogen donor this system required a particle fraction, soluble proteins and a heat-stable dialysable factor for maximum activity. The particle fraction apparently contained an electron carrier which could be replaced by any of a number of redox dyes.

Kaji & McElroy (1959) partly purified a soluble thiosulphate reductase from bakers' yeast which required reduced glutathione for activity. Stoichiometric amounts of sulphite and sulphide were produced in the presence of a glutathione regenerating system (e.g. glucose-6-phosphate plus G-6-P dehydrogenase plus NADH plus glutathione reductase). Kaji and McElroy proposed that thiosulphate reacts with a disulphide group on the enzyme

266

to form a sulphenyl thiosulphate which then reacts with glutathione to form sulphide and sulphite as depicted in the following scheme:

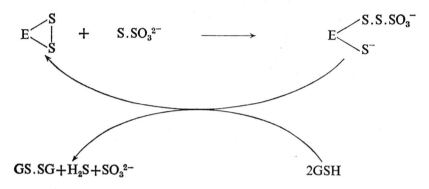

$$GS.SG + H_2S + SO_3^{2-} \qquad\qquad 2GSH$$

Sulphite competitively inhibited the reaction with thiosulphate and was thought to do so by itself reacting with the disulphide groups. Cyanide, however, which also reacts with disulphide groups, did not inhibit the enzyme nor did the thiol-binding reagent, $p$-chloromercuribenzoate. The optimum pH for thiosulphate reduction was about 8·6 and the $K_m$ for the thiosulphate was 6 mM. Homocysteine and cysteine, but not ascorbate, substituted for glutathione.

Leinweber & Monty (1963) demonstrated the presence of a thiosulphate-reducing system in extracts of *S. typhimurium* with very similar properties to those of the yeast enzyme. They made the following observations regarding the relationship between the enzyme and thiosulphate metabolism in the intact bacteria.

1. Extracts of several sulphide-requiring mutants of *S. typhimurium* which were unable to utilize thiosulphate for growth had normal glutathione-dependent thiosulphate reductase activity.

2. With one strain of *S. typhimurium*, cysteine suppressed thiosulphate reduction by intact cells but had no effect on the formation of the GSH-dependent reductase.

3. Sulphite, but not sulphide, inhibited the cell-free enzyme whereas the reverse was true for thiosulphate metabolism by intact cells.

For these reasons Leinweber & Monty suggested that the GSH-dependant reductase played no role in thiosulphate reduction in the intact cell. An alternative mechanism for thiosulphate reduction catalysed by rhodanese is discussed in chapter 8.

### 10.1.5  Uptake of sulphate by micro-organisms

Sulphate is transported into a number of micro-organisms by an 'active' process rather than by passive diffusion. Cowie, Bolton & Sands (1950) reported that *E. coli* is freely permeable to sulphate ions but recent evidence suggests that this is not so (Ellis, 1964 *b*). Segel & Johnson (1961) showed that *Penicillium chrysogenum* is able to concentrate sulphate to levels considerably above those of the external medium and later Yamamoto & Segel (1966) reported that sulphate transport in *P. chrysogenum* is temperature-, pH-, and concentration-dependent, is suppressed by inhibitors of energy metabolism such as 2,4-dinitrophenol or azide, and is independent of sulphate reduction. An energy- and pH-dependent uptake of sulphate has been demonstrated in *Euglena gracilis* (Abraham & Bachhawat, 1964) while in *Chlorella* sulphate accumulation is stimulated by light and inhibited by 2,4-dinitrophenol (Wedding & Black, 1960). Dreyfuss (1964) studied an energy- and temperature-dependent transport of sulphate into *S. typhimurium*: the uptake depended upon the integrity of the three cistrons of the cys-A region of the *Salmonella* linkage map (Mizobuchi, Demerec & Gillespie, 1962) and mutants in this region were unable to grow on sulphate although they possessed all the enzymes necessary for the assimilation of intracellular sulphate into cysteine (Dreyfuss & Monty, 1963 *a*). The sulphate transport in *Salmonella* had a $K_m$ of 36 $\mu$M sulphate, a value very similar to that reported by Yamamoto & Segel (1966) for the transport system in *P. chrysogenum*. The sulphate-transport system in *Salmonella* appears also to catalyse thiosulphate transport (Dreyfuss, 1964). Sulphate-transport negative mutants of *Neurospora* (Metzenberg & Parson, 1966) and *Aspergillus* (Hussey, Orsi, Scott & Spencer, 1965; Spencer, Hussey, Orsi & Scott, 1968) have also been reported.

### *Binding of sulphate by* Salmonella

Mutants of *S. typhimurium* which are unable to transport sulphate (transport-negative mutants) are nevertheless able to bind sulphate and the rate of binding is comparable to the rate of sulphate-transport in transport-positive bacteria (Dreyfuss & Pardee, 1965; Pardee, Prestidge, Whipple & Dreyfuss, 1966). At saturation about $10^4$ molecules of sulphate are bound per bacterium and the binding system has a $K_m$ for sulphate of the order of 4 $\mu$M. Energy is not required for the process which is inhibited by anions structurally similar to sulphate such as selenate, sulphite and chromate. The following observations suggest that the binding is a specific process and may be part of the active transport system for sulphate (Pardee *et al.* 1966):

1. Both sulphate transport in transport-positive cells (Dreyfuss & Pardee, 1965) and binding in transport-negative mutants are absent in spheroplasts, indicating that structural integrity of the cell membrane is essential for both activities.

2. The binding and active transport of sulphate are repressed by the growth of the bacteria on cysteine (see also § 10.1.6).

3. Both binding and transport activities are lost simultaneously in a class of chromate-resistant mutants and are regained by phage-transduction or reversion.

*S. typhimurium* cells which have been subjected to osmotic shock lose their ability to bind sulphate and the binding activity is then found in solution (Pardee *et al.* 1966). Pardee & Prestidge (1966) showed that cell-free extracts of *S. typhimurium* contain a soluble, high molecular weight compound with sulphate-binding properties which is absent in bacteria grown on cysteine. This compound was later purified (Pardee, 1966, 1967) and shown to be a protein of molecular weight 32,000 which had a typical amino acid composition except that it lacked the sulphur-containing amino acids. When saturated with sulphate it bound one ion of the latter per molecule of protein. The protein is located in the cell wall-membrane region of the bacterial cell (Pardee & Watanabe, 1968).

## 10.1.6 Control of assimilatory sulphate reduction

### Repression

Assimilatory sulphate reduction is under tight control and there is now much evidence to suggest that suppression of sulphate incorporation by cysteine (Roberts *et al.* 1955; Chaste & Pierfitte, 1965) and by other sulphur compounds (see § 10.1.1) is, in part, the result of repression of the enzymes of the sulphate reduction pathway.

Pasternak (1961, 1962) showed that the suppression of sulphate incorporation in *E. coli* and *B. subtilis* by cysteine is accompanied by repression of the sulphate-activating systems in these bacteria: glutathione also repressed sulphate activation in *B. subtilis* but not in *E. coli*. Both ATP-sulphurylase and APS-kinase are repressed by cyst(e)ine in *E. coli* and probably also in *B. subtilis* (Wheldrake & Pasternak, 1965). PAPS formation in *E. coli* is repressed by sulphite, thiosulphate, sulphide, alanine

3-sulphinic acid, methionine (in the presence of sulphate) and, apparently, sulphate since the activity of the sulphate-activating system of bacteria grown on sulphate is below that of bacteria grown on glutathione (Ellis & Pasternak, 1962; Ellis, Humphries & Pasternak, 1964). Since sulphate, sulphite and alanine 3-sulphinic acid do not repress sulphate activation in mutants of *E. coli* which are blocked between sulphite and sulphide, Ellis, Humphries & Pasternak (1964) suggested that repression by these compounds in wild strains is due to a metabolic product, probably cysteine. (The different responses of different organisms to glutathione might then be explained in terms of the rates of hydrolysis of glutathione to cysteine). Support for this hypothesis was obtained by Wheldrake (1967) who found an inverse relationship between the intracellular concentration of cysteine and the activity of the sulphate-activating enzymes in *E. coli* grown on a variety of sulphur sources. De Vito & Dreyfuss (1964) reported that the ATP-sulphurylase in yeast is repressed by methionine; the activity of this enzyme in yeast grown on cysteine, however, was about four times that of yeast grown on sulphate.

Sulphite reductase is repressed by cysteine in *E. coli* (Mager, 1960; Kemp, Atkinson, Ehret & Lazzarini, 1963), in *Neurospora* (Metzenberg & Parson, 1966) and in yeast (De Vito & Dreyfuss, 1964). Dreyfuss & Monty (1963 b) demonstrated coincident repression of PAPS-reductase, sulphite reductase and thiosulphate reductase in *S. typhimurium* by cysteine; a partial repression of sulphite reductase by alanine 3-sulphinate but not by thiosulphate was also shown. The activity of sulphite reductase in *S. typhimurium* grown on djenkolic acid is considerably above that of sulphate-grown bacteria indicating that sulphate represses this enzyme (Dreyfuss & Monty, 1963 b).

Repression of the enzymes converting PAPS to cysteine in *E. coli* and *B. subtilis* was studied by Pasternak, Ellis, Jones-Mortimer & Crichton (1965). PAPS-reductase and sulphite reductase are strongly repressed by cysteine and sulphide and to a much lesser extent by sulphate and sulphite. None of the sulphur compounds repressed serine sulphhydrase, although *O*-acetylserine sulphhydrase in *S. typhimurium* has been reported to be repressed by cysteine (Kredich & Tomkins, 1966).

Repression of the sulphate- and thiosulphate-transporting system of *S. typhimurium* by cysteine was reported by Dreyfuss (1964). In *A. nidulans*, however, sulphate transport appears to be regulated by the level of the internal sulphur pool and is not repressed by cysteine, methionine, taurine, djenkolic acid nor choline sulphate (Scott & Spencer, 1965).

Jones-Mortimer, Wheldrake & Pasternak (1968) made the interesting observation that the *O*-acetylserine is required for the synthesis of PAPS reductase, sulphite reductase and the sulphate-activating enzymes in *E. coli*. Cyst(e)ine-requiring mutants were obtained which lacked the structural gene for serine transacetylase, and which were able to grow on sulphate only in the presence of *O*-acetylserine. The activities of the sulphate-activating enzymes and sulphite reductase were negligible in the mutants grown on cystine (see also Jones-Mortimer, 1968) but synthesis of the enzymes was induced by the addition of *O*-acetylserine to the growth medium. The intracellular concentrations of cyst(e)ine of the mutants, when grown on limiting cystine, were below that required for substantial repression of synthesis of the sulphate-activating enzymes (Wheldrake, 1967). Thus the possibility that *O*-acetylserine prevents repression of the synthesis of sulphate-activating and other enzymes by interfering with cysteine biosynthesis appears to be eliminated. Kredich & Tomkins (1966, 1967) reported that L-cysteine at concentrations as low as 2 $\mu$M strongly inhibits synthesis of serine transacetylase; the inhibition is competitive with respect to acetyl-CoA. This feedback inhibition of *O*-acetylserine synthesis may thus regulate the level, not only of the immediate precursor of cysteine, but also of the inducer of the early enzymes of the sulphate-reduction pathway. Indeed repression of synthesis of the latter enzymes by cysteine may possibly be due solely to the inhibition of *O*-acetylserine synthesis by cysteine.

Pasternak *et al.* (1965) pointed out that some of the findings on the repression of the cysteine biosynthetic enzymes are incompatible with the 'coordinate repression' hypothesis of Ames & Garry (1959) whereby all enzymes of a biosynthetic pathway are considered to be repressed equally by the end-product. For example, sulphite, which appears to act by its metabolic conversion to cysteine (see above), almost completely represses sulphate activation in *E. coli* but has a much smaller effect on PAPS-reductase and sulphite reductase; cysteine, on the other hand, represses strongly all of these enzyme activities. Pasternak *et al.* proposed an alternative hypothesis, 'differential repression', whereby increasingly higher concentrations of the end-product are required to inhibit the synthesis of successive enzymes along the pathway of cysteine biosynthesis. The hypothesis requires that added sulphate, sulphite and sulphide give rise to increasingly higher concentrations of intracellular cysteine. That this is so is indicated not only by the fact that sulphide is a better repressor of the sulphate-activating enzymes in *E. coli* than sulphite, which is in turn better than sulphate (Ellis, Humphries & Pasternak, 1964), but also by direct determinations of the levels of intracellular cysteine in bacteria grown on sulphate, sulphite or sulphide (Wheldrake, 1967).

### Feedback inhibition

The addition of cysteine to *E. coli* grown in the presence of sulphate results in an immediate and almost complete inhibition of sulphate incorporation (Roberts *et al.* 1955; Ellis, Humphries & Pasternak, 1964). This effect cannot be accounted for by the repression of the enzymes involved in the biosynthesis of cysteine but indicates end-product inhibition of one or more of these enzymes. Ellis (1964 *b*) suggested that the site of the end-product inhibition in *E. coli* is the sulphate-uptake mechanism since cysteine inhibits sulphate accumulation in intact cells but has no effect on cell-free PAPS-reductase, sulphite reductase, or sulphate activation (Ellis, Humphries & Pasternak, 1964): sulphate activation is also insensitive to sulphite, thiosulphate and sulphide.

Sulphate transport in *S. typhimurium* is inhibited by sulphite and thiosulphate but not, apparently, by cysteine (Dreyfuss, 1964): transport is also deficient in mutants which accumulate PAPS, perhaps indicating a feed-back inhibition by a 'high energy' compound formed during sulphate transport (Dreyfuss & Pardee, 1966). In *Penicillium chrysogenum* the uptake of sulphate is inhibited by thiosulphate and also by cysteine and methionine which are apparently converted to feed-back inhibitors (Segel & Johnson, 1961; Yamamoto & Segel, 1966).

In yeast, inhibition of ATP-sulphurylase by low concentrations of ATP, PAPS and sulphide (Wilson *et al.* 1961; De Vito & Dreyfuss, 1964) suggests that the intracellular levels of these compounds may regulate sulphate reduction: cysteine, methionine, sulphite and thiosulphate do not inhibit yeast ATP-sulphurylase (De Vito & Dreyfuss, 1964). Thiosulphate reduction in *Salmonella* may be controlled by the intracellular level of sulphide (Leinweber & Monty, 1963). Inhibition of serine transacetylase by low levels of cysteine in *Salmonella* has been reported by Kredich & Tomkins (1966, 1967).

### 10.1.7 Miscellaneous reductions catalysed by 'assimilatory' sulphate-reducing organisms

#### Tetrathionate reductase (*Tetrathionase*)

Many facultative anaerobes reduce tetrathionate to thiosulphate (equation 10.17):

$$S_4O_6^{2-} + 2e \rightarrow 2S_2O_3^{2-} \tag{10.17}$$

Pollock, Knox & Gell (1942) first reported this activity in *Salmonella* and *Proteus* and later Le Minor & Pichinoty (1963) showed that it is widely distributed amongst Enterobacteriaceae, *Aeromonas* and *Pasteurella*. Although these bacteria are incapable of dissimilatory sulphate reduction, tetrathionate reductase should be classed as a dissimilatory enzyme since it appears to function in anaerobic respiration in an analogous manner to dissimilatory nitrate reductase (Knox, Gell & Pollock, 1943; Knox, 1945).

The formation of tetrathionate reductase is induced by tetrathionate (Knox & Pollock, 1944) and suppressed by oxygen (Pichinoty & Bigliardi-Rouvier, 1962, 1963; Pichinoty, 1965): thiosulphate is neither an inducer nor a repressor of tetrathionate reductase synthesis.

Tetrathionate reduction is inhibited by oxygen and in the intact cell it is supported by a variety of hydrogen donors such as glucose, mannitol, lactose and formate (Pollock & Knox, 1943). Cell-free tetrathionate reductase does not oxidize reduced pyridine nucleotides but catalyses the oxidation of $FMNH_2$ by tetrathionate (Pichinoty & Bigliardi-Rouvier, 1963). Whether the latter reaction has physiological significance is uncertain. Pichinoty & Bigliardi-Rouvier (1963) reported the following properties of tetrathionate reductase from *Escherichia intermedia*: optimum pH, 7·5; $K_m$, 0·5 mM tetrathionate; and activation energy, 29,600 calories. Pichinoty (1963) described an assay method for tetrathionate (and other) reductases based on the facts that reduced dyes such as benzyl viologen support tetrathionate reduction and that benzyl viologen is reduced by hydrogen in the presence of hydrogenase.

$$H_2 \xrightarrow{\text{hydrogenase}} \text{benzyl viologen} \xrightarrow{\text{reductase}} \text{tetrathionate}$$

An excess of hydrogenase (prepared from *D. desulfuricans*) is added to the system to be assayed and the reduction followed by measuring the hydrogen consumption manometrically.

Tetrathionate reductase, and also thiosulphate reductase, are particulate enzymes in facultative anaerobes and appear to be associated with the cell membrane (Pichinoty & Bigliardi-Rouvier, 1963; Puig, Azoulay & Pichinoty, 1967). Piéchaud, Puig, Pichinoty, Azoulay & Le Minor (1967) obtained mutants of *Proteus vulgaris* which were unable to synthesize tetrathionate reductase, thiosulphate reductase and the 'membrane bound' enzymes, nitrate reductase and hydrogenlyase. This multiple enzyme deficiency appears to be due to a single mutation and the hypothesis was advanced that this resulted in an alteration either to the structure of the elements constituting the membrane particles or to the mechanism responsible for the assembling of these elements. Later, it was indeed shown by sedimentation analysis of the particulate material of *P. vulgaris* that mutants lacking the membrane-bound reductases also lacked a particle fraction which was associated with the activity of these enzymes in the wild-type bacteria (Azoulay, Puig & Pichinoty, 1967).

## Tetrathionate reduction in aerobic bacteria

Tetrathionate reduction by the strict aerobe *Pseudomonas aeruginosa* was reported by Pollock & Knox (1943). Trudinger (1967 *a*) also demonstrated tetrathionate reduction by a strict aerobe isolated from soil. In the latter case the enzyme was induced by thiosulphate and tetrathionate but the enzyme differed from that found in facultative anaerobes both in being formed aerobically and in its affinity for tetrathionate. No function for tetrathionate reductase in aerobic bacteria has yet been found.

## Reduction of elemental sulphur

The reduction of elemental sulphur to sulphide is a wide-spread reaction among bacteria, yeasts, fungi, actinomycetes, plants and animal tissues (Dumas, 1874; Pollacci, 1875; Cugini, 1876, Selmi, 1876; Miquel, 1879; de Rey-Pailhade, 1888 *a, b*, 1898; Beijerinck, 1901; Heffter, 1908; Chowrenko, 1912; Sasaki & Otsuka, 1912; Tanner, 1917, 1918; Barker, 1929, 1930; McCallan & Wilcoxon, 1931; Tarr, 1933; Starkey, 1937; Braun & Silberstein, 1942; Ahlström, von Euler, Gernow & Hägglund, 1944; Miller, McCallan & Weed, 1953; Woolfolk, 1962; Rankine, 1963, 1964). The production thereby of the toxic hydrogen sulphide by fungi has been proposed on a number of occasions to account for the fungicidal effects of sulphur but this hypothesis has not met with general acceptance (see discussions by McCallan, 1964; Horsfall, 1956; Miller, McCallan & Weed, 1953; Martin, 1964). In many instances sulphur is reduced by a thermostable system which may involve chemical interaction between sulphur and thiol groups (e.g. glutathione) in the cell (Heffter, 1908; Barker, 1930; Buchanan & Fulmer, 1930; McCallan & Wilcoxon, 1931) as represented by equation 10.18.

$$2R.SH + S^0 \rightarrow R.S.S.R. + H_2S \qquad (10.18)$$

Sulphite and arsenite, which react with disulphide and with thiol groups respectively, strongly inhibit sulphur reduction in yeast and *Neurospora* (Miller, McCallan & Weed, 1953).

Nevertheless an enzymatic reduction of sulphur is indicated in a few cases. Ahlström *et al.* (1944) reported that the ability of yeast and of animal tissues to reduce sulphur was destroyed by heat whereas the reaction shown in equation 10.18 should not be so affected. They therefore suggested that succinic dehydrogenase or triose phosphate dehydrogenase might be involved in the reduction of sulphur. Earlier Sciarini & Nord (1943) had reported that sulphur stimulates the fermentation of isopropanol, glycerol, glucose and xylose by *Fusarium*, apparently by acting as a hydrogen acceptor for various dehydrogenation reactions. Tweedy & Turner (1966, see also Tweedy, 1964) studied the effects of sulphur on the metabolism of conidia of *Monilinia fructicola*. They found that sulphur increased the respiratory quotient, ($CO_2$ released/$O_2$ absorbed), of respiring conidia and reduced the rate and extent of cytochrome-*c* reduction by NADH in extracts of conidia. Respiration and sulphur reduction by intact conidia were affected equally by amytal and antimycin A. These results were interpreted as indicating that sulphur accepts electrons at a stage between NADH and cytochrome-*c*, thus competing with oxygen in the electron transport system.

Whether the reduction of sulphur is chemical or enzymic the indications are that it is a non-specific process. It is questionable, therefore, whether sulphur reduction plays any significant role in the normal metabolism of the cell.

## 10.2  Dissimilatory sulphate reduction

Dissimilatory sulphate reduction is the property of a few specialized bacteria. The original type species *Desulfovibrio desulfuricans* (*Spirillum desulfuricans*) was described by Beijerinck (1895) and isolated in pure culture by van Delden (1903). A comprehensive study of its physiology and metabolism was made by Baars (1930). The dissimilatory sulphate reducing bacteria are widely distributed in sea water, marine muds, fresh water, soil and oil-bearing environments. They play a major role in anaerobic corrosion processes and other processes of economic importance (see chapter 13).

Campbell & Postgate (1965) and Postgate & Campbell (1966) distinguished two genera: *Desulfovibrio* (Kluyver & van Niel, 1936); mesophilic, non-sporulating, sulphate-reducing bacteria and *Desulfotomaculum*; spore-forming, sulphate-reducing bacteria with both mesophilic and thermophilic representatives. Typical species of both genera are listed in table 10.3.

TABLE 10.3  *Classification of dissimilatory sulphate-reducing bacteria*

Genus *Desulfovibrio* (Kluyver & van Niel, 1936; Postgate & Campbell, 1966)

Non sporulating, Gram-negative vibrios, sometimes sigmoid or spirilloid; occasionally straight. Obligate anaerobes with polar flagella showing progressive motility. Mesophilic, sometimes halophilic. Contain cytochrome-$c_3$ and desulphoviridin and generally hydrogenase. Facultative or obligate sulphate reducers; sulphate reduction is their respiratory dissimilatory process. Habitats, sea water, marine mud, fresh water and soil

Species

*D. desulfuricans* (type species); *D. vulgaris*; *D. salexigens* (salt water species); *D. gigas*; *D. africanus*

Genus *Desulfotomaculum* (Campbell & Postgate, 1965)

Gram-negative, straight or curved rods, usually single or sometimes in chains; thermophilic types show lenticulate and otherwise swollen forms. Terminal or subterminal sporulation, slightly swelling the cells. Motile, with peritrichous flagella. Obligate anaerobes which reduce sulphate to sulphide. Habitats, fresh water, soils, geothermal regions, certain spoiled foods, intestines of insects and in rumen contents of ruminant animals

Species

*D. nigrificans* (thermophilic, opt. temp. 55°); *D. orientis* (mesophilic, opt. temp. 30–37°); *D. ruminus* (mesophilic, opt. temp. 37°)

The current classification is supported to some extent by the infrared spectral properties of sulphate-reducing bacteria (Booth, Miller, Paisley & Saleh, 1966). The thermophilic bacterium, *Desulfotomaculum nigrificans*, formerly known as *Clostridium nigrificans*, is clearly distinguished from species of *Clostridium* on the basis of its DNA-base composition (Saunders, Campbell & Postgate, 1964; Saunders & Campbell, 1966). The genus, *Desulfovibrio* appears to consist of three main groups on the basis of DNA-base composition (Sigal, Senez, Le Gall & Sebald, 1963; Saunders, Campbell & Postgate, 1964; Miller, Hughes, Saunders & Campbell, 1968) and on tolerance to *bis-p*-chlorophenyldiguanidinehexane diacetate (Saleh, 1964).

All dissimilatory sulphate-reducing bacteria are strict anaerobes and, in general, media must be poised at a low $E_h$ (0 to $-100$ mV) for rapid growth to occur (Grossman & Postgate, 1953; Abd-el-Malek & Rizk, 1958, 1960; Postgate, 1959 *a*; Alico & Liegey, 1966). Postgate (1959 *b*, 1965 *b*, 1966) has reviewed procedures for the enrichment, isolation and mass culture of these bacteria and a technique for the continuous culture of *Desulfovibrio* has been reported by Leban, Edwards & Wilke (1966). Methods for the enumeration of sulphate-reducing bacteria have been discussed by Postgate (1963 *a*). Simple three-carbon and four-carbon

276

compounds are generally used by sulphate-reducing bacteria as hydrogen donors; growth is, however, stimulated by complex organic material such as yeast extract (Miller, 1949; Butlin, Adams & Thomas, 1949; Postgate, 1951 *a*). Some strains of *D. desulfuricans* utilize oxamate (Postgate, 1963 *c*) and choline (Hayward & Stadtman, 1959; Senez & Pascal, 1961; Baker, Papiska & Campbell, 1962). Numerous other carbon compounds including carbohydrates, petroleum hydrocarbons, long chain fatty acids and alcohols have been claimed to be utilized by *Desulfovibrio*. The validity of many of these claims has been questioned by Postgate (1959 *a*, 1965 *a*) although it is clear that the utilization of organic compounds may be influenced by factors such as the $E_h$ of the medium (Grossman & Postgate, 1953). Molecular hydrogen is also a hydrogen donor for sulphate reduction by some species because of the presence of the enzyme, hydrogenase, in these organisms (Stephenson & Stickland, 1931; Sisler & ZoBell, 1950, 1951; Senez & Volcani, 1951; Sadana & Jagannathan, 1954; Senez, 1955; King & Winfield, 1955; Littlewood & Postgate, 1956; Coleman, 1960; Akagi & Campbell, 1961). In general, sulphate, or one of a number of other inorganic sulphur compounds (table 10.4), is an essential requirement for the growth of sulphate-reducing bacteria but some strains are able to grow on pyruvate (Postgate, 1952 *a*, 1963 *b*) or fumarate (Miller & Wakerley, 1966) in 'sulphate-free' media.

TABLE 10.4   *Sulphur substrates for* Desulfovibrio desulfuricans
(*Postgate, 1959a*)

| Substrate | Reduction | Comments |
|---|---|---|
| $SO_4{}^{2-}$ | + | . |
| $SO_3{}^{2-}$ | + | . |
| $S_2O_3{}^{2-}$ | + | . |
| $S_4O_6{}^{2-}$ | + | not by *D. orientis* |
| $S_2O_4{}^{2-}$, $S_2O_5{}^{2-}$ | + | decompose into one or more of above ions at physiological pH values |
| $S_3O_6{}^{2-}$, $S_5O_6{}^{2-}$ | − | . |
| colloidal S | + | partly non-enzymic |
| commercial S | − | impurities may be reduced |
| $SO_3NH_2{}^-$, $CH_3CH_2SO_3{}^-$ | − | . |
| $C_6H_5SO_3{}^-$, $CH_3SO_3{}^-$ | − | . |
| $CH_3C_6H_4SO_2{}^-$, $C_6H_5SO_2{}^-$ | − | . |

Carbon dioxide is incorporated during the growth of *D. desulfuricans* (Sorokin, 1954 *a, b*, 1960; Mechalas & Rittenberg, 1960; Postgate, 1960 *a*) by mechanisms which appear to be similar to those present in heterotrophic organisms. Earlier reports that *D. desulfuricans* is capable of autotrophic growth on $H_2$, $CO_2$ and sulphate (Wight & Starkey, 1945; Butlin & Adams, 1947; Butlin, Adams & Thomas, 1949; Sisler & ZoBell, 1951; Senez & Volcani, 1951; Adams, Butlin, Hollands & Postgate, 1951; Senez, 1954; Sorokin, 1954 *a*) appear to be in error and growth in these cases was probably due to organic impurities in the reagents used (Postgate, 1965 *a*).

### 10.2.1  Pigments of sulphate-reducing bacteria

*D. desulfuricans* contains two characteristic pigments, cytochrome-$c_3$ and a porphyroprotein, desulphoviridin (Postgate, 1954, 1956 *a*, 1961; Ishimoto, Koyama & Nagai, 1954, 1955; Ishimoto, Koyama, Yagi & Shiraki, 1957). These pigments have also been detected in *Desulfovibrio gigas* (Le Gall, 1963) and *Desulfovibrio africanus* (Campbell, Kasprzycki & Postgate, 1966). Some properties of cytochrome-$c_3$ are shown in table 10.5; the pigment from *D. gigas* has similar properties except that its isoelectric point is at pH 5·2 (Le Gall, Mazza & Dragoni, 1963). Cytochrome-$c_3$ is an integral part of the sulphate-reducing system in *Desulfovibrio* and is unusual in having an extremely low redox potential. The protein moiety has a high histidine and cysteine content and low amounts of aromatic amino acids (Takahashi, Titani & Minakami, 1959; Coval, Horio & Kamen, 1961; table 10.6). Cytochrome-$c_3$ synthesis by *D. desulfuricans* is regulated by the iron content of the medium (Postgate, 1956 *b*, *c*).

Desulphoviridin dissociates in alkaline solution to yield a photo-sensitive chromophoric group with a strong red fluorescence in ultra-violet light at 365 m$\mu$; this provides a sensitive diagnostic test for organisms containing this pigment (Postgate, 1959 *c*). No function has yet been assigned to desulphoviridin; its absence in one strain of *D. desulfuricans* (Miller & Saleh, 1964) suggests that it plays no part in sulphate reduction.

Cytochrome-$c_3$ and desulphoviridin are not present in *Desulfotomaculum* (Postgate, 1956 *b*; Adams & Postgate, 1959). Instead, these organisms contain a protohaem pigment, possibly cytochrome-*b* (Campbell, Frank & Hall, 1957; Postgate & Campbell, 1963).

The presence of ferredoxin in *D. desulfuricans* was reported by Tagawa & Arnon (1962). Ferredoxin from *D. nigrificans* was partly purified by Akagi (1965) and the pigment from *D. gigas* was crystallized by Le Gall & Dragoni (1966). *D. gigas* ferredoxin shows absorption peaks in the oxidized state at 305 and 390 m$\mu$, contains 0·9 $\mu$mole of non-haem iron

TABLE 10.5    *The properties of cytochrome-$c_3$ from*
Desulfovibrio desulfuricans

|  | $\alpha$ | $\beta$ | $\gamma$ | Reference† |
|---|---|---|---|---|
| absorption maxima (m$\mu$) of reduced | 553 | 523 | 419 | 1 |
| form at pH 7 | 553 | 525 | 419 | 2 |
|  | 552 | 522 | 418 | 3* |
|  | 552 | 525 | 419 | 5 |
| $E_{553}^{1\%}$ | . | 42 | . | 2 |
| molecular weight (ultracentrifuge) | . | 13,000 | . | 2 |
| „        „    (amino acid analysis) | . | 13,605 | . | 4 |
| „        „    (ultracentrifuge) | . | 11,300 | . | 3 |
| mesohaem per mole | . | 2 | . | 2,3 |
| isoelectric point | . | 10·30– 10·66 | . | 2 |
| $E_0'$, pH 7, 30° | . | −205 mV | . | 2 |

\* Crystalline cytochrome-$c_3$.
† *References:* 1, Ishimoto, Koyama & Nagai (1954). 2, Postgate (1956 *a*). 3, Horio & Kamen (1961). 4, Coval, Horio & Kamen (1961). 5, Ishimoto, Koyama, Yagi & Shiraki (1957).

TABLE 10.6    *Amino acid composition of cytochrome-$c_3$*

| Amino acid | Residues per mole | |
|---|---|---|
|  | (1) | (2) |
| aspartic acid | 12·1 | 13 |
| glutamic acid | 6·3 | 6 |
| threonine | 10·8 | 5 |
| serine | 7·3 | 5 |
| proline | 6·5 | 4 |
| glycine | 15·6 | 10 |
| alanine | 13·4 | 11 |
| valine | 8·1 | 9 |
| methionine | 0 | 3 |
| isoleucine | 0·9 | 0 |
| leucine | 4·3 | 3 |
| tyrosine | 0 | 3 |
| phenylalanine | 3·0 | 2 |
| lysine | 23·4 | 22 |
| histidine | 6·1 | 9 |
| arginine | 1·8 | 1 |
| cystine | 0 | . |
| ½-cysteine | . | 8 |
| tryptophan | . | . |
| ammonia | 8·3 | 11 |

(1)  Takahashi, Titani & Minakami (1959) (mol.wt 12,400).
(2)  Coval, Horio & Kamen (1961).

and 0·8 $\mu$mole of labile sulphide per milligram of protein and has an $E'_0$ value in the order of $-310$ mV. It thus appears to resemble ferre-doxins from other bacterial sources (cf. Valentine, 1964). *D. gigas* con-tains a flavoprotein (Le Gall & Hatchikian, 1967) which resembles in some of its properties the flavodoxin of *Clostridium pasteurianum* (Knight & Hardy, 1966). The purified flavoprotein has a molecular weight of approximately 14,000 and shows an unusual response to ultra-violet irradiation. On such irradition under anaerobic conditions in the presence of ascorbate the colour changes from yellow (absorption maxima 375, 485 m$\mu$; shoulder at 480 m$\mu$) to purple (absorption maxima 350, 585, 620 m$\mu$; shoulder at 380 m$\mu$). The original spectrum is regained when the irradiated flavoprotein is left in darkness under aerobic conditions. It is not yet known whether this property is significant from the point of view of the physiological action of the flavoprotein.

In addition to cytochrome-$c_3$, ferredoxin and flavodoxin, *D. gigas* contains a pink-coloured protein (Le Gall & Dragoni, 1966) with a spectrum similar to that of rubredoxin of *Cl. pasteurianum* (Lovenberg & Sobel, 1965). The rubredoxin of *D. gigas* mediates electron transfer between NADH and mammalian cytochrome-$c$ (and, less effectively, the cytochrome-$c_3$ of *D. gigas*) in the presence of a purified rubredoxin oxido-reductase from *D. gigas* (Le Gall, 1968). Rubredoxin has also been detected in *Desulfovibrio vulgaris* (Haschke & Campbell, 1967) and in a nitrogen-fixing strain of *Desulfovibrio* (Postgate, 1967).

## 10.2.2 Reduction of sulphate

From the comparative rates of reduction of sulphate and sulphite by hydrogen by *D. desulfuricans* (Postgate, 1951 *b*) and from the fact that sulphite is a non-competitive antagonist of the selenate inhibition of sulphate reduction (Postgate, 1949, 1952 *b*), Postgate deduced that sulphite is an intermediate in the reduction of sulphate to sulphide by this organism. This view was supported by the finding of Koyama, Tamiya, Ishimoto & Nagai (1954) that sulphite inhibits the reduction of [35]S-sulphate. Later Millet (1955, 1956) showed directly that [35]S-sulphite is formed from [35]S-sulphate during sulphate reduction by *D. desulfuricans*.

Cell-free systems of *D. desulfuricans*, catalysing the reduction of sulphate to sulphide by hydrogen, were first obtained by Peck (1959) and Ishimoto (1959). The systems required ATP and either methyl viologen or cytochrome-$c_3$ as an electron mediator. At pH 7·8 sulphite was the main product of sulphate reduction by Peck's preparation. Peck further showed that approximately one mole of ATP is utilized for each mole of $H_2$ consumed and that sulphate reduction is completely inhibited by molybdate, tungstate and chromate. Chromate (Postgate, 1952 *b*) and tungstate (Ishimoto, Koyama, Omura & Nagai, 1954) had previously been shown to inhibit sulphate reduction by whole bacteria and Wilson & Bandurski (1958 *b*) had shown that molybdate, tungstate and chromate react with the ATP-sulphurylase of yeast (see chapter 5): selenate, which also inhibits sulphate reduction by *Desulfovibrio* (Postgate, 1949, 1952 *b*), also inhibits ATP-sulphurylase. Peck (1959) suggested, therefore, that a similar enzyme in *Desulfovibrio* may catalyse the formation of APS as the primary step in sulphate reduction (equation 10.19).

$$ATP + SO_4^{2-} \rightleftharpoons APS + PP_i \qquad (10.19)$$

An [35]S-labelled nucleotide having the same electrophoretic mobility as APS was detected when crude extracts of *Desulfovibrio* were incubated with ATP and [35]S-sulphate: PAPS, the characteristic intermediate in assimilatory sulphate reduction (§ 10.1.2) was not detected (Peck, 1959).

Peck's hypothesis has received substantial support from succeeding investigations.

### *ATP-sulphurylase*

The formation of APS during sulphate reduction by extracts of *Desulfovibrio* was confirmed by Ishimoto & Fujimoto (1959), by Baker, Papiska & Campbell (1962) and by Peck (1962 *a, b*) who also demonstrated that APS, but not PAPS, is reduced by these preparations (see also Peck, 1966; Campbell, Kasprzycki & Postgate, 1966). ATP-sulphurylases have been isolated and partly purified from *D. desulfuricans* (Baliga, Vartak &

TABLE 10.7  *Properties of ATP-sulphurylases from* D. desulfuricans *and* D. nigrificans *(After Akagi & Campbell, 1962 b)*

|  | *D. desulfuricans* | *D. nigrificans* |
|---|---|---|
| nucleotide specificity | ATP active; ITP, UTP, deATP, CTP, GTP inactive | . |
| magnesium requirement | $> 3 \cdot 3$ mM | $1 \cdot 7$ mM |
| pH optimum | $8 \cdot 0$–$9 \cdot 5$ | $7 \cdot 0$–$7 \cdot 5$ |
| equilibrium constant | $6 \cdot 2 \times 10^{-9}$ (pH $8 \cdot 0$) | $1 \cdot 8 \times 10^{-8}$ (pH $7 \cdot 2$) |
| thermostability at 60 °C | complete inactivation in 3 min (no ATP present) 40% inactivation in 15 min (in presence of ATP) | no loss in 2 hr (no ATP) . |

Jagannathan, 1961; Akagi & Campbell, 1962 *a, b*) and from *D. nigrificans* (Akagi & Campbell, 1962 *a, b*). Both enzymes require magnesium ions, are specific for ATP and catalyse the formation of unstable nucleotide anhydrides with molybdate, chromate and tungstate (cf. yeast ATP-sulphurylase; chapter 5). The enzyme from the thermophilic organism *D. nigrificans* is considerably more thermostable than that from *D. desulfuricans*. Some properties of the ATP-sulphurylases from the sulphate reducing bacteria are shown in table 10.7.

## Inorganic pyrophosphatase

The equilibrium constant of the ATP-sulphurylase catalysed reaction is in the order of $10^{-8}$ in favour of ATP and $SO_4^{2-}$ ions (Robbins & Lipmann, 1958 *b*; Wilson & Bandurski, 1958 *b*; see table 10.7). The synthesis of APS is aided by the action of inorganic pyrophosphatase which 'pulls' the reaction in the direction of synthesis by removing pyrophosphate (Peck, 1959, 1962 *a*).

In *D. desulfuricans* both soluble and insoluble pyrophosphatases have been detected: these differ in some of their physical properties and in their requirements for metal ions. According to Baliga, Vartak & Jagannathan (1961) the soluble enzyme requires magnesium ions whereas the insoluble enzyme specifically requires cobalt. Akagi & Campbell (1963) purified the soluble pyrophosphatase from *D. desulfuricans* and found that either magnesium or manganese ions, or to a lesser extent cobalt ions, were activators of the enzyme.

## APS-reductase

An enzyme, APS-reductase, which catalyses the reduction of APS to sulphite by hydrogen in the presence of hydrogenase and methyl viologen (equation 10.20) has been extracted from *D. desulfuricans* and partly

$$APS + H_2 \rightarrow AMP + SO_3^{2-} + 2H^+ \qquad (10.20)$$

purified (Ishimoto & Fujimoto, 1959, 1961; Egami, Ishimoto & Taniguchi, 1961; Peck, Deacon & Davidson, 1965). According to Egami, Ishimoto & Taniguchi (1961) cytochrome-$c_3$ can replace methyl viologen as an electron acceptor in the reaction. Peck (1962 *b*), however, was unable to demonstrate a stimulation of APS-reductase by cytochrome-$c_3$ (see also Peck, Deacon & Davidson, 1965) although a role for this pigment in the reduction of sulphate by crude extracts of *Desulfovibrio* had been reported earlier (Peck, 1959). However, Guarraia, Laishley,

Forget & Peck (1968) reported briefly that a single pathway of electron transfer functions between hydrogen and sulphite reductase, thiosulphate reductase and APS-reductase. A functional role for cytochrome-$c_3$ in the reactions catalysed by the first two enzymes is well documented (see below).

The synthesis of APS from AMP and sulphite (equation 10.21),

$$AMP + SO_3^{2-} \rightarrow APS + 2e \qquad (10.21)$$

catalysed by APS-reductase, was demonstrated by Ishimoto & Fujimoto (1961) and by Peck (1961 *b*). Using ferricyanide as an electron acceptor, Peck, Deacon & Davidson (1965) developed an assay for APS-reductase based on this reaction. A number of other electron acceptors such as methylene blue, 2,6-dichlorophenolindophenol, benzyl viologen, phenazine methosulphate, FMN, FAD, NADP$^+$, NAD$^+$, mammalian cytochrome-$c$, cytochrome-$c_3$, ferredoxin and $O_2$ were unable to support the formation of APS by reaction 10.21 (Peck *et al.* 1965).

APS-reductase also catalyses the reduction of guanylyl sulphate, uridylyl sulphate, and cytidylyl sulphate in the presence of hydrogenase and hydrogen (Ishimoto & Fujimoto, 1961); AMP, but not guanylic acid, cytidylic acid nor uridylic acid, inhibits the reduction of all the nucleoside sulphatophosphates. The synthesis of nucleoside sulphatophosphates from sulphite occurs only with AMP, deoxy-AMP, inosinic acid and guanylic acid: uridylic acid, cytidylic acid, thymidylic acid, deoxyguanylic acid, deoxycytidylic acid, adenosine 5'-sulphate, adenosine 2',5'-diphosphate, adenosine 3',5'-diphosphate, ADP and ATP are all inactive (Peck *et al.* 1965).

APS-reductase is activated by 2-mercaptoethanol and inhibited by a number of thiol-binding reagents (Ishimoto & Fujimoto, 1961; Peck *et al.* 1965) which suggests that thiol groups are essential for its activity. The fact that arsenite does not inhibit, however, indicates that vicinal thiol groups are not involved (cf. PAPS-reductase; p. 257).

Purified APS-reductase contains FAD and iron in the ratio of 1:4–5 (Peck, Deacon & Davidson, 1965). The enzyme-bound FAD is reduced by sulphite in the absence of AMP (Peck & Davidson, 1967). Peck and Davidson also reported that [35]S-labelled protein is formed on the incubation of oxidized APS-reductase with [35]$SO_3^{2-}$, or of the reduced enzyme with [35S]APS. The chemical reactivity of the enzyme-bound sulphur suggests that it is covalently bound to two specific sites on the enzyme and it is therefore of interest that Aubert (1956) had obtained evidence that in intact *D. desulfuricans* organically-bound sulphite is formed by the reduction of sulphate.

APS-reductase is present in all the dissimilatory sulphate-reducing bacteria which have so far been studied and it appears to be absent from assimilatory sulphate-reducing organisms (Peck, 1961 *a*, 1962 *a*). The enzyme in dissimilatory sulphate-reducing organisms is very similar to that involved in the terminal stages of sulphur oxidation by thiobacilli and by photosynthetic sulphur bacteria (see chapter 9).

### ADP-sulphurylase

Extracts of *D. desulfuricans* contain the enzyme ADP-sulphurylase which catalyses the phosphorolysis of APS to ADP and sulphate ions as in equation 10.22 (Peck, 1962 *b*).

$$APS + P_i \rightarrow ADP + SO_4^{2-} \tag{10.22}$$

Its function is unknown. It is apparently not affected by molybdate, which completely inhibits sulphate activation, so suggesting that the enzyme plays no essential role in this process.

### 10.2.3 Reduction of sulphite

Millet (1956) prepared cell-free extracts of *D. desulfuricans* which reduced sulphite to sulphide in the presence of hydrogen: three moles of hydrogen were consumed per mole of sulphide produced (equation 10.23).

$$SO_3^{2-} + 3H_2 \rightarrow S^{2-} + 3H_2O \tag{10.23}$$

The enzyme responsible for this reaction, sulphite reductase, is present in both the soluble and the particulate fractions obtained after high-speed centrifugation of extracts of *Desulfovibrio* (Postgate, 1959 *a*, 1961). Haschke & Campbell (1967, 1968) have briefly reported the purification of sulphite reductase from *D. vulgaris*.

The nature of the intermediates between sulphite and sulphide is uncertain. Iverson (1966) noted that under certain conditions, the growth of *D. desulfuricans* on simple agar media (for example, lactate plus inorganic salts) is stimulated when cultures of the same organism on trypticase soy agar plus sulphate are present in the same growth chamber. This stimulation was attributed to the formation, by the bacteria growing on the rich medium, of a gaseous substance which was later (Iverson, 1967) tentatively identified as disulphur monoxide (S₂O). The formation of small amounts of elemental sulphur by *D. desulfuricans* was also reported. Iverson suggested that disulphur monoxide may be an intermediate in sulphite reduction but the possibility that the gas arose by the degradation of, for example, an enzyme-bound intermediate containing sulphur with an oxidation state intermediate between sulphite and sulphide must obviously be considered. Suh, Nakatsukasa & Akagi (1968) reported that thiosulphate and dithionite are formed during sulphite reduction by crude extracts of *D. vulgaris* and *D. nigrificans*. The enzyme responsible for thiosulphate formation was, however, distinct from sulphite reductase and could be separated from the latter enzyme by ammonium sulphate fractionation.

Postgate (1956 *a*) reported that cytochrome-$c_3$ stimulates sulphite reduction by detergent-treated cells or by extracts of *D. desulfuricans*. The participation of cytochrome-$c_3$ in the transfer of electrons from hydrogen to sulphite was confirmed by Ishimoto and his colleagues (Ishimoto, Kondo, Kameyama, Yagi & Shiraki, 1958; Ishimoto & Yagi, 1961) who showed that crude extracts of *D. desulfuricans* lose the ability to reduce sulphite with hydrogen when they are treated with Amberlite IRC-50 to remove cytochrome-$c_3$: subsequent addition of the cytochrome to the resin-treated extracts restores the activity. With partly purified preparations of sulphite reductase, however, cytochrome-$c_3$ failed to couple sulphite reduction to hydrogenase (Ishimoto & Yagi, 1961) and an artificial electron carrier such as methyl viologen was necessary. Reconstitution experiments with fractions of the bacterial extracts indicated that two or more components other than cytochrome-$c_3$ and hydrogenase were required for the reduction of sulphite by hydrogen. Postgate (1961) also found evidence for the requirement of two cofactors in addition to cytochrome.

Akagi (1965) briefly reported that ferredoxin stimulates sulphite reduction coupled to the pyruvate phosphoroclastic reaction in 'ferredoxin-free' extracts of *D. nigrificans* but since ferredoxin was required for the phosphoroclastic reaction itself (see Suh & Akagi, 1966; Akagi, 1967) it is not clear whether ferredoxin is an essential part of the sulphite reductase reaction. Further evidence for a functional role for this pigment in sulphite reduction was provided by Le Gall & Dragoni (1966) who showed that in extracts of *D. gigas* the rate of sulphite reduction by hydrogen is a function of the ferredoxin concentration. Recently Le Gall & Hatchikian (1967) and Guarraia, Laishley, Forget & Peck (1968) have shown that flavodoxin from *D. gigas* can replace ferredoxin in the sulphite-hydrogen reaction: rubredoxin, however, is not active (Le Gall & Dragoni, 1966). The precise functions of cytochrome-$c_3$, ferredoxin and flavodoxin

in the electron transport chain associated with sulphite reduction therefore remains to be elucidated.

Electron transport phosphorylation coupled to sulphite reduction by
hydrogen in *D. gigas* has been demonstrated by Peck (1966): $P/H_2$ ratios
of 0·18 were obtained. The phosphorylation was inhibited by uncoupling
agents such as gramicidin, pentachlorophenol and dinitrophenol. Guarraia
& Peck (1967) have reported the presence of DNP-stimulated ATPases in
the soluble and particulate components of *D. gigas* extracts. Both enzymes
were stimulated by magnesium ions, and the particulate enzyme exhibited
two pH maxima (at pH 6·8 and 8·0) and was stimulated by calcium ions
at pH 6·5. The existence of DNP-stimulated ATPase activity is consistent
with the presence of a system for electron-transport phosphorylation in
the extracts but the significance of the apparent multiplicity of enzymes
remains to be elucidated.

### 10.2.4   Reduction of thiosulphate and tetrathionate

Thiosulphate and tetrathionate support the growth of *D. desulfuricans*
(table 10.4). ZoBell & Rittenberg (1948) suggested that these compounds
might be intermediates in the reduction of sulphite to sulphide but
kinetic evidence (Postgate, 1951 *b*) does not support this contention:
moreover, Ishimoto & Yagi (1961) have achieved a partial separation of
the enzyme systems catalysing the formation of $H_2S$ from sulphite and
thiosulphate. Thiosulphate and tetrathionate are reduced quantitatively
under hydrogen by cell suspensions of *D. desulfuricans* (equations 10.24 and
10.25) (Postgate, 1951 *b*; see also Ishimoto, Koyama, Omura & Nagai,
1954).

$$S_2O_3{}^{2-} + 4H_2 \rightarrow 2HS^- + 3H_2O \qquad (10.24)$$

$$S_4O_6{}^{2-} + 9H_2 \rightarrow S^{2-} + 3H_2S + 6H_2O \qquad (10.25)$$

The reduction of thiosulphate under hydrogen by detergent-treated
bacteria (Postgate, 1956 *a*) and by extracts of *Desulfovibrio* (Ishimoto,
Koyama & Nagai, 1955; Postgate, 1956 *a*; Ishimoto, Koyama, Yagi &
Shiraki, 1957; Ishimoto, Kondo, Kameyama, Yagi & Shiraki, 1958) is
stimulated by cytochrome-$c_3$: methyl viologen can replace cytochrome
as an electron mediator but $NAD^+$, FAD, FMN, riboflavin and mammalian cytochrome-*c* are inactive.

During thiosulphate reduction small amounts of sulphite are formed
which indicates that the initial reaction is a reduction of thiosulphate to
sulphide and sulphite (Ishimoto, Kondo, Kameyama, Yagi & Shiraki,
1958). A thiosulphate reductase which could catalyse this reaction is

associated mainly with a particulate fraction from *D. desulfuricans* and it has been prepared virtually free of hydrogenase by Ishimoto & Koyama (1957).

### 10.2.5   Miscellaneous reductions catalysed by *Desulfovibrio*

Butlin, Adams & Thomas (1949) reported that elemental sulphur supports the growth of *Desulfovibrio*. Postgate (1951 *a*), using purified preparations of sulphur, was unable to confirm this result although cell suspensions of *Desulfovibrio* reduce colloidal sulphur under hydrogen (Postgate, 1951 *b*; Ishimoto, Koyama, Omura & Nagai, 1954). Extracts of *Desulfovibrio* also reduce elemental sulphur to sulphide under hydrogen in the presence of cytochrome-$c_3$ and hydrogenase but the reaction is of doubtful physiological significance because reduced cytochrome-$c_3$ is oxidized chemically by sulphur (Ishimoto, Yagi & Shiraki, 1957; Ishimoto, Kondo, Kameyama, Yagi & Shiraki, 1958). Oxygen (Postgate, 1956 *a*), nitrite (Senez, Pichinoty & Konavaltchikoff-Mazoyer, 1956) and hydroxylamine (Ishimoto, Yagi & Shiraki, 1957; Senez & Pichinoty, 1958 *a, b*; Pichinoty & Senez, 1958) are also reduced by *Desulfovibrio* through simple chemical interaction between these compounds and reduced cytochrome-$c_3$.

### 10.2.6   The path of sulphur in dissimilatory sulphate reduction

The pathway of sulphur in dissimilatory sulphate reduction and its associated electron transport by *Desulfovibrio* is summarized in the following scheme (fig. 10.2).

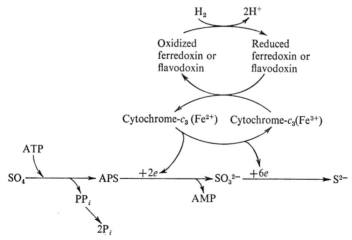

Fig. 10.2   Pathway of dissimilatory sulphate reduction in *Desulfovibrio*.

287

### 10.2.7    The uptake of sulphate by *Desulfovibrio*

Littlewood & Postgate (1957) demonstrated that *D. desulfuricans* is not freely permeable to sulphate and they proposed that either sulphate reduction occurred outside the osmotic barrier (cell membrane?) or that the uptake of sulphate into the cell is so regulated that the internal concentration of this ion remains below the level detectable by their analytical procedure (titration with $Ba^{2+}$ ions). Support for the latter proposal was obtained by Furusaka (1961) who, by using low concentrations of $^{35}SO_4^{2-}$, showed that *D. desulfuricans* is able to accumulate sulphate to concentrations beyond that of the external medium. The accumulation was apparently dependent upon sulphate reduction and was abolished by inhibitors of this process such as selenate or oxygen. The rate of sulphate transport at 37° in 0·5 mM $SO_4^{2-}$ was calculated to be in the order of 4 n-moles $SO_4^{2-}$ per milligram dry weight per minute but the rate of accumulation of sulphate was about one tenth of this value. These results indicated that most of the sulphate transported into the cell is rapidly reduced and could account for Littlewood & Postgate's failure to detect internal sulphate by their less sensitive methods.

# 11

---

## THE METABOLISM IN ANIMALS OF
## INORGANIC SULPHUR COMPOUNDS

It has long been obvious that in the higher animals the anions of the different oxyacids of sulphur are readily interconvertible and that in particular $SO_4^{2-}$ ions and $S_2O_3^{2-}$ ions are quite general constituents of mammalian urines. As these anions are not normally present to any great extent in an animal's diet they must be produced *in vivo* from other sulphur compounds, the only quantitatively important of which are the sulphur-containing amino-acids. Further, inorganic compounds of sulphur can be incorporated into organic substances. The interconversions of these various sulphur-containing compounds in animals are by no means clearly understood and many gaps are only too apparent in our knowledge. The situation is not helped by the great reactivity of many sulphur compounds and the extreme ease with which many of the reactions undergone by them occur even in the absence of any catalyst.

### 11.1   The oxidation of cysteine to sulphate ions
### 11.1.1   The formation of alanine 3-sulphinic acid
Although the reactions leading to the production of alanine 3-sulphinic acid are often presumed to be some of the most important in the metabolism of cysteine in animal tissues detailed knowledge of this process is scanty indeed: theories, however, abound. It is not even clear whether cysteine or cystine is the starting point for the oxidative processes: the key role of cysteine seems to be generally accepted but this view probably stems only from the classical observations of Pirie (1934 *a*) who showed that slices of rat liver and kidney would form $SO_4^{2-}$ ions only so long as the reaction mixtures gave a nitroprusside reaction, presumably indicating the presence of cysteine. Only recently (Wainer, 1965; Sörbo & Ewetz, 1965) has more direct evidence for the participation of cysteine been obtained.

It is usually stated that the first stable intermediate in the oxidation of cysteine is alanine 3-sulphinic acid (cysteine sulphinic acid). This view again stems from the work of Pirie (1934 *a*) but direct evidence for the statement is scanty. Certainly Medes (1937) showed that alanine 3-sulphinic acid could give rise to $SO_4^{2-}$ ions *in vivo* but its occurrence in animal tissues was shown only in 1954 by Bergeret & Chatagner who isolated it from brain. Alanine 3-sulphinic acid has since been shown to be produced from cysteine both *in vivo* (Chapeville & Fromageot, 1955) and *in vitro* (Awapara & Doctor, 1955) but it has not been detected in any enzyme system actually producing $SO_4^{2-}$ ions from cysteine.

If it be assumed for the moment that alanine 3-sulphinic acid is in fact the key intermediate in the oxidation of cysteine to $SO_4^{2-}$ ions then the mode of formation of this compound is of obvious interest. Unfortunately, facts are few. The most commonly held view is that cysteine is oxidized to the unstable alanine 3-sulphenic acid which undergoes a dismutation to alanine 3-sulphinic acid and cysteine as in reaction 11.1.

$$
\begin{array}{cccc}
\mathrm{CH_2SH} & \mathrm{CH_2.SOH} & \mathrm{CH_2.SO_2H} & \mathrm{CH_2.SH} \\
| & | & | & | \\
2\mathrm{H.C.NH_2} \rightarrow 2\mathrm{H.C.NH_2} \rightarrow \mathrm{H.C.NH_2} & +\mathrm{H.C.NH_2} \\
| & | & | & | \\
\mathrm{COOH} & \mathrm{COOH} & \mathrm{COOH} & \mathrm{COOH}
\end{array} \quad (11.1)
$$

Another scheme which has been often postulated is the hydrolytic cleavage of cystine to alanine 3-sulphenic acid and cysteine shown in reaction 11.2

$$
\begin{array}{cccc}
\mathrm{CH_2.S-S.CH_2} & & \mathrm{CH_2.SOH} & \mathrm{CH_2.SH} \\
| & | & | & | \\
\mathrm{H.C.NH_2} & \mathrm{H.C.N.H_2} \rightarrow \mathrm{H.C.NH_2} & +\mathrm{H.C.NH_2} \\
| & | & | & | \\
\mathrm{COOH} & \mathrm{COOH} & \mathrm{COOH} & \mathrm{COOH}
\end{array} \quad (11.2)
$$

which is again followed by the dismutation of the former compound to alanine 3-sulphinic acid and cysteine. It is perhaps of interest that Cecil (1950) has shown that this hydrolysis may be catalysed by acetate ions. A rather less favoured route for the oxidation of cyst(e)ine involves the formation of the so-called cysteine disulphoxide which is in reality a thiosulphinate ester (*11.1*) (Sweetman, 1959).

$$
\begin{array}{cc}
\mathrm{CH_2.SO_2.S.CH_2} & \\
| & | \\
\mathrm{H.C.NH_2} & \mathrm{H.C.NH_2} \\
| & | \\
\mathrm{COOH} & \mathrm{COOH}
\end{array} \quad (\mathit{11.1})
$$

This readily reacts with cysteine to give cystine and alanine 3-sulphinic acid (Toennies & Lavine, 1936; Sweetman, 1959) and such reactions have recently been suggested to be of importance in soil biochemistry (Freney, 1958, 1960).

Actual experimental evidence for the pathway between cysteine and alanine 3-sulphinic acid is lacking and only recently has a start been made to the isolation of the enzymes involved. Wainer (1965) and Sörbo & Ewetz (1965) have shown that an enzyme, or enzyme system, in the soluble fraction of a rat liver homogenate can oxidize cysteine to alanine 3-sulphinic acid. Both groups showed that the sulphinic acid was rapidly further metabolized and only accumulated if the enzyme system were partially purified (Wainer, 1965) or inhibitors such as hydroxylamine were present (Sörbo & Ewetz, 1965). The latter workers (Ewetz & Sörbo, 1966) have shown that the formation of alanine 3-sulphinic acid in such preparations requires the presence of oxygen and is greatly stimulated by liver microsomes, $Fe^{2+}$ ions and NADPH. The detailed mechanism is not clear but the oxidation is not due to a 'mixed function' oxidase and it cannot be excluded that the NADPH simply prevents inactivation of the enzyme. With regard to the properties of the enzyme, it is inhibited by mercuribenzoate and iodoacetate, but not by arsenite, and it is apparently quite specific because neither D-cysteine, L-cystine, glutathione nor cysteamine were oxidized. This specificity is perhaps slightly surprising because the rat can *in vivo* produce $SO_4^{2-}$ ions equally well from D- and from L-cysteine (Cavallini, De Marco & Mondovi, 1958). The situation at present, then, remains that enzymes present in rat liver—but not rat heart, kidney, spleen nor brain—can oxidize L-cysteine to alanine 3-sulphinic acid by some unknown pathway. The absence of the system from kidney is surprising in view of the early work of Pirie (1934 a).

It is perhaps of interest to compare this oxidation to that of cysteamine to hypotaurine which is catalysed by an enzyme present in horse kidney. The enzyme has been purified some 300-fold (Cavallini, Scandurra & De Marco, 1963) and it catalyses the aerobic oxidation of cysteamine in a reaction which shows an absolute requirement for $S^0$ or $S^{2-}$ ions as cofactors. The $S^{2-}$ ions are purely catalytic in their function and at high concentrations (>7 mM) they are in fact inhibitory. Unfortunately it is not possible to decide which particular molecular species of sulphur is active in this reaction because under the experimental conditions there must be a rapid interconversion of $S^0$, $S^{2-}$ and $S_n^{2-}$. It had previously been suggested (De Marco, Mondovi, Scandurra & Cavallini, 1962) that the persulphide of cysteamine, thiocysteamine, $(H_2N.C_2H_4.S.SH)$ was the true substrate of the enzyme but the later work showed that this was not so and if the persulphide were present, it could only be as a transient intermediate. Thiotaurine $(H_2N.C_2H_4.SO_2SH)$ was another product of

291

the oxidation but this appeared to be formed secondarily by a reaction between hypotaurine and a polysulphide (De Marco, Coletta & Cavallini, 1961). The oxidation of cysteamine catalysed by this enzyme is not inhibited by $CN^-$ (1 mM) which might suggest that the reaction does not involve a heavy metal but the very strong chelating properties of cysteamine perhaps make this conclusion dangerous. Certainly the properties of this kidney enzyme seem to be quite different from those of the liver enzyme which oxidizes cysteine; in particular, the oxidation of cysteamine is not stimulated by NADPH. This effect, however, may reflect only the different stabilities of the two systems.

Finally, one further oxidative reaction undergone by cysteine must be mentioned because it places in considerable doubt the physiological significance of all the above reactions. Wainer (1964) in a preliminary, report, showed that mitochondrial preparations from rat liver could oxidize cysteine to $SO_4^{2-}$ ions by a route which did not involve free alanine 3-sulphinic acid. Addition of alanine 3-sulphinic acid to this mitochondrial system did not alter the rate of production of $SO_4^{2-}$ ions, nor was it itself utilized. Wainer therefore concluded that the role of alanine 3-sulphinic acid in the formation of $SO_4^{2-}$ ions was obscure and that if this acid did occur it must be tightly bound to the enzyme, despite the fact that all previous evidence for its participation in the production of $SO_4^{2-}$ ions was based on studies of the free compound. More recently Wainer (1967) has extended these observations and has shown that under certain conditions there is an unexplained dependence of the reaction on glutathione, either oxidized or reduced, which can be replaced by no other SH compound. The mechanism of the reactions is still quite obscure: no co-factor requirement has been shown and according to Wainer neither free mercaptopyruvate nor free $H_2S$ can be intermediates although the participation of a 'bound' form of the latter cannot be excluded. Pyruvate is probably the immediate product from the carbon skeleton of cysteine. These findings certainly make the role of alanine 3-sulphinate in the formation of $SO_4^{2-}$ ions very obscure and it is to be hoped that further studies of this interesting system will soon be forthcoming. It may well be that the alanine 3-sulphinic acid which undoubtedly can be produced by liver, as Wainer himself has shown (1965), is utilized for the production of taurine rather than $SO_4^{2-}$ ions although Wheldrake & Pasternak (1967) claim, on the contrary, that in cultured neoplastic mast cells alanine 3-sulphinate is the precursor of $SO_4^{2-}$ ions.

## 11.1.2 The metabolism of alanine 3-sulphinic acid

Here again it is clear that alanine 3-sulphinic acid can be actively meta-bolized by animal tissues but the exact pathway is still problematical. There are certainly two major routes, one leading to taurine and isethionic acid, the other leading to $SO_4^{2-}$ ions. The first of those pathways need not be considered in detail here and it can be summarized in the reactions shown below (11.3).

$$
\begin{array}{ccc}
\underset{\displaystyle |}{CH_2.SO_2H} & \underset{\displaystyle |}{CH_2.SO_3H} & \\
H.\overset{\displaystyle |}{\underset{\displaystyle |}{C}}.NH_2 & \to H.\overset{\displaystyle |}{\underset{\displaystyle |}{C}}.NH_2 & \\
COOH & COOH & \\
\downarrow & \downarrow & \\
\underset{\displaystyle |}{CH_2.SO_2H} & \underset{\displaystyle |}{CH_2.SO_3H} & \underset{\displaystyle |}{CH_2.SO_3H} \\
CH_2.NH_2 & \to \quad CH_2.NH_2 & \to CH_2.OH
\end{array} \qquad (11.3)
$$

The route through hypotaurine (i.e. decarboxylation before oxidation) seems to be the dominant one (Awapara, 1953; Awapara & Wingo, 1953) in rat and in dog liver. There are, however, large species differences in the activity of the decarboxylase (or decarboxylases) involved in these re-actions: for instance, cat liver and horse liver are virtually devoid of the enzyme while in both rabbit and in guinea pig liver it is certainly present in very small amounts (Hope, 1955; Sörbo & Heyman, 1957). More recently Jacobsen & Smith (1963) showed that human liver could not decarboxylate either alanine 3-sulphinate or alanine 3-sulphonate and they suggested that in this tissue, in contrast to human brain, taurine may be formed through some quite different pathway. Such an alternate route could be that involving the oxidation of cysteamine (see § 11.1.1) whose occurrence in man was demonstrated by Eldjarn (1954).

With regard to the formation of $SO_4^{2-}$ ions, Fromageot, Chatagner & Bergeret (1948) showed that these were derived from $SO_3^{2-}$ ions coming from alanine 3-sulphinic acid, the other reaction product being alanine. A long series of investigations carried out by Fromageot and his school showed that α-oxoglutarate and pyridoxal phosphate were involved in this desulphinication and it was proposed that the reaction involved a transamination of alanine 3-sulphinic acid to give 3-sulphinopyruvic acid which spontaneously decomposed—by analogy with oxaloacetic acid—to give $SO_2$, and so $SO_3^{2-}$ ions. The other product of the final reaction, pyruvic acid, underwent a transamination to give alanine so that the overall process for the formation of $SO_2$ from alanine 3-sulphinic acid could be represented as follows:

$$\text{Alanine 3-sulphinate} + \alpha\text{-oxoglutarate} \rightleftharpoons \text{sulphinopyruvate} + \text{glutamate}$$

$$\text{Sulphinopyruvate} \rightarrow \text{pyruvate} + SO_2$$

$$\text{Pyruvate} + \text{glutamate} \rightleftharpoons \alpha\text{-oxoglutarate} + \text{alanine}$$

$$\text{Alanine 3-sulphinate} \rightarrow \text{alanine} + SO_2$$

Much of this early work has been summarized by Chatagner, Bergeret, Séjourné & Fromageot (1952).

These reactions were investigated in some detail by Singer & Kearney (1954) who showed that various mitochondrial preparations could catalyse the pyridoxal-dependent transamination between alanine 3-sulphinic acid and either $\alpha$-oxoglutarate or oxaloacetate. They also showed the occurrence in rat liver mitochondria of a relatively slow $NAD^+$-dependent oxidative deamination of alanine 3-sulphinic acid which likewise lead to the formation of sulphinopyruvic acid. This oxidative pathway, unlike the transaminative one, does not occur in micro-organisms. Kearney & Singer (1953) further investigated the breakdown of sulphinopyruvic acid and showed that $Mn^{2+}$ ions very effectively catalysed the liberation of $SO_2$ from this compound in a reaction quite analogous to the similarly catalysed breakdown of oxaloacetate. Certainly there seems no reason to postulate a requirement for an enzyme to catalyse this rather rapid spontaneous decomposition.

Unfortunately the situation has become more complicated in recent years. In particular, Sumizu (1961) has shown that preparations from rat liver can bring about the direct formation of $SO_3^{2-}$ ions and alanine from alanine 3-sulphinic acid by some pyridoxal-dependent process. The evidence that the alanine was formed directly seems conclusive: the addition of [$^{14}$C]pyruvate to the reaction mixture did not lead to the formation of labelled alanine, thus excluding the transamination postulated by Fromageot and his group. The mechanism of this reaction is unknown, but it may be significant that Soda, Novogrodsky & Meister (1964) have shown that the purified aspartate $\beta$-decarboxylase of *Aspergillus faecalis* can bring about the direct desulphinication of alanine 3-sulphinic acid in a reaction stimulated by $\alpha$-oxo acids and by pyridoxal phosphate. It seems not impossible that a similar reaction could occur in rat liver.

One further point should be noted. It is possible that the $SO_2$ liberated from the alanine 3-sulphinic acid—by whatever mechanism—may not appear as $SO_3^{2-}$ ions but as some more complex compound. De Marco & Coletta (1961 $b$) have, for instance, shown that the sulphur from alanine 3-sulphinic acid can, by a reaction with cystine or with cystamine, appear as $S$-sulphocysteine or $S$-sulphocysteamine respectively. These reactions were detected in preparations from either rat liver mitochondria (where transamination was occurring) or pig kidney (where oxidative deamination was occurring).

### 11.1.3  The oxidation of $SO_3^{2-}$ ions

The oxidation of $SO_3^{2-}$ ions to $SO_4^{2-}$ ions can, of course, proceed spontaneously, especially in the presence of metal ions, or of certain haem derivatives such as protohaematin (Fromageot & Chapeville, 1961). Nevertheless, this oxidation is inhibited by proteins, presumably by their ability to bind metal ions, and it seems certain that $in$ $vivo$ the reaction must be an enzymatic one (Heimberg, Fridovich & Handler, 1953). Further evidence for this was given by Hunter & Ford (1954) who, in a brief communication, reported that mitochondria from rat liver could oxidize $SO_3^{2-}$ ions to $SO_4^{2-}$ ions with the simultaneous esterification of phosphate, the P/O ratio being unity.

Fridovich & Handler (1956 $a, b$) showed that extracts from dog liver could oxidize $SO_3^{2-}$ to $SO_4^{2-}$ ions by what was obviously a complex reaction, apparently involving a flavoprotein, lipoic acid and hypoxanthine, the latter functioning as a cofactor. The preparations were devoid of xanthine oxidase activity when tested in the usual way with purine substrates. It was suggested that the first stage of the oxidation involved the reaction of $SO_3^{2-}$ ions with lipoic acid to give the thiosulphate ester of dihydrolipoic acid which in turn yielded $SO_4^{2-}$ ions. Unfortunately there was no direct evidence for such a compound being formed.

More recently MacLeod, Farkas, Fridovich & Handler (1961) have prepared a soluble sulphite oxidase from ox, dog and rat liver. This enzyme is a haem protein, very similar in spectroscopic properties to cytochrome-$b_5$, which occurs in the microsomal fractions of liver, heart and kidney. Oxygen, cytochrome-$c$ and various dyes could act as acceptors for the oxidation and the reaction could be represented by equation 11.4.

$$SO_3^{2-} + H_2O + 2cyt\text{-}c^{3+} \rightarrow SO_4^{2-} + 2H^+ + 2cyt\text{-}c^{2+} \qquad (11.4)$$

The production of hydrogen peroxide when oxygen was used as the acceptor suggested the participation of a flavin in the reaction but no spectroscopic evidence for this could be obtained. The enzyme was saturated at low concentrations of reactants, the $K_m$ values being approximately 10 $\mu$M $SO_3^{2-}$ ions and 0·1 $\mu$M cytochrome-$c$.

Howell & Fridovich (1968) have now obtained from ox liver a sulphite oxidase which is, to judge by spectroscopic evidence, free from any haem component. This preparation has little ability to transfer electrons to oxygen but it will rapidly oxidize sulphite in the presence of cytochrome-$c$. Kinetic studies suggested that the reaction had a ping-pong mechanism with the $K_s$ values for the substrates (in the forward direction) being 22 $\mu$M $SO_3^{2-}$ and 0·021 $\mu$M cytochrome-$c$. The latter value was considerably influenced by the presence of certain anions, including sulphate (a product of the reaction), which competed with cytochrome-$c$. The $K_i$ for sulphate was 3·5 mM. Polymeric forms of cytochrome-$c$ were also powerful inhibitors of the reaction. This enzyme shows obvious affinities with the sulphite oxidase of *Thiobacillus novellus* (see p. 215) and it now seems doubtful whether a sulphite oxidase capable of directly utilizing oxygen as an acceptor in fact exists.

The apparent participation of hypoxanthine in the oxidation of $SO_3^{2-}$ ions lead Fridovich & Handler (1957, 1958, 1961) to an investigation of milk xanthine oxidase and other oxidative enzymes which can, under abnormal conditions, initiate a free radical chain between $SO_3^{2-}$ ions and $O_2$ in the presence of a substrate such as hypoxanthine, so leading to the formation of $SO_4^{2-}$ ions. This process is very sensitive to the presence of any free-radical trapping agent, even to an excess of xanthine oxidase itself, and it requires quite high concentrations of $SO_3^{2-}$ ions (approximately 1 mM) for efficient chain initiation: it is therefore apparent that this reaction can have little significance *in vivo* (MacLeod, Fridovich & Handler, 1961).

From these discussions it should be obvious that although there is no doubt that the oxidation of the sulphur-containing amino-acids leads to the formation of $SO_4^{2-}$ ions there is an unfortunate lack of information about the details and it seems that several interlocking pathways may well be involved.

## 11.2 The oxidation of inorganic sulphur compounds

### 11.2.1 The oxidation of $S^{2-}$ ions

It is well known that $S^{2-}$ ions can be produced in the animal body, although the pathways involved in this overall reaction are far from clear (see §§ 11.3.2 and 11.4.2), and that these ions can certainly be oxidized by animal tissues to give $SO_4^{2-}$ and $S_2O_3^{2-}$ ions (Dziewiatkowski,

1945; Baxter, van Reen & Pearson, 1956). The oxidation of $S^{2-}$ ions to $SO_4^{2-}$ ions involves the transfer of eight electrons and so is likely to involve, at the very least, four separate steps. These steps are by no means well-defined at present. Until the work of Sörbo (1958b) it had been assumed that this oxidation of $S^{2-}$ ions was a typical enzymatic reaction but Sörbo pointed out that the 'enzyme' was remarkably heat stable, being able to withstand boiling for two hours. Sörbo (1958b) also showed that the oxidation could be catalysed non-enzymatically by many iron-containing compounds, including haem derivatives, and therefore suggested that the reaction in tissues was normally non-enzymatic. This conclusion has been essentially confirmed by the work of Baxter and his co-workers (Baxter, van Reen, Pearson & Rosenberg, 1958; Baxter & van Reen, 1958) although they claim both a heat-stable and a heat-labile system to occur in rat liver and kidney. The most active catalyst of the oxidation of $S^{2-}$ ions was ferritin and as the latter is present in high concentrations in the gut, where $S^{2-}$ ions are produced by the intestinal flora, this type of oxidation may well be of considerable physiological importance (Baxter & van Reen, 1958). The mechanism of this oxidation is unknown; Baxter & van Reen (1958) claimed that a 'protein-bound sulphite' ($R.S.SO_3^-$, i.e. a thiosulphate ester) was an intermediate stage in the oxidation but this has been disputed by Sörbo (1960) who implicated polysulphides in the reaction.

The oxidation of $S^{2-}$ by liver extracts is stimulated by hypoxanthine (Ichihara, 1959) so that here, as in the oxidation of $SO_3^{2-}$, a free-radical chain reaction may operate under suitable conditions. Again it is improbable that such a pathway could operate in vivo (see MacLeod, Fridovich & Handler, 1961).

In the higher animals it seems certain that at least from a purely quantitative point of view the oxidation of $S^{2-}$ ions must be a relatively unimportant process.

The very high toxicity of $H_2S$, which is due to its direct action on nerve centres, must be kept in mind: the $LD_{50}$ in cats, for instance, is as low as 25 $\mu g/kg$ body weight (Lovatt Evans, 1967). This toxic action is so rapid that the 'detoxication' of $H_2S$ by its oxidation to $S^0$ with oxyhaemoglobin is of little practical significance.

## 11.2.2 The oxidation of $S_2O_3^{2-}$ ions

Although it has long been known that $S_2O_3^{2-}$ ions can given rise to $SO_4^{2-}$ ions both in vivo (Nyiri, 1923; Zörkendörfer, 1935) and in vitro (Pirie,

1934 *b*) it is now clear that the reaction is a complex one depending upon the prior conversion of $S_2O_3^{2-}$ ions to $S^{2-}$ and $SO_3^{2-}$ ions, as is described in § 11.3.2. The direct oxidation of $S_2O_3^{2-}$ ions almost certainly cannot occur.

### 11.2.3 The oxidation of SCN⁻ ions

This reaction, the oxidation of SCN⁻ ions to $SO_4^{2-}$ and CN⁻ ions, can be catalysed both by myeloperoxidase (Sörbo & Ljunggren, 1958) and by a microsomal peroxidase present in the thyroid gland, but not the liver and kidney, of sheep, pig and ox (Maloof & Soodak, 1964). The enzyme from the thyroid gland has been solubilized and it seems to require ascorbic acid as a component of the reaction mixture: this may, however, simply be as a source of hydrogen peroxide which is utilized in the oxidation.

### 11.3 The reduction of inorganic sulphur compounds

### 11.3.1 The reduction of $SO_4^{2-}$ ions to $SO_3^{2-}$ ions

As far as is known the direct reduction of sulphate to sulphite does not occur in animal tissues. Indeed this reaction does not appear to be brought about by any living organism. As discussed in chapter 10, reduction of sulphate takes place indirectly in micro-organisms and plants after its conversion to APS or PAPS by reactions with ATP. There is some evidence that a similar indirect reduction of sulphate ions can be catalysed by animal tissues, although only to a very slight extent.

The possibility of this reduction occurring in mammals was first demonstrated by Wortman (1963) who showed that extracts of beef corneal epithelium could reduce PAPS with the production of a protein-bound derivative of $SO_3^{2-}$ ions. The enzyme responsible for this reduction could utilize either NADH or NADPH as reductant and required as a constituent of the reaction mixture a heat-stable, non-dialysable substance which was present in the corneal extract. This reaction, and its cofactor requirement, is very reminiscent of that which leads to the reduction of PAPS in yeast (Torii & Bandurski, 1964).

More recently a similar type of reduction of PAPS has been noted in the intestinal mucosa of the rat: in this case the reaction was NADPH-dependent and $SO_3^{2-}$ ions were produced (Robinson, 1965). Neither NADH nor glutathione could replace NADPH. When cystine or oxidized glutathione was present in the system the sulphur of PAPS appeared as the sulpho residue of *S*-sulphocysteine or of *S*-sulphoglutathione. It seems possible that this reaction could be one at least of the physiological routes leading to the formation of $S$-sulphoglutathione which has been isolated from calf lens (Waley, 1959) and from rat intestinal mucosa (Robinson & Pasternak, 1964).

It is therefore now clear that $SO_4^{2-}$ ions can be reduced, albeit indirectly, in some mammalian tissues but it must be stressed that the physiological significance of this reduction is quite unknown, although it is not without interest that the cornea can incorporate the sulphur of $SO_4^{2-}$ ions into cystine and methionine (Dohlman, 1957).

The route whereby the chick embryo can reduce $SO_4^{2-}$ ions is quite unknown (Chapeville & Fromageot, 1957).

## 11.3.2 The reduction of $S_2O_3^{2-}$ ions

It has been known for some time that the two sulphur atoms of $S_2O_3^{2-}$ ions have different metabolic fates in animal tissues (Skarżyński, Szczepkowski & Weber, 1960), as they also have in micro-organisms; the sulphane sulphur is incorporated into proteins, while the sulphonate sulphur appears as $SO_4^{2-}$ ions. This situation could arise through the action of a thiosulphate reductase similar to that which occurs in yeast and which catalyses a reaction (11.5) between $S_2O_3^{2-}$ ions and glutathione, as follows (Kaji & McElroy, 1959):

$$S_2O_3^{2-}+2G.SH \rightleftharpoons G.S.S.G.+SO_3^{2-}+S^{2-}+2H^+ \qquad (11.5)$$

The thiosulphate reductase of liver (Sörbo, 1964) catalyses a similar reaction in homogenates of rat liver fortified with glutathione or certain other SH compounds, dihydrolipoate, mercaptoethanol and cysteine being decreasingly effective as reductants. This may simply reflect a relatively low specificity towards the SH compound but the utilization of dihydrolipoate is of particular interest because Villarejo & Westley (1963 a) have shown that crystalline beef liver rhodanese (see chapter 8), which cannot utilize glutathione as a substrate, catalyses the reduction of $S_2O_3^{2-}$ ions to $SO_3^{2-}$ ions by dihydrolipoate or dihydrolipoamide, lipoate persulphide (L.SH.SSH) probably being an intermediate in this reaction. However, Koj (1968) has now separated the rhodanese and thiosulphate reductase activities of beef liver and has conclusively shown the existence in this tissue of a true thiosulphate reductase catalysing reaction 11.5. It is therefore possible that there are two pathways for thiosulphate reduction in liver: the first catalysed by thiosulphate reductase using glutathione as reductant and the second catalysed by rhodanese using dihydrolipoate as a reductant. In liver, where the final product is of course $SO_4^{2-}$ ions formed by the action of sulphite oxidase, it is likely that the thiosulphate reductase pathway is the more important because the powerful inhibition of thiosulphate reduction by arsenite (70% at 0·1 mM) is reversed to an equal extent by monothiols and by dithiols whereas only the latter would be expected to be effective if dihydrolipoate were involved in the reaction. Nevertheless, this once again stresses the complexity, and multiplicity, of the reactions of sulphur-containing compounds in animal tissues.

Recently Koj, Frendo & Janik (1967) have pointed out that during the oxidation of $S_2O_3^{2-}$ ions by the mitochondria of rat liver more $SO_4^{2-}$ ions are produced than can be accounted for by the oxidation of only the sulphonate group and they have shown that the sulphane sulphur atom must also yield some $SO_4^{2-}$ ions. This could occur by the $S^{2-}$ ions being oxidized to $SO_4^{2-}$ ions, either directly or through $S_2O_3^{2-}$ ions, by those ill-understood reactions considered in § 11.2.1. Whatever be the exact route, the above authors have certainly drawn attention to the existence in animal tissues of a 'thiosulphate cycle' which can be represented as follows:

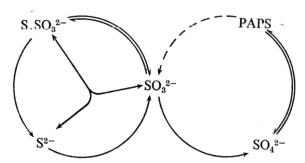

Sörbo (1964) has also suggested, despite his previous observations (Sörbo, 1958 a), that thiosulphate reductase may be responsible for the formation of $SO_3^{2-}$ ions from aminoethyl thiosulphuric acid ($H_2N.C_2H_4S.SO_3^-$) ions and so indirectly for the conversion of this compound to $SO_4^{2-}$ ions by rats.

The reduction of $S_2O_3^{2-}$ ions by SH compounds is reminiscent of the similar (but non-enzymatic) reduction of tetrathionate to thiosulphate (Gilman, Philips, Koelle, Allen & St John, 1946) by reaction 11.6.

$$S_4O_6^{2-} + 2R.SH \rightarrow R.S.S.R. + 2S_2O_3^{2-} + 2H^+ \qquad (11.6)$$

As tetrathionate is rapidly converted to $S_2O_3{}^{2-}$ ions *in vivo* a similar reaction may be involved here and so be responsible for the highly nephrotoxic action of tetrathionate. In a dog, a dose of 500 mg/kg will cause complete anuria within one hour of its administration and this anuria is accompanied by drastic changes in the levels of the SH compounds present in the kidney. Another reaction which could possibly be responsible for the toxicity of tetrathionate is that described by Pihl & Lange (1962) in which it reacted with glyceraldehyde 3-phosphate dehydrogenase to give a sulphenyl thiosulphate:

$$R\text{—}SH + S_4O_6{}^{2-} \rightarrow R\text{—}S.S_2O_3{}^{-} + S_2O_3{}^{2-} + H^+ \qquad (11.7)$$

This reaction (11.7) presumably represents a quite general one between tetrathionate and cysteine residues in proteins.

### 11.3.3   The reduction of $SO_3{}^{2-}$ ions

Although micro-organisms can bring about the reduction of $SO_3{}^{-}$ to $S^{2-}$ ions, as far as is known this reaction hardly occurs in animal tissues. It is of some interest that the NADPH-linked sulphite reductase of *E. coli*, which catalyses this reaction, also shows nitrite reductase and hydroxyl-amine reductase activities, all apparently due to a single protein molecule which can bring about the 6-electron reduction of $SO_3{}^{2-}$ or $NO_2{}^{-}$ ions (see chapter 9).

## 11.4   The interconversions of sulphur compounds

Most of the reactions which have been considered in the previous sections have involved a change in the valency of the sulphur but there are many other reactions undergone by cysteine, and its immediate derivatives, in which no such valence change occurs. In many cases the significance of these reactions is not as clear as that of the oxidative or reductive reactions and it is possible that at least some of them may simply be artefacts caused by the great reactivity of many inorganic compounds of sulphur.

### 11.4.1   The reactions of 3-mercaptopyruvate

The formation of this important metabolite of cysteine need hardly be considered here because it arises by a normal transamination reaction with $\alpha$-oxoglutarate or oxaloacetate (Meister, Fraser & Tice, 1954). The further reactions of this compound are, however, a matter of great interest.

Some of the most important reactions of 3-mercaptopyruvate are the so-called transsulphurase reactions, an example of which is the reaction, catalysed by several rat tissues, between 3-mercaptopyruvate and $SO_3^{2-}$ ions to give pyruvate and $S_2O_3^{2-}$ ions (Sörbo, 1957 a). The enzyme responsible for this reaction is the so-called mercaptopyruvate sulphurtransferase which has been intensively studied by Kun. As has already been discussed in chapter 8, mercaptopyruvate sulphurtransferase can catalyse the transfer of the sulphur atom from mercaptopyruvate to a wide range of acceptors and Fanshier & Kun (1962) have suggested that the reaction involves the formation of an enzyme-persulphide which is the actual sulphur donor so that the reactions catalysed by the enzyme can be represented by reactions 11.8 and 11.9.

$$
\begin{array}{ccc}
CH_2SH & & CH_3 \\
| & & | \\
CO + E.SH & \to & CO + E.S.SH \\
| & & | \\
COOH & & COOH
\end{array}
\tag{11.8}
$$

$$
E.S.SH +
\left\{
\begin{array}{cc}
SO_3^{2-} & S_2O_3^{2-} \\
CN^- & SCN^- \\
R.SO_2^- & R.SO_2S^- \\
R.SII & R.S.SH
\end{array}
\right\}
\to \; + E.SH
\tag{11.9}
$$

The last example in reaction 11.9 is of considerable interest because it may account in part for the so-called cysteine desulphydrase reaction which leads to the elimination of $H_2S$ from cysteine. This over-all reaction could take place by the formation of mercaptopyruvate followed by the transfer of the sulphur from this to cysteine (R.SH above) yielding cysteine persulphide (R.S.SH above) which is turn could react with a further molecule of cysteine to give cystine and $H_2S$. The sequence of reactions from mercaptopyruvate can therefore be written as in (11.10).

$$
\begin{array}{llll}
CH_2SH & CH_2SH & CH_3 & CH_2S.SH \\
| & | & | & | \\
CO \;+\; & H.C.NH_2 \to & CO \;+\; & H.C.NH_2 \\
| & | & | & | \\
COOH & COOH & COOH & COOH
\end{array}
$$

$$
\begin{array}{llll}
CH_2S.SH & CH_2SH & CH_2.S.S.CH_2 & \\
| & | & | \quad\quad | & \\
H.C.NH_2 + & H.C.NH_2 \to & H.C.NH_2 \; H.CNH_2 + H_2S & \\
| & | & | \quad\quad | & \\
COOH & COOH & COOH \quad COOH &
\end{array}
\tag{11.10}
$$

So far the formation of a persulphide has not been proven to occur during the transsulphurase reaction but it is a very likely intermediate, especially in view of the possible formation of persulphides during the reactions catalysed by rhodanese and by cystathionase. Certainly there is ample evidence that persulphides of the general type R.S.SH are very reactive compounds, especially at pH values below about 9 (Hylin & Wood, 1959; Flavin, 1962).

As already pointed out, $S_2O_3^{2-}$ ions can be formed directly by the reaction of mercaptopyruvate and $SO_3^{2-}$ ions: they can also be formed indirectly by a second route from mercaptopyruvate. Alanine 3-sulphinate can act as an acceptor in the transsulphurase reaction, with the formation of alanine thiosulphonate as represented in reaction 11.11 (Sörbo, 1957 a). It has been reported (Zgliczyński & Stelmaszyńska, 1961) that in human erythrocytes alanine thiosulphonate can also be formed by a reaction between $S_2O_3^{2-}$ ions and alanine 3-sulphinate.

$$
\begin{array}{cccc}
\mathrm{CH_2SH} & \mathrm{CH_2SO_2H} & \mathrm{CH_3} & \mathrm{CH_2SO_2SH} \\
| & | & | & | \\
\mathrm{CO} + \mathrm{H.\overset{|}{C}.NH_2} & \rightarrow & \mathrm{CO} + \mathrm{H.\overset{|}{C}.NH_2} \\
| & | & | & | \\
\mathrm{COOH} & \mathrm{COOH} & \mathrm{COOH} & \mathrm{COOH}
\end{array} \qquad (11.11)
$$

(Reaction 11.11, incidentally, probably accounts for the inhibition by mercaptopyruvate of the production of $SO_4^{2-}$ ions from alanine 3-sulphinate which has been reported by Tigert & Smith (1965).) Alanine thiosulphonate can then undergo the simultaneous transamination (with α-oxoglutarate or oxaloacetate) and desulphuration shown in reaction 11.12, which is catalysed by rat liver mitochondria, to give $S_2O_3^{2-}$ ions (De Marco & Coletta, 1961 a), thus:

$$
\begin{array}{ccc}
\mathrm{CH_2.SO_2SH} & \mathrm{CH_3} & \\
| & | & | \\
\mathrm{H.\overset{|}{C}.NH_2} + \mathrm{CO} & \rightarrow \mathrm{CO} + \mathrm{CH.NH_2} + S_2O_3^{2-} \\
| & | & \\
\mathrm{COOH} & \mathrm{COOH} &
\end{array} \qquad (11.12)
$$

The detailed mechanism is not known: the thiosulphate could arise directly on analogy with the corresponding reaction of alanine 3-sulphinate (§ 11.1.2), or the transamination could result in the liberation of $S^0$ and of $SO_3^{2-}$ ions which would react to give $S_2O_3^{2-}$ ions. A similar reaction can probably occur *in vivo* because De Marco, Coletta, Mondovi & Cavallini (1960) have shown that alanine thiosulphonate is an efficient precursor of urinary thiosulphate.

303

*S*-Sulphocysteine can also give rise to $S_2O_3^{2-}$ ions when it is incubated with rat liver slices or homogenates (Sörbo, 1958 *a*) as well as giving $SO_3^{2-}$ ions by a non-enzymatic reaction with SH groups.

## 11.4.2   The desulphydration of cysteine

The desulphydration of cysteine, a reaction which leads to the production of $H_2S$ from cysteine, is perhaps one of the most confused aspects of cysteine metabolism. The occurrence of this pathway in animal tissues was first noted by Fromageot, Wookey & Chaix (1939) who postulated the existence of the enzyme cysteine desulphydrase. The early and complex history of this enzyme has been ably summarized by Smythe (1945) and by Fromageot (1947), and the reaction (11.13) supposedly catalysed by it can be written

$$
\begin{array}{ccc}
CH_2SH & CH_2 & CH_3 \\
| & \parallel & | \\
H.C.NH_2 \xrightarrow{-H_2S} & C.NH_2 \xrightarrow{-NH_2} & CO \\
| & | & | \\
COOH & COOH & COOH
\end{array}
\qquad (11.13)
$$

A chemical model for this reaction, based on catalysis by pyridoxin and a metal ion, was suggested by Snell (Metzler & Snell, 1952; Metzler, Ikawa & Snell, 1954) but nevertheless evidence for its occurrence *in vivo* is slight. Certainly current ideas seem to favour the view that the formation of $H_2S$ by animal tissues is not due to a cysteine desulphydrase but to the action of several other known enzymes. One possible route, involving a transsulphuration between mercaptopyruvate and cysteine, had already been discussed above (§ 11.4.1, reaction 11.10) and another involves the enzyme cystathionase.

Probably the first suggestion that cysteine desulphydrase and cystathionase activities are due to a single enzyme was made by Binkley & Okeson (1950) but only more recently has rather direct evidence for this suggestion been obtained. Flavin (1962) studied in great detail the reactions catalysed by a highly purified cystathionase from *Neurospora* and showed that this enzyme could catalyse a pyridoxal-dependent desulphydration of cystine. During this reaction the persulphide analogue of cysteine was formed. Flavin suggested that the desulphydration of cystine took place by the following series of reactions (11.14, 11.15)

$$
\begin{array}{cccc}
CH_2-S-S-CH_2 & CH_3 & CH_2.S.SH \\
| & | & | & | \\
H.C.NH_2 & H.C.NH_2 \rightarrow CO+H.C.NH_2+NH_3 \\
| & | & | & | \\
COOH & COOH & COOH\ COOH
\end{array}
\qquad (11.14)
$$

$$\begin{array}{cc} \underset{|}{CH_2.S.SH} \quad \underset{|}{CH_2SH} & \underset{|}{CH_2}\!-\!S\!-\!S.\underset{|}{CH_2} \\[4pt] H.\underset{|}{C}.NH_2 + H.\underset{|}{C}.NH_2 \;\rightarrow\; H.\underset{|}{C}.NH_2 \quad H.\underset{|}{C}.NH_2 + H_2S \\[4pt] COOH \qquad COOH & COOH \qquad COOH \end{array} \quad (11.15)$$

which gives for the over-all reaction the classical desulphydration of cysteine, 11.16.

$$\begin{array}{cc} \underset{|}{CH_2.SH} & \underset{|}{CH_3} \\[4pt] H.\underset{|}{C}.NH_2 & \rightarrow \underset{|}{C}O + NH_3 + H_2S \\[4pt] COOH & COOH \end{array} \qquad (11.16)$$

A scheme such as this in which the true substrate for the desulphydration is cystine, not cysteine, makes easily understandable the observation by Cavallini, Mondovi, De Marco & Scioscia-Santoro (1962 *a*) that the desulphuration of cysteine by rat liver cystathionase was inhibited by mercaptoethanol and by hypotaurine: mercaptoethanol would remove the substrate by converting the cystine to the unreactive cysteine and hypotaurine would act as a sulphur-acceptor from cysteine persulphide, so preventing the cyclic formation of cystine. As pointed out above, the scheme suggested by Flavin implies that the true substrate for the desulphydration reaction is cystine, a possibility which had previously been suggested by Cavallini, Mondovi, De Marco & Scioscia-Santoro (1962 *b*), but it should be noted that Loiselet & Chatagner (1964) claim to have shown from kinetic studies that both cysteine and cystine are substrates for cystathionase. In fact, under certain conditions cysteine can inhibit the desulphydration of cystine. The situation is obviously complex but the conclusion that cystathionase can catalyse the desulphydration of the sulphur-containing amino-acids seems inescapable.

The distribution of cystathionase in animal tissues has recently been investigated by Mudd, Finkelstein, Irreverre & Laster (1965) and the enzyme is present in considerable amounts in the liver, pancreas and kidney of several species although it is almost absent from many other tissues (e.g. brain and muscle). Eagle, Washington & Friedman (1966) have made the interesting observation that ten diploid human cell lines require to be supplied with cystine when maintained in tissue culture, presumably because of their demonstrated lack of cystathionase, whereas a number of polyploid lines showed enhanced cystathionase activity and could grow in the absence of cystine. It may therefore be that the cystathionase activity of liver, for example, is due to the presence therein of polyploid cells.

305

Further complications have been added to an already difficult situation by the report that crystalline tryptophanase (Newton & Snell, 1964) and the B protein of tryptophan synthetase (Crawford & Ito, 1964) from *E. coli* can catalyse the deamination and desulphydration of cysteine. Both of these enzymes are pyridoxal phosphate-dependent.

It should therefore be clear that $H_2S$ can be produced from cysteine, or cystine, by a number of apparently quite different pathways and at present it is impossible to decide which, if any, of them are of physiological importance in higher animals. From a quantitative point of view desulphydration must be relatively unimportant.

## 11.5 The incorporation of inorganic sulphur compounds into organic compounds

The formation of organic compounds of sulphur from inorganic compounds is a process well-known to occur *in vitro* but details of the incorporation are poorly known indeed. Examples of such transformations are: the formation of taurine from $SO_4^{2-}$ ions (Boström & Åqvist, 1952); of cysteine from $S^{2-}$ ions (Dziewiatkowski, 1946) or from $S_2O_3^{2-}$ ions (Schneider & Westley, 1963); of alanine 3-sulphinate from $SO_3^{2-}$ ions (Chapeville, Fromageot, Brigelhuber & Henry, 1956) and of alanine 3-sulphonate from $SO_3^{2-}$ ions (Chapeville & Fromageot, 1957, 1960). A useful summary of much of this work has recently been provided by Dodgson & Rose (1966).

### 11.5.1 Reactions *in vitro*

Chapeville & Fromageot (1958, 1961) have suggested that many of the reactions mentioned above can be explained on the basis of reactions of an activated $\beta$-carbon atom of cysteine, or more correctly of some enzyme-bound desulphurated form thereof. This activated intermediate is postulated to react as follows:

1. Add $S^{2-}$ ions to give cysteine.
2. Add $SO_3^{2-}$ ions to give alanine 3-sulphonate.
3. Add cysteine to give lanthionine.
4. React with water to give pyruvate and ammonia.

The enzyme believed to be responsible for this series of reactions has been named 'cysteine lyase': it is present in the yolk sac and yolk of embryonated hens' eggs but it has not been significantly purified (only about 20 times). Despite this lack of purification it has been claimed to

306

be a pyridoxal-dependent enzyme, the apo-enzyme being significantly reactivated by $0.1 \mu$M pyridoxal although not completely reactivated at concentrations as high as $10 \mu$M. It is certainly to be hoped that further studies of this potentially interesting enzyme will be forthcoming although it should perhaps be pointed out that it itself obviously cannot bring about the *de novo* synthesis of cysteine. Even if it be assumed that the $S^{2-}$ ion liberated in the initial stage of the reaction is quantitatively utilized by serine sulphhydrase (see below) to give cysteine there is no net synthesis of the latter and the overall process can, at the best, only be that represented in reaction 11.17.

$$
\begin{array}{c}
\begin{array}{c}
\text{CH}_2.\text{SH} \\
| \\
\text{H.C.NH}_2 \\
| \\
\text{COOH}
\end{array}
\longrightarrow
\left[
\begin{array}{c}
\text{CH}_2\text{-X} \\
| \\
\text{H.C.NH}_2 \\
| \\
\text{COOH}
\end{array}
\right]
\quad + \quad S^{2-}
\end{array}
$$

$$
\begin{array}{cc}
\downarrow \text{SO}_3^{2-} & \qquad \downarrow \raisebox{0.5ex}{$\diagdown$}\text{Serine} \qquad (11.17) \\
\begin{array}{c}
\text{CH}_2.\text{SO}_3\text{H} \\
| \\
\text{H.C.NH}_2 \\
| \\
\text{COOH}
\end{array}
&
\begin{array}{c}
\text{CH}_2.\text{SH} \\
| \\
\text{H.C.NH}_2 \\
| \\
\text{COOH}
\end{array}
\end{array}
$$

In view of this it is difficult not to suggest that 'cysteine lyase' may only be yet another component of the 'cysteine desulphydrase' which has already been considered in § 11.4.2.

It is of considerable interest that these reactions can be catalysed non-enzymatically by pyridoxal or pyridoxal phosphate in the presence of any of several metal ions, including iron, copper, vanadium, etc. (Ratsisalovanina, Chapeville & Fromageot, 1961). Under such conditions there is a facile substitution of the —SH, —OH and —OPO$_3^{2-}$ groups of cysteine, serine and phosphoserine by $S^{2-}$ or $SO_3^{2-}$ ions, to give cysteine and alanine 3-sulphonate or even by cysteine and homocysteine to give lanthionine and cystathionine respectively. More recently Fromageot, Roderick & Pouzat (1963) showed that similar reactions were catalysed by pyruvate in the presence of copper, but not of other metals, and they have discussed possible reasons for this specificity.

Another route leading to the incorporation of $S^{2-}$ ions is that catalysed by the pyridoxal phosphate dependent enzyme serine sulphhydrase, first detected in yeast by Schlossmann & Lynen (1957), which catalyses the interconversion of L-serine and L-cysteine shown in reaction 11.18.

$$\underset{\overset{|}{\text{COOH}}}{\overset{\overset{\text{CH}_2\text{OH}}{|}}{\text{H.C.NH}_2}} + \text{H}_2\text{S} \rightarrow \underset{\overset{|}{\text{COOH}}}{\overset{\overset{\text{CH}_2.\text{SH}}{|}}{\text{H.C.NH}_2}} + \text{H}_2\text{O} \qquad (11.18)$$

This reaction obviously does give a *de novo* synthesis of cysteine. The existence of serine sulphhydrase in animal tissues has been shown by Brüggemann, Schlossmann, Merkenschlager & Waldschmidt (1962) who found it to be present in several tissues of rats and of hens, hen brain being a particularly rich source. Its distribution in the chick embryo and associated tissues has been described by Sentenac & Fromageot (1964). The enzyme from chicken liver has been purified about 100-fold (Brüggemann & Waldschmidt, 1962) and distinguished from the 'cysteine desulphydrase' which was also present. Confusion has again been caused by the claim (Kato, Ogura, Kimura, Kawai & Suda, 1966) that the homoserine dehydratase, cystathionase, cysteine desulphydrase and cysteine desulphurase activities of rat liver are all associated with a single protein.

Sentenac, Chapeville & Fromageot (1963) have described what may be a related enzyme, phosphoserine phospholyase, in the vitelline sac of chick embryo which is believed to catalyse the reaction of $S^{2-}$ ions with phosphoserine (reaction 11.19), serine itself not being a substrate.

$$\underset{\overset{|}{\text{COOH}}}{\overset{\overset{\text{CH}_2.\text{O.PO}_3{}^{2-}}{|}}{\text{H.C.NH}_2}} + \text{H}_2\text{S} \rightarrow \underset{\overset{|}{\text{COOH}}}{\overset{\overset{\text{CH}_2.\text{SH}}{|}}{\text{H.C.NH}_2}} + \text{HPO}_4{}^{2-} \qquad (11.19)$$

Again this enzyme has not been purified but it has been claimed that it is not pyridoxal dependent, a finding which is perhaps surprising in view of the non-enzymatic catalysis of the same reaction by pyridoxal (Ratsisalovanina *et al.* 1961). It is perhaps of some significance that this activity could not be detected in any other tissue of the embryonic chick.

Another interesting reaction is that catalysed by the lysed mitochondria of rat liver in which the sulphane sulphur of an $S_2O_3{}^{2-}$ ion is transferred to serine with the production of cysteine (Schneider & Westley, 1963). Once again the details have not been clarified and there are some puzzling features of the reaction, especially the facts that both D- and L-serine are utilized and that the reaction mixture must contain cysteine which, it is claimed, protects the newly-formed cysteine from enzymatic destruction. Schneider & Westley (1963) claim that $S^{2-}$ ions cannot be an intermediate in this transfer, which excludes the possible participation of serine sulphydrase, and they suggest that a lipoate persulphide is formed during the reactions. No dependence of the reaction on added lipoate could, however, be shown.

This reaction, in which only the sulphane sulphur of $S_2O_3^{2-}$ ions is transferred, should perhaps be compared with that taking place in *Aspergillus nidulans* in which the intact $S_2O_3^{2-}$ ion is utilized to give *S*-sulphocysteine which can then give rise to cysteine (Nakamura & Sato, 1963 *a*). There is as yet no evidence for the occurrence of such a process in animal tissues.

## 11.5.2   Reactions *in vivo*

The problem of the incorporation *in vivo* of inorganic forms of sulphur into organic compounds is a difficult one to investigate but, at least in the majority of mammals, there is little evidence that the process can occur to any significant extent. The situation in ruminants is of course different and will be considered separately, as will that in birds.

Even when $S^{2-}$ ions are used as a sulphur source the extent of the incorporation of those ions into the sulphur-containing amino-acids is small; their utilization by rats for the synthesis of cysteine *in vivo* is minute (Dziewiatkowski, 1946) but significant (Huovinen & Gustafsson, 1967).

The situation which holds with $SO_3^{2-}$ ions is more difficult to evaluate with certainty. According to Chapeville, Fromageot, Brigelhuber & Henry (1956), when $^{35}SO_3^{2-}$ ions are administered to eviscerated rabbits, or to rabbits whose gastro-intestinal tracts have been carefully sterilized, traces of alanine 3-[$^{35}S$]sulphinate could be isolated from kidney tissue. Furthermore, the tissues of those animals yielded [$^{35}S$]cysteine and [$^{35}S$]-taurine both of which were apparently metabolic products of alanine 3-sulphinate. Certainly the latter can readily give rise to taurine, but how it could be reduced to cysteine is not clear. Perhaps it is pertinent that Bennet (1937) long ago showed that alanine 3-sulphinate could not replace cysteine in the diet of the rat. Whatever be the immediate source of the cysteine there was no doubt that its sulphur was ultimately derived from $SO_3^{2-}$ ions, under conditions in which $SO_4^{2-}$ ions were not utilized, but the process certainly did not occur to any great extent. Further, Huovinen

& Gustafsson (1967) have shown that in germ-free rats $SO_3^{2-}$ ions cannot be utilized for the formation of cysteine: this is obviously contrary to the findings of Chapeville *et al.* (1956) in the rabbit.

Chapeville & Fromageot (1957) have also shown that the living chick embryo can incorporate $SO_3^{2-}$ ions into alanine 3-sulphonate but, as already discussed (§ 11.5.1), this reaction probably involves cysteine as one of the reactants so that it does not represent a *de novo* synthesis of organically bound sulphur.

With $SO_4^{2-}$ ions the situation is simpler and it seems clear that those ions cannot to any significant extent be incorporated into organic compounds by the intact animal: where incorporation into cysteine does occur the micro-organisms of the gastro-intestinal tract play a major role (Chapeville *et al.* 1956; Dziewiatkowski & Di Ferrante, 1957; Huovinen & Gustafsson, 1967). Dziewiatkowski & Di Ferrante have also pointed out that although considerable amounts of radioactivity appear bound to the plasma proteins after the administration of $^{35}SO_4^{2-}$ ions to rats, only about 5% of this bound activity is present as cysteine or methionine, the remainder still occurring as sulphate, probably in the form of a mucopolysaccharide. In fact, it is now clear that quite simple sulphate esters (aryl sulphates and steroid sulphates) are firmly bound to plasma proteins (Plager, 1965; Wang & Bulbrook, 1967) and it seems that the formation of compounds of this type may account for a considerable proportion of the so-called incorporation of $SO_4^{2-}$ ions into proteins. Such binding may, of course, be of considerable physiological importance especially in connection with steroid metabolism. John & Miller (1965) have also stressed the association of $^{35}SO_4^{2-}$ ions and protein: following the addition of such ions to blood perfusing an isolated rat liver there was a considerable amount of radioactivity associated with the albumin and $\alpha$-globulin of the serum. This radioactivity was apparently mainly present as small molecules which were non-covalently bound to the protein and only traces, if any, were present in covalent linkage. There is therefore little, if any, evidence to suggest that $SO_4^{2-}$ ions can be utilized to any significant extent for the synthesis of the sulphur-containing amino-acids by the higher animals. One exception to this statement must be made: Rambaut & Miller (1965) have claimed that adult cats can be maintained on a synthetic diet lacking in the sulphur-containing amino-acids and that in such animals $SO_4^{2-}$ ions are extensively used for the synthesis of cysteine and methionine in the blood proteins. The cat, then, appears to be unique among the monogastric animals in that it does not require an exogenous source of cysteine and methionine. Certainly further investigation of this point at the enzymatic level would be most valuable.

The situation in the ruminant is, of course, quite different to that in other mammals. Ruminants can utilize inorganic forms of sulphur for the extensive synthesis of the sulphur-containing amino-acids but this process is basically a function of the micro-organisms of the rumen, not of the mammal itself. The reduction of $SO_4^{2-}$ ions by rumen micro-organisms has been studied by Lewis (1954) while the work of Block & Stekol (1950) and of Block, Stekol & Loosli (1951) has amply shown that $SO_4^{2-}$ ions can be utilized by the ruminant for the synthesis of cysteine and methionine. Some aspects of this work have recently been briefly reviewed by Dodgson & Rose (1966).

In the hen the situation is again different from that in mammals because Machlin, Pearson, Denton & Bird (1953) have shown that the laying hen can incorporate the sulphur of injected $^{35}SO_4^{2-}$ ions into the cystine present in the egg proteins. Again, of course, the participation of intestinal micro-organisms cannot be completely excluded. In the chick embryo $^{35}SO_4^{2-}$ ions are also incorporated into organic compounds although here the principal product is not cystine but taurine: no incorporation into cystine, methionine or glutathione could be detected (Machlin, Pearson & Denton, 1955; Lowe & Roberts, 1955). This difference between the pathway in the essentially aseptic conditions of the chick embryo and that in the adult hen suggests that the utilization by the latter of $SO_4^{2-}$ ions for the production of cystine may well be a function of the intestinal flora. Unfortunately there is no information on the way in which these ions are utilized by either the chick embryo or by the hen and it is therefore interesting that Mason, Hansen & Jakobsen (1965) have suggested that taurine may be a precursor of the sulphur-containing amino-acids in this species. If this be the case it is difficult to rationalize the finding by Miraglia, Martin, Spaeth & Patrick (1966) that the incorporation of sulphate-sulphur into taurine by the hen is increased when the diet is supplemented with cysteine or methionine, and indeed with $SO_3^{2-}$ or $S_2O_3^{2-}$ ions. Further information is needed before these results can be explained.

### 11.5.3 $SO_4^{2-}$ ions in nutrition

The above discussion has stressed that higher animals can only very poorly utilize $SO_4^{2-}$ ions for the synthesis of the sulphur-containing amino-acids. Further, these ions (and $S_2O_3^{2-}$ ions) are only poorly absorbed from the gastro-intestinal tract, hence the use of $Na_2SO_4$ and $MgSO_4$ as saline cathartics. The mechanism of this absorption of $SO_4^{2-}$ ions is unknown: Deyrup (1963) has shown that in the rat it occurs only in the lower ileum and that although it was not dependent upon the presence of exogenous substrates it was inhibited by dinitrophenol, cyanide and fluoride so that it was presumably an energy-dependent process. Dodgson & Rose (1966) have suggested that the uptake of $SO_4^{2-}$ ions in the intestine involves the formation of PAPS but there is no evidence for or against this postulate.

Despite this poor absorption of $SO_4^{2-}$ ions, it is clear that they do play a significant role in the diet of the animal. Wellers & Chevan (1959) and Wellers, Boelle & Chevan (1960), for example, have shown that about 30% of the daily sulphur requirement of the rat can be supplied by dietary sodium sulphate: this amount is sufficient to satisfy all bodily requirements apart from the anabolism of protein. Michels & Smith (1965) have reached an essentially similar conclusion and have shown that if no $SO_4^{2-}$ ions are present in the diet then the requirement for methionine is increased. Gordon & Sizer (1955) showed a similar 'sparing' action of sulphate ions in the nutrition of the chick.

It should be noted that other dietary factors could conceivably alter the requirements for sulphate although no such effect has yet been detected. For example, as has already been discussed in §§ 5.1.4 and 6.8, a hypovitaminosis A leads to diminished levels of sulphate activation and transfer. Hypervitaminosis A also influences sulphate metabolism, as was shown originally by Fell, Mellanby & Pelc (1956) and more recently by Mukherji & Bachhawat (1966). The latter effects at least seem to be indirect, due to the action of vitamin A on the lysosomes (Lucy, Dingle & Fell, 1961; Dingle, 1961; Fell & Dingle, 1963), and a direct effect on the metabolism of $SO_4^{2-}$ ions seems most unlikely. Vitamin D also can effect the metabolism of $SO_4^{2-}$ ions because the uptake of these by the long bones of the chick is diminished in hypovitaminosis D (Sahashi, Suzuki, Nishikawa, Tanaka, Takahashi, Inaba & Miyazawa, 1962; Tanaka, Takahashi, Miyazawa, Higaki, Inaba, Nishikawa, Suzuki & Sahashi, 1963). This effect is reversed by the administration of vitamin D, vitamin $D_3$ having the expected greater activity than vitamin $D_2$, but again there is nothing to show that this action is a direct one on the metabolism of $SO_4^{2-}$ ions.

In ruminants there exist even further complications: for instance, the complex, ill-understood but practically very important inter-relationships of sulphate, copper and molybdenum. For a discussion of these problems reviews by Underwood (1956), De Renzo (1962) and Adelstein & Vallee (1962) should be consulted.

### 11.5.4    The active transport of $SO_4^{2-}$ ions

It has long been known (see Smith, H.W. 1951) that $SO_4^{2-}$ ions are re-absorbed as the glomerular filtrate passes down the kidney tubules but little is known of the mechanism of this process. Deyrup (1956) showed that slices of kidney cortex could concentrate $SO_4^{2-}$ ions *in vitro* by a process which was independent of added ATP or substrates. More recently Deyrup (1964) has shown that this uptake of $SO_4^{2-}$ ions is inhibited by $S_2O_3^{2-}$ ions and she suggests that both types of ion bind to common sites in the kidney, as had of course previously been suggested by Berglund, Helander & Howe (1960) from their studies in dogs. Deyrup also showed that the arylsulphatase substrates p-nitrophenyl sulphate and nitrocatechol sulphate (see § 7.1) inhibited the uptake of $^{35}SO_4^{2-}$ ions by kidney slices, but only when they were present in concentrations at least 1000-fold greater than that of the $^{35}SO_4^{2-}$ ions, and there was nothing to suggest the participation of arylsulphatases in this absorption of sulphate.

Mitochondria from rat kidney cortex can also take up sulphate ions from their environment (Winters *et al.* 1962), effecting a several hundred-fold concentration. Rasmussen, Sallis, Fang, DeLuca & Young (1964) have shown that this uptake of $SO_4^{2-}$ ions by mitochondria, both from rat liver and rat kidney, is stimulated by parathormone. The concentration of parathormone which was used was very high and the authors themselves point out that the effect may be pharmacological rather than physiological. The accumulation of phosphate and arsenate ions was similarly stimulated by parathormone and it was suggested that one site was responsible for the uptake of all three ions: if this be the case it would appear that it is the divalent form of the phosphate ion which is concentrated.

Quite recently Ulrich & Körmendi (1967) have uncovered another facet of the metabolism of sulphate ions: they have shown that the loss of $K^+$ ions from mitochondria is greatly increased if sulphate ions are present in the medium. The interpretation of the results was difficult but it appeared that $SO_4^{2-}$ ions altered in some way the permeability of the mitochondrial membrane.

That still further complications in the active transport of $SO_4^{2-}$ ions are likely to be discovered is suggested by the observation that the uptake of these ions from sea water into the plasma of the ascidian *Ciona intestinalis* is stimulated by vanadate whereas chromate has no effect (Bielig, Pfleger, Rummel & Seifen, 1961). Although no interpretation of these results is possible, at first sight they might be taken to suggest that ATP-sulphurylase (which is inhibited by these anions, see § 5.1.1), and therefore PAPS, can play no part in this transport of $SO_4^{2-}$ ions.

### 11.5.5    The formation of sulphate esters *in vivo*

As has already been discussed in § 6.9, Wellers has suggested that *in vivo* the sulphur of cysteine can be utilized directly for the formation of sulphate esters. There is, however, little evidence for this view and it seems certain that the route *in vivo* is, as it is *in vitro*, that involving the transfer of sulphate from PAPS.

There are, of course, added complications in the study of the formation of sulphate esters *in vivo*: for example, there is the problem of the role of vitamin A which has already been considered in chapters 5 and 6, and in § 11.5.3 above. Vitamin C also plays a role in the formation of sulphate esters *in vivo* and Kodicek & Loewi (1955) showed that granulation tissue from scorbutic guinea pigs could incorporate much less $^{35}SO_4^{2-}$ ions into mucopolysaccharides than could the corresponding tissue from normal animals. There is, however, no suggestion that this effect of vitamin C is a direct one because the addition of ascorbic acid *in vitro* made no difference to the rate of incorporation in preparations from scorbutic animals.

Another factor to be considered in studies of the formation of sulphate esters *in vivo* is the 'serum sulphation factor' discovered by Salmon & Daughaday (1957) in their work on the formation of mucopolysaccharide sulphates by the costal cartilage of the rat *in vitro*. The nature of this factor is still not clear: it has been stated (Daughaday & Parker, 1965) that it is not identical with growth hormone but this seems not to have been proven and certainly the action of the factor appears to be mediated through protein synthesis and not directly through sulphate metabolism. For example, Salmon, Von Hagen & Thompson (1967) have shown that when the incorporation of $^{35}SO_4^{2-}$ ions into rat cartilage *in vitro* is inhibited by puromycin or actinomycin the stimulating effect of the serum sulphation factor is no longer apparent. This finding would be in keeping with the original observation (Salmon & Daughaday, 1957) that at least part of the activity of the sulphation factor was due to the presence of essential amino acids in the serum and also with the finding of Telser, Robinson &

Dorfman (1965) that mucopolysaccharides are probably built up on a core of protein. Unfortunately there is as yet no information on the role of this factor *in vivo* although there seems no reason to suppose that it is not functional under such conditions. Likewise there has been no report of the action of the serum sulphation factor on the formation of other types of sulphate ester: if its action indeed be on protein synthesis then any effect on the formation of simpler esters would seem unlikely.

Corticosteroids also influence the incorporation of sulphate into mucopolysaccharides as was first noted by Layton (1951 *a*) who showed that cortisone inhibited this reaction in chick cartilage and regenerating muscle tissue *in vitro*. Layton suggested that this effect could be related to the well known anti-rheumatic action of cortisone and showed that a similar inhibition of the sulphate incorporation into mucopolysaccharides occurred *in vivo* in rats (Layton, 1951 *b*). Boström & Odeblad (1953) confirmed the existence of this effect *in vitro* but Kodicek & Loewi (1955) on the other hand could not. The problem was investigated in greater detail by Lash & Whitehouse (1961) who showed that the uptake of sulphate into mucopolysaccharides was inhibited not only by cortisone but also by a wide range of anti-inflammatory steroids and by salicylate. It is now clear that salicylate acts by uncoupling oxidative phosphorylation and the anti-inflammatory steroids by inhibiting oxidative reactions in the mitochondria (Boström, Berntsen & Whitehouse, 1964). Again, therefore, although these compounds can undoubtedly depress the incorporation of sulphate ions into sulphate esters (not only mucopolysaccharides but also simple urinary sulphate esters) *in vivo*, the effect is an indirect one mediated through the energy-producing reactions of the cell. It should perhaps be again pointed out here that the esterification of a sulphate ion is an expensive process: two molecules of ATP are required for the esterification of each sulphate ion so that this reaction may be particularly sensitive to changes in the level of oxidation *in vivo*.

### 11.5.6 The metabolism of sulphate esters *in vivo*

It is not intended to discuss this subject in detail but a few general points must be made. Roy (1960 *c*) pointed out that there was little evidence that sulphatases, in particular arylsulphatases, were active *in vivo*: this situation certainly no longer holds. Probably the first convincing demonstration of the (presumably) sulphatase-catalysed hydrolysis of a steroid sulphate was that of Roberts, VandeWiele & Lieberman (1961) who showed that [7α-³H]androstenolone sulphate was converted to androsterone and aetiocholanolone glucuronides by human subjects. Similarly there can be little doubt that arylsulphatases are active *in vivo*: for example, nitrocatechol sulphate (Flynn, Dodgson, Powell & Rose, 1967), biphenylyl 4-sulphate (Hearse, Powell, Olavesen & Dodgson, 1967) and cyclohexylphenyl 4-sulphate (Hearse, Powell, Olavesen & Dodgson, 1968) have been shown to give rise to $SO_4^{2-}$ ions *in vivo* so that the action of an arylsulphatase must be presumed. There is therefore now little reason to doubt that sulphatases can show the same action *in vivo* as *in vitro* but the factors which control the extent of this action are by no means clearly understood. Why, for example, is biphenylyl 4-sulphate hydrolysed *in vivo* (Hearse *et al.* 1967) while 2-naphthyl sulphate is not (Hawkins & Young, 1954)?

A very valuable contribution which may help to solve this problem has recently been made by Powell (Powell, Curtis & Dodgson, 1967) who has introduced the technique of whole-body autoradiography for the study of the distribution of [$^{35}$S]sulphate esters in mice and rats. Using this technique Powell has shown that only those sulphate esters which are metabolized *in vivo* accumulate in cells and that different esters have very different distributions in different organs. For example, serine *O*-sulphate and threonine *O*-sulphate which are desulphated *in vivo* (see § 7.6) are concentrated in brown adipose tissue, an organ which is not normally considered to be involved in sulphur metabolism, while tyrosine *O*-sulphate is concentrated in a well-defined zone of the kidney, an organ in which an aminotransferase acting on tyrosine *O*-sulphate is known to occur (Rose, Flanagan & John, 1966). It is already obvious, therefore, that this technique can provide information not otherwise readily accessible and further results are awaited with interest.

Finally, it should again be pointed out that it is now obvious that many steroid sulphates can undergo extensive metabolic changes *in vivo* without removal of the sulphate group (see Lieberman & Roberts, 1969) but this topic cannot be further considered here.

# 12

# THE CLINICAL CHEMISTRY OF SOME INORGANIC SULPHUR COMPOUNDS

Until recently there has been little direct investigation of the clinical chemistry of the inorganic forms of sulphur despite the considerable interest in this aspect of the biochemistry of cysteine and other compounds containing divalent sulphur. Much of this has recently been summarized by Eldjarn (1965) and by Berlow (1967). Of the compounds containing sulphur in a higher oxidation state only one, taurine, has been extensively studied and this substance scarcely enters into the scope of the present discussion. Interest in taurine was stimulated by the observation (Hempelmann, Lisco & Hoffman, 1952) that its urinary excretion was greatly increased following radiation injury in man, a finding which was confirmed in experimental animals by Kay, Harris & Entenman (1956). Sörbo (1962 c) showed that the taurine which was excreted under such conditions was not newly synthesized but was preformed taurine coming from the damaged lymphatic and muscular tissues. This observation is of course in keeping with the fact that a hypertaurinuria occurs in other diseases involving cellular damage such as, for instance, muscular dystrophy (Hurley & Williams, 1955) and paroxysmal myoglobinuria (Bowden, Fraser, Jackson & Ford Walker, 1956). Other aspects of the clinical chemistry of taurine have been reviewed by Sörbo (1965) and they need not be considered further at present.

The major metabolite of the sulphur-containing amino-acids is the sulphate ion of which the daily urinary excretion in man amounts to about 25 m-moles. This, it must be presumed, is derived mainly from the oxidation of cysteine and it is therefore of some interest that Patrick (1962) has shown that liver samples obtained from patients with cystinosis (in which a massive deposition of cystine occurs in the tissues) can oxidize cysteine to $SO_4^{2-}$ ions at the normal rate. Cystinosis cannot therefore be due to an inability to carry out this oxidation. It should perhaps be pointed out that the oxidation of each molecule of cysteine to an $SO_4^{2-}$ ion is equivalent to the production of two equivalents of acid (Lemann,

317

Relman & Connors, 1959), a situation of some importance in the endogenous acid-base metabolism of the human. Any decrease in the efficiency of the excretion of $SO_4^{2-}$ ions in the urine must lead to an increased level of serum sulphate and hence to an acidosis: an increased level of serum sulphate might therefore prove useful as an indication of renal insufficiency although it should perhaps be noted that in the normal kidney there is considerable tubular reabsorption of $SO_4^{2-}$ ions so that the situation is not necessarily so simple (see Smith, H.W. 1951). Standard methods are available for the determination of serum sulphate but perhaps one of the most reliable is that of Berglund & Sörbo (1960).

Only very recently has there been a description of a condition in which the body is apparently unable to produce $SO_4^{2-}$ ions. Mudd, Irreverre & Laster (1967) and Irreverre, Mudd, Heizer & Laster (1967) have described a male infant showing at birth neurological abnormalities which progressed until he was virtually decorticate at the age of nine months. The urine of this patient contained large amounts of $SO_3^{2-}$ ions, of $S_2O_3^{2-}$ ions and of S-sulphocysteine but virtually no $SO_4^{2-}$ ions. At autopsy the liver and kidney contained no detectable sulphite oxidase activity (see § 11.1.3) although this could readily be detected in control samples. The levels of several other enzymes involved in the metabolism of sulphur-containing compounds were unchanged in the patient's tissues. The biochemical findings in this condition were easily explicable on the basis of a lack of sulphite oxidase leading to the accumulation of $SO_3^{2-}$ ions in the body and thence to $S_2O_3^{2-}$ ions and to S-sulphocysteine. It is to be hoped that further information on this condition, which should be detectable by spot tests for $SO_3^{2-}$ ions in the urine, may become available so that the obviously great severity of the symptoms may be explained. At first sight a deficiency of $SO_4^{2-}$ ions would have been expected to have manifested itself in gross changes in the connective tissue rather than in the central nervous system and the neurological defects are not easily explicable. Two possibilities seem to be obvious: firstly, the deficiency of $SO_4^{2-}$ ions could have caused a defective synthesis of cerebroside sulphate or, secondly, the sulphite ions could have inhibited cerebroside sulphatase (or arylsulphatase, see §§ 7.1.8 and 7.3.6) and so caused an accumulation of cerebroside sulphate comparable to that which occurs in metachromatic leucodystrophy.

As has already been discussed in § 7.1.8, the latter condition is the only one in which a deficiency of a sulphatase has been well authenticated. The congenital condition of metachromatic leucodystrophy is associated with an asulphatasia (or hyposulphatasia) and is characterized clinically by profound neurological changes associated with the deposition of cerebroside sulphates in several tissues, including the kidney (Austin, 1965). The relation between the enzymatic defect and the symptoms is obscure but the very low level of urinary arylsulphatase activity in metachromatic leucodystrophy (Austin, McAfee & Shearer, 1965) has provided a useful tool for the differential diagnosis of the disease. A detailed description of a rapid screening test for this purpose has been published (Austin, Armstrong, Shearer & McAfee, 1966).

A further neurological condition involving a deranged metabolism of cerebroside sulphate has recently been studied. Bachhawat, Austin & Armstrong (1967) have shown that in globoid leucodystrophy, a disease which involves massive destruction of the myelin, there was a lack of cerebroside sulphotransferase activity (see § 6.7) in the brain. This enzyme defect might well have been predicted from the clinical findings of the greatly decreased cerebroside sulphate level and the increased cerebroside level which occurs in nervous tissue of patients suffering from this condition. In globoid leucodystrophy there was no change in the level of the arylsulphatase activity.

It certainly seems of interest that in all of the conditions so far known in which there is some enzymatic defect in the metabolism of $SO_4^{2-}$ ions the principal symptoms are neurological. The obvious conclusion might be that it is in the metabolism of nervous tissue that $SO_4^{2-}$ ions are playing their major role and that this role is in the metabolism of cerebroside sulphates. This type of compound may well have an importance which has not so far been ascribed to it.

Some other aspects of the pathology of the arylsulphatases have already been considered in § 7.1.8 and only two further points need be mentioned here. The first involves the use of these enzymes in the differential diagnosis of the leukaemias. It has been known since the work of Austin & Bischel (1961) and of Tanaka, Valentine & Fredricks (1962) that the cytoplasm of human white cells show a pronounced arylsulphatase activity, this being greatest in the granulocytes (especially the eosinophils) and least in the lymphocytes. The same general picture was obtained either by the classical enzymological assays used by Tanaka or by the histochemical method used by Austin. Lawrinson & Gross (1964) later pointed out that by using the same technique as Austin & Bischel (1961)—i.e. a histochemical test with 6-benzoyl-2-naphthyl sulphate as substrate—a strong *nuclear* arylsulphatase activity could be detected in lymphocytes and lymphoblasts but not in myeloid cells, monocytes or their blast cells. This observation has been used in a differential diagnosis of acute leukaemia by Ekert & Denett (1966). They showed that marrow smears from four children suffering from an acute undifferentiated leukaemia gave the same picture of nuclear arylsulphatase activity as did smears from patients with acute lymphatic leukaemia: from a practical point of view this was of some importance because these undifferentiated leukaemias responded to treatment as did the lympatic leukaemias. Therefore, whatever be the biochemical basis of this method it may prove to be one of considerable practical importance. It should be recalled that previous biochemical and histochemical investigations (§ 7.1.7) have given no suggestion of the existence of a nuclear arylsulphatase in any other tissue. It may be that the nuclear arylsulphatase of lymphocytes is an artefact of staining but even if this be the case it is of no significance in the present connection.

The second use of arylsulphatase determinations is in the characterization of acid-fast bacteria and, in particular, of the mycobacteria. Kubica & Vestal (1961) and Tarshis (1963) have made much use of phenolphthalein sulphate for this purpose.

The clinical chemistry of other inorganic forms of sulphur has received virtually no attention despite the rather important metabolic role of, for instance, thiosulphate. Thiocyanate has been investigated to some extent both in connection with its inhibition of the uptake of iodine by the thyroid gland and with the elevated level of this ion which occurs in the plasma of smokers (Maliszewski & Bass, 1955). In view of the obvious ill-effects of smoking on at least some vascular diseases it may not be without importance that Clemedson, Sörbo & Ullberg (1960) have shown a quite important uptake of $[^{35}S]CNS^-$ ions by arterial walls.

Investigations of the metabolism of thiocyanate have brought to light a further possible relationship, albeit indirect, between the metabolism of inorganic compounds of sulphur and the central nervous system. As has already been mentioned, the level of thiocyanate is higher in the plasma of smokers than of non-smokers, values of 0·10 mM and 0·058 mM respectively having been quoted by Wilson (1965) who also made the interesting observation that this increased level of thiocyanate was not found in smokers suffering from Leber's optic atrophy. He therefore suggested that this rare hereditary disease was caused by an inborn error of metabolism which allowed the accumulation of cyanide by preventing its conversion to thiocyanate. The diffuse neurological changes occurring in Leber's disease are certainly comparable to the neurotoxic effects of the chronic administration of cyanide to experimental animals (Smith, Duckett & Waters, 1963). Unfortunately no biochemical evidence for any enzyme defect has been forthcoming and certainly in the tissues of the only patient so examined there was no drop in the level of rhodanese activity. Other enzymes which could be involved in the removal of cyanide were not investigated.

Wilson (1965) found no difference in the levels of cyanide in the plasma of smokers and of non-smokers. This suggests that any oxidation of thio-cyanate to cyanide by peroxidases (see § 11.2.3) or by the so-called thio-cyanate oxidase of erythrocytes (Goldstein & Rieders, 1953) is negligible, despite the claim that an equilibrium exists between these two ions in the body (Boxer & Rickards, 1952).

Interest in thiocyanate has been further stimulated by the finding that in patients suffering from a chronic degenerative neuropathy (nutritional ataxic neuropathy) which is common in parts of Nigeria there is an increased concentration of thiocyanate in plasma and in urine (Osuntokun, 1968). An acute retrobulbar neuritis which is probably a manifestation of the same condition in children is accompanied by similar changes in the levels of thiocyanate. Both these diseases are associated with the long-standing consumption of large amounts of cassava (manioc tuber) which contains considerable quantities of the cyanogenic glycoside, linamarin. Hospitalization of such patients, with the removal of cassava from the diet, is accompanied by a drop in the plasma thiocyanate concentration from the high value of 0·085 mM towards normal levels but this trend is reversed when a return is made to the normal diet. Osuntokun, Durowoju, McFarlane & Wilson (1968) have also shown that this high level of thio-cyanate in plasma is associated with a very low level of cysteine + cystine in the plasma and they suggested that this was due to the utilization of these amino acids for the detoxication of the cyanide derived from the cassava. Such a detoxication could take place through the formation of thiocyanate by transulphuration from cysteine (see chapter 8 and § 11.4.1) or through the formation of 2-amino-4-thiazolidine carboxylic acid from cystine (Wood & Cooley, 1956). Osuntokun et al. (1968) therefore pro-posed to use the sulphur-containing amino-acids in the therapy of these neurological conditions which could be the result of a chronic intoxication with cyanide derived from cassava and perhaps from other foods. The use of such therapy would certainly be justified in view of the old observa-tion (Voegtlin, Johnson & Dyer, 1926) that cystine effectively reduces the toxicity of cyanide. However, it must be stressed that the daily intake of cyanide by these patients is small, probably of the order of 8 mg (0·3 m-mole) per day (Osuntokun, 1968), so that the amount of cyst(e)ine required for its detoxication must be negligible in comparison with the amounts oxidized to produce the 25 m-moles of sulphate normally excreted each day. It is therefore hard to believe that the daily ingestion of 0·3 m-mole of cyanide would cause a significant drop in the concentration of cyst(e)ine in plasma unless a severe protein deficiency were also present: such is not the case to judge by the results of Osuntokun (1968).

These few examples obviously show, even with the rather scanty attention the subject has so far attracted, that disorders of the metabolism of the inorganic compounds of sulphur have important consequences in several fields of medicine. Doubtless many other important effects remain to be discovered.

# 13

# ECONOMIC ASPECTS OF INORGANIC
# SULPHUR METABOLISM

The role of biological activity in modifying the distribution of sulphur in nature was outlined briefly in chapter 1. Large-scale transformations by living organisms have important economic significance. In agriculture they influence to a considerable extent soil fertility while in geochemistry such transformations may result in the production of forms of sulphur which are amenable to economic exploitation. Further, secondary effects of inorganic sulphur metabolism, such as the formation of acid and alkali, may result in mobilization and immobilization of cations in natural situations and may contribute to various forms of corrosion.

Much effort, particularly of a technical nature, has been applied to a number of these economic aspects but it is not the purpose of this chapter to review this work in detail. Rather we wish to highlight briefly some of the more important of these aspects in order to emphasize that research on inorganic sulphur metabolism is not simply an interesting academic exercise but has much application to primary and secondary industry. Many of the references cited in this chapter are review articles to which the reader is referred for details.

## 13.1 Soil fertility

It is now generally accepted that the regeneration of sulphate from the reduced sulphur present in the organic residues in soils is brought about by the activities of the soil microflora (Starkey, 1950; Waksman, 1952; Alexander, 1961). The microflora, therefore, play a vital role in maintaining a supply of sulphate for the growth of plants. The exact mechanism(s) by which the organic sulphur is oxidized is not fully understood. Although it is possible that oxidation of hydrogen sulphide, released from cysteine by cysteine desulphydrase, may occur to a certain extent, there is a body of evidence that much of the oxidation takes place while sulphur remains bound in organic forms (Freney & Stevenson, 1966; Freney, 1967 b). The following pathway was proposed by Freney (see also § 11.1.1).

$$\begin{array}{ccccc}
CH_2SH & CH_2.S{-}S.CH_2 & & CH_2.S{-}SO_2.CH_2 \\
| & | \quad\quad | & & | \quad\quad\quad | \\
H.\overset{|}{C}.NH_2 \longrightarrow H.\overset{|}{C}.NH_2 & H.\overset{|}{C}.NH_2 \longrightarrow & H.\overset{|}{C}.NH_2 & H.\overset{|}{C}.NH_2 \\
| & | \quad\quad | & & | \quad\quad\quad | \\
COOH & COOH \quad COOH & & COOH \quad\quad COOH
\end{array}$$

Cysteine            Cystine                 Cystine 'disulphoxide'

$$\begin{array}{ccc}
CH_2.SO_2H & CH_2.SO_3H & \\
| & | & \\
H.\overset{|}{C}.NH_2 \longrightarrow & H.\overset{|}{C}.NH_2 \longrightarrow & SO_4^{2-} \\
| & | & \\
COOH & COOH &
\end{array}$$

Alanine              Alanine
3-sulphinic acid     3-sulphonic acid

Inorganic forms of sulphur such as elemental sulphur and thiosulphate are oxidized by soils with the intermediate formation of polythionates (Gleen & Quastel, 1953). The oxidation is probably due to the autotrophic sulphur-bacteria, thiobacilli, and to the combined action of heterotrophic bacteria, actinomycetes and fungi which have a limited ability to oxidize inorganic sulphur compounds (see chapter 9). The use of elemental sulphur as an artificial fertilizer (Burns, 1967) depends upon the action of the sulphur-oxidizing organisms and considerable increases in the populations of thiobacilli have been noted following the application of elemental sulphur to soils (e.g. Moser & Olson, 1953).

The reduction of sulphate to sulphide occurs in waterlogged soils where anaerobic conditions are created. The reaction becomes particularly important in situations such as rice-paddy fields where the formation of sulphide may give rise to a browning disease of rice which is prevented by addition of inhibitors of sulphate reduction (e.g. nitrate) to the soil (see Starkey, 1966).

Sulphate reduction in poorly drained or waterlogged soils results in the reduction of iron and its precipitation as ferrous sulphide with the production of alkaline conditions as is shown by equation 13.1 where H represents the source of reducing agent (e.g. organic matter; Starkey, 1966).

$$Na_2SO_4 + Fe(OH)_3 + 9H \rightarrow FeS + 2NaOH + 5H_2O \qquad (13.1)$$

Fixation of carbon dioxide by the alkali appears to be one of the major mechanisms for the formation of carbonates in poorly drained soils (e.g. Whittig & Janitzky, 1963). In some situations the formation of hydrated sodium carbonate (Natron) by this means reaches commercially exploitable proportions (Gubin & Tzechomskaja, 1930; Verner & Orlovskii, 1948; Abd-el-Malek & Rizk, 1963 a). The infertility of soils in the Nile Delta has been tentatively attributed to the alkalinity developed as the result of large-scale sulphate reduction by bacteria (Gracie, Rizk, Moukhtar & Moustafa, 1934).

The precipitation of insoluble calcium and magnesium carbonates is a frequently observed consequence of sulphate reduction in waterlogged soils (Whittig & Janitzky, 1963; Abd-el-Malek & Rizk, 1963 b; Ogata & Bower, 1965). By contrast, the oxidation of sulphur increases the acidity of soils and is generally accompanied by increases in soluble calcium, magnesium and iron. Soluble phosphorus also increases during sulphur oxidation and attempts have been made to exploit this fact by introducing rock-phosphate mixed with elemental sulphur to soils as a source of phosphate fertilizer (Nimgade, 1968).

## 13.2 Corrosion

Sulphur-metabolizing organisms are responsible, at least in part, for several types of corrosion. At times their activities have caused serious industrial problems until suitable remedies were developed. The production of sulphuric acid by oxidation of sulphide by *Thiobacillus* spp. is particularly prevalent and has been reported to cause corrosion of concrete sewers (Parker, 1945) and water cooling towers (Taylor & Hutchinson, 1947), of stone (Pochon, Coppier & Tchan, 1951) and of metal pipes (Postgate, 1959 a). In some of these instances evidence was obtained that the sulphide was generated by the action of sulphate-reducing organisms. An interesting example of corrosion by thiobacilli, reported by Thaysen, Bunker & Adams (1945), was that of rubber fire hoses which was caused by the oxidation of the vulcanized rubber lining by the sulphur-oxidizing bacteria.

The corrosion of buried iron pipes is characterized by the presence of sulphide in the corrosion products and it is now fairly generally accepted that the sulphate-reducing bacteria play a major role in this type of anaerobic corrosion process (Postgate, 1960 b). A widely held hypothesis for the mechanism is that of cathodic depolarization advanced by von Wolzogen Kühr & van der Vlugt (1934) in which the bacteria act as the depolarizing agents by utilizing the hydrogen available at the cathodic areas of the metal for the reduction of sulphate to sulphide (equations 13.2 to 13.8).

$$4Fe \rightarrow 4Fe^{2+} + 8e \text{ (anodic reaction)} \tag{13.2}$$

$$8H_2O \rightarrow 8H^+ + 8OH^- \left.\right\} \text{ cathodic reaction} \tag{13.3}$$
$$8H^+ + 8e \rightarrow 8H \tag{13.4}$$

$$SO_4^{2-} + 8H \rightarrow S^{2-} + 4H_2O \text{ (cathodic depolarization, bacterial)} \tag{13.5}$$

$$Fe^{2+} + S^{2-} \rightarrow FeS \text{ (corrosion product)} \tag{13.6}$$

$$3Fe^{2+} + 6OH^- \rightarrow 3Fe(OH)_2 \text{ (corrosion product)} \tag{13.7}$$

$$\text{Sum: } 4Fe + SO_4^{2-} + 4H_2O \rightarrow FeS + 3Fe(OH)_2 + 2OH^- \tag{13.8}$$

Although this mechanism has been challenged from time to time it has received convincing support from recent investigations (see Booth, 1964).

According to Iverson (1968) iron phosphide is also one of the corrosion products as the result of an interaction between ferrous ions and phosphine, produced by the sulphate-reducing bacteria by reduction of phosphate.

## 13.3 Recovery of base metals by leaching of sulphides

Considerable interest has been taken in recent years in the use of microorganisms in the recovery of base metals from low-grade sulphide ores (Woodcock, 1967). The acidophilic iron and sulphur-oxidizing thiobacilli, *Thiobacillus ferro-oxidans* and *Thiobacillus thio-oxidans*, are frequently found in natural environments in which oxidative leaching of sulphide minerals is taking place, and the oxidation of pyrite ($FeS_2$), for example, is thought to occur by a combination of biological and chemical reactions as illustrated by equations 13.9 to 13.12 (Temple & Delchamps, 1953).

$$\text{Chemical: } 2FeS_2 + 2H_2O + 7O_2 \rightarrow 2FeSO_4 + 2H_2SO_4 \tag{13.9}$$

$$\text{Biological: } 4FeSO_4 + O_2 + 2H_2SO_4 \rightarrow 2Fe_2(SO_4)_3 + 2H_2O \tag{13.10}$$

$$\text{Chemical: } Fe_2(SO_4)_3 + FeS_2 \rightarrow 3FeSO_4 + 2S \tag{13.11}$$

$$\text{Biological: } 2S + 3O_2 + 2H_2O \rightarrow 2H_2SO_4 \tag{13.12}$$

Several non-ferrous sulphide minerals, for example chalcocite ($Cu_2S$), covellite ($CuS$), enargite ($3Cu_2S.As_2S_5$) galena ($PbS$), millerite ($NiS$), molybdenite ($MoS_2$), orpiment ($As_2S_3$), sphalerite ($ZnS$) and tetrahedrite ($Cu_8Sb_2S_7$) are apparently attacked directly by *T. ferro-oxidans* (Silverman & Ehrlich, 1964), but in a number of cases the rate of release of the metal is greatly increased by the addition of pyrite or other iron salts.

It would appear that this stimulation is due to the chemical leaching of the metal by ferric ions (equation 13.13; Ehrlich & Fox, 1967).

$$MS + 2Fe^{3+} \rightarrow M^{2+} + S^0 + 2Fe^{2+} \qquad (13.13)$$

This reaction occurs readily under acidic conditions and the principle function of the iron- and sulphur-oxidizing organisms is to regenerate the acidic, ferric sulphate leaching liquor. The inclusion of elemental sulphur in biological leaching mixtures to increase the production of acid has been found to improve considerably the leaching efficiency (Napier, personal communication).

A number of metals have been successfully extracted from natural ore material in laboratory-scale experiments while the extraction of copper has been developed on a commercial scale which competes economically with the standard metallurgical methods.

## 13.4   Sulphide and sulphur deposits

### 13.4.1   Sulphide deposits

The idea that bacterial sulphate reduction may have played a role in the genesis of some natural base-metal sulphide deposits is a relatively old one. Siebenthal (1915) suggested that the zinc sulphide deposits of Missouri, Oklahoma and Kansas arose in this way and a similar suggestion was made by Bastin (1926) for the origin of copper, lead and zinc deposits of the type found in the Mississippi Valley. Over the last ten to fifteen years renewed interest has been taken in the possibility of 'biogenic' sulphide deposits. This interest has arisen, in part, from the realization that many stratiform sulphide ore bodies may not be magmatic or hydro-thermal in origin, as had previously been widely assumed by geologists, but may instead be sedimentary in nature and laid down under moderate temperatures and pressures (Roberts, 1967; King, 1967).

It cannot be said that a major role for sulphate-reducing bacteria in the formation of economic sulphide deposits has been proven. Several observations, however, indicate that this could be a theoretical possibility. There is no doubt that at the present time sulphate reduction is a wide-spread activity and is responsible for extensive precipitation of iron sulphides (see Silverman & Ehrlich, 1964). Sulphides of metals other than iron have not been shown to form under natural conditions by biological action but under laboratory conditions sulphides of copper, silver, lead, zinc, antimony, cobalt, cadmium and nickel have been produced in cultures of sulphate-reducing bacteria (Miller, 1950; Baas Becking &

Moore, 1961; Suckow & Schwartz, 1963). A number of the sulphides showed distinct mineralogical characteristics (Baas Becking & Moore, 1961). These experiments demonstrated unequivocally that sulphate-reducing bacteria can exist and be active in situations in which heavy metal sulphides may arise. They belie the criticism of Davidson (1962) and Booth & Mercer (1963) that sulphate-reducing bacteria could play no role in the formation of, for example, copper sulphides since copper is extremely toxic to these organisms. These criticisms took no account of the fact that in natural situations the bulk of the copper is nonionic (Cheney & Jensen, 1962) and is consequently very much less toxic than free copper ions (Sadler & Trudinger, 1967) nor did they allow for the detoxifying effect of the hydrogen sulphide produced by sulphate reduction (Temple & Le Roux, 1964).

### 13.4.2 Sulphur

The role of sulphur-metabolizing organisms in the production of sulphur deposits is now undisputed (Ivanov, 1968). Butlin & Postgate (1954, see also Butlin, 1953) demonstrated that certain warm, saline lakes in Libya, notably Ain-ez-Zania, are heavily infected with *Desulfovibrio desulfuricans* and they obtained convincing evidence that sulphur formation in these lakes is due to oxidation of hydrogen sulphide produced by *D. desulfuricans* by the photosynthetic, sulphide-oxidizing bacteria *Chromatium* and *Chlorobium*. The photosynthetic bacteria also provide the organic matter necessary to support sulphate reduction. The lakes yielded 100–200 tons of crude sulphur per annum. In other situations (e.g. Ljunggren, 1960; Ivanov, 1957, 1960; Sokolova, 1962) hydrogen sulphide of volcanic or fumarolic origin is oxidized to sulphur by photosynthetic bacteria, the chemosynthetic thiobacilli or *Beggiatoa*.

Major economic sulphur deposits for which a biogenic origin has been suggested include deposits in Sicily (Hunt, 1915; Dessau, Jensen & Nakai, 1962), France (Schneegans, 1935), Texas and Louisiana (Starkey & Wight, 1945; Jones, Starkey, Feely & Kulp, 1956) and the Carpathian sulphur deposits (Kuznetsov, Ivanov & Lyalikova, 1963).

### 13.4.3 Fractionation of stable sulphur isotopes

Although not in itself of economic significance the ability of sulphate-reducing organisms to discriminate partly between the stable isotopes of sulphur has been used in speculations on the origin of natural sulphide and sulphur deposits.

Sulphur has four stable isotopes whose average relative abundance is as follows: $^{32}$S, 95·1%; $^{33}$S, 0·74%; $^{34}$S, 4·2%; $^{36}$S, 0·016%

Natural sulphur-containing materials, however, vary widely in isotope composition, the $^{32}$S/$^{34}$S ratio ranging from about 21·3 to 23·2. Meteoritic sulphur is of almost constant composition with a $^{32}$S/$^{34}$S ratio of about 22·2 which is often assumed to be that of 'primordial' terrestial sulphur. Variations from this value are taken as evidence of isotope fractionation during natural transformations of sulphur and are generally expressed relative to the meteoritic value.

$$\delta^{34}S\ (‰) = \frac{^{34}S/^{32}S\ \text{sample} - {^{34}S/^{32}S}\ \text{meteoritic}}{^{34}S/^{32}S\ \text{meteoritic}} \times 1000$$

At equilibrium, the isotope exchange reaction (equation 13.14) has been calculated to give an enrichment of $^{34}$S in sulphate with a $\delta^{34}$S of 71‰ at 25° with respect to the initial sulphate (Tudge & Thode, 1950).

$$^{32}SO_4{}^{2-} + H_2{}^{34}S \rightleftharpoons {^{34}SO_4{}^{2-}} + H_2{}^{32}S \qquad (13.14)$$

This most favoured isotope distribution has not been realized in laboratory experiments although it is approached in waters of Green Lake, New York (Deevey, Nakai & Stuiver, 1963; Nakai & Jensen, 1964).

The chemical reduction of sulphate (when sulphate is in excess) at 18–50° results in an enrichment of $^{32}$S in sulphide ($\delta^{34}$S $= -22‰$) due to a kinetic discrimination against the heavier $^{34}$S atom (Harrison & Thode, 1957).

The bacterial reduction of sulphate is, however, more complicated and in laboratory experiments enrichments of $^{32}$S in hydrogen sulphide give $\delta^{34}$S values ranging from zero to $-46‰$ (see Harrison & Thode, 1958; Jones & Starkey, 1957, 1962; Kaplan & Rittenberg, 1962a, 1964). The extent of sulphur isotope fractionation by sulphate-reducing bacteria appears to depend partly on the rate of reduction, lower rates tending to result in greater fractionation (Harrison & Thode, 1958; Kaplan & Rittenberg, 1964). Other factors, including the nature of the hydrogen donor, also control the fractionation in unknown ways (Kemp & Thode, 1968). An examination of the isotope composition of natural sulphide deposits (table 13.1) reveals that the so-called sedimentary deposits show the widest spread, and have a tendency to be enriched to a considerable extent in $^{32}$S relative to meteoritic sulphur and to magmatic or hydro-thermal sulphide. These results would be consistent with a 'biogenic' origin of the sedimentary sulphides.

In certain sulphur deposits the sulphur is considerably enriched in $^{32}$S relative to sulphate in the same environments, a result which has been

TABLE 13.1  *Distribution of sulphur isotopes in natural sulphides*
(*Adapted from Ault, 1959; Ault & Kulp, 1960*)

| Source of sulphide | $\delta^{34}S$ (‰) relative to meteoritic ($^{32}S/^{34}S = 22\cdot21$) | |
|---|---|---|
| | range | mean |
| hydrothermal (including ore deposits) | $+13\cdot3$ to $-36\cdot4$ | $+ 4\cdot5$ |
| silicic igneous rocks | $+10\cdot0$ to $- 0\cdot9$ | $+ 4\cdot7$ |
| all mafic igneous rocks | $+ 7\cdot7$ to $- 1\cdot8$ | $+ 2\cdot7$ |
| large bodies of mafic igneous rocks | $+ 6\cdot3$ to $- 0\cdot9$ | $+ 4\cdot5$ |
| sedimentary | $+44\cdot2$ to $-42\cdot7$ | $-12\cdot0$ |

taken to indicate that the sulphur arose by biological activity (see Thode, Wanless & Wallouch, 1954; Feely & Kulp, 1957; Jensen, 1962).

Isotope fractionation, to varying extents, also occurs during sulphite reduction by *Desulfovibrio*, during sulphate and sulphite reduction by assimilatory sulphate-reducing organisms, and during sulphide oxidation by chemosynthetic and photosynthetic bacteria (see Kaplan & Rittenberg, 1962 a, 1964; Nakai & Jensen, 1964; Mekhtieva, 1964; Mekhtieva & Kondrat'eva, 1966; Kondrat'eva & Mekhtieva, 1966; Krouse, McCready, Husain & Campbell, 1967).

These processes undoubtedly contribute to the natural distribution of sulphur isotopes and help complicate the interpretations of such distribution in geological terms.

### 13.5  Petroleum technology

The invariable association between sulphate-reducing bacteria and oil-bearing waters and oil deposits has given rise to frequent speculations that these bacteria are involved in the genesis of oil (Bunker, 1936; Davis, 1967). Oil-like materials have been reported in cultures of sulphate-reducing bacteria (Jankowski & ZoBell, 1944; Sisler & ZoBell, 1951; Hvid-Hansen, 1951) but whether the production of these materials is significant from the point of view of natural oil formation is an open question.

Rather better established is a role for sulphate-reducing organisms in the secondary recovery of oil from shales. ZoBell (1947 a,b) reported that cultures of sulphate-reducing bacteria, grown in oil and sand mixtures, released the oil which remained absorbed to the sand in their absence. The oil was not utilized for growth. This phenomenon was confirmed by Dostálek & Spurný (1957) and appears to be applicable to industrial operations (Dostálek & Spurný, 1958; Dostálek, 1961). It would appear that the release of oil is due to one or more of the following mechanisms (ZoBell, 1950; Stone & ZoBell, 1952).

1. The production of gases within the pores of the absorbent.
2. Displacement of oil from the absorbing particles by the bacteria which have a tendency to adhere to such particles.
3. Formation of surface active agents by the bacteria.
4. Bacterial oxidation of the oil which might reduce its viscosity.
5. In natural situations, the utilization of mineral sulphates or carbonates to release oil absorbed to such materials.

Some support for the third mechanism was obtained by La Rivière (1955) who showed that sulphate-reducing bacteria produce a material which increases the mobility of oil. This was not, however, a unique property of the sulphate-reducing organisms since several other species of bacteria also excreted detergent-like materials.

### 13.6 Miscellaneous

The pollution of natural waters as the result of the activities of sulphur-metabolizing organisms now appears to be well established. On the one hand it may be caused by the liberation of sulphuric acid and toxic metals due to the oxidation of metal sulphides in mine wastes and in bituminous coal deposits (see § 13.3) and, on the other, to the formation of toxic hydrogen sulphide by sulphate-reducing bacteria in anaerobic environments (Butlin, 1949). Additional deleterious effects of biological sulphate reduction include the blackening of paper pulp due to the precipitation of FeS generated from sulphates and sulphites; the contamination of town gas with $H_2S$ (Senez, Geoffray & Pichinoty, 1956; Pankhurst, 1968) and the spoilage of canned foods (Werkman, 1929) and of brines for preserving olives (Levin, Ng, Nagel & Vaughn, 1959; Levin & Vaughn, 1966). Lastly, mention must be made of the possible role of sulphate-reducing bacteria and both photosynthetic and chemo-synthetic bacteria in the maturation of thermal 'medicinal' muds (Starka, 1951; Laporte, Laurent & Kaiser, 1965).

This is by no means a complete list of the economic effects of inorganic sulphur metabolism and, no doubt, many others will be uncovered in the future. These examples, however, serve to illustrate the wide-ranging implications of the particular branch of biochemistry and to stress the importance, to other disciplines, of research in sulphur biochemistry.

# BIBLIOGRAPHY

Abbott, E. V. (1923). *Soil Sci.* **16**, 207.

Abbott, L. D. (1947). *Archs. Biochem.* **15**, 205.

Abd-el-Malek, Y. & Rizk, S. G. (1958). *Nature, Lond.* **182**, 538.

Abd-el-Malek, Y. & Rizk, S. G. (1960). *Nature, Lond.* **185**, 635.

Abd-el-Malek, Y. & Rizk, S. G. (1963*a*). *J. appl. Bact.* **26**, 20.

Abd-el-Malek, Y. & Rizk, S. G. (1963*b*). *J. appl. Bact.* **26**, 14.

Abegg, R., Auerbach, Fr. & Koppel, I. (1927). *Handbuch der Anorganischen Chemie.* Vol. 4, Part I, 1st half. Leipzig: S. Hirzel.

Abelson, P. H. (1966). *Proc. natn. Acad. Sci. U.S.A.* **55**, 1365.

Abraham, A. & Bachhawat, B. K. (1963). *Biochim. biophys. Acta* **70**, 104.

Abraham, A. & Bachhawat, B. K. (1964). *Indian J. Biochem.* **1**, 192.

Abrahams, S. C. (1955). *Acta Cryst.* **8**, 661.

Adair, F. W. (1966). *J. Bact.* **92**, 899.

Adair, F. W. (1968). *J. Bact.* **95**, 147.

Adair, F. W. & Umbreit, W. W. (1965). *Bact. Proc.* p. 84.

Adams, J. B. (1960). *Biochem. J.* **76**, 520.

Adams, J. B. (1962). *Biochim. biophys. Acta* **62**, 17.

Adams, J. B. (1963). *Archs. Biochem. Biophys.* **101**, 478.

Adams, J. B. (1964*a*). *Biochim. biophys. Acta* **82**, 572.

Adams, J. B. (1964*b*). *J. clin. Endocr. Metab.* **24**, 988.

Adams, J. B. (1964*c*). *Biochim. biophys. Acta* **83**, 127.

Adams, J. B. (1967). *Biochim. biophys. Acta* **146**, 522.

Adams, J. B. & Chulavatnatol, M. (1967). *Biochim. biophys. Acta* **146**, 509.

Adams, J. B. & Edwards, A. M. (1968). *Biochim. biophys. Acta* **167**, 122.

Adams, J. B. & Meaney, M. F. (1961). *Biochim. biophys. Acta* **54**, 592.

Adams, J. B. & Poulos, A. (1967). *Biochim. biophys. Acta* **146**, 493.

Adams, J. B. & Rienits, K. G. (1961). *Biochim. biophys. Acta* **51**, 567.

Adams, M. E., Butlin, K. R., Hollands, S. J. & Postgate, J. R. (1951). *Research, Lond.* **4**, 245.

Adams, M. E. & Postgate, J. R. (1959). *J. gen. Microbiol.* **20**, 252.

Adelstein, S. J. & Vallee, B. L. (1962). In *Mineral Metabolism*, Vol. 2 B, p. 32. Ed. by Comar, C. L. & Bronner, F. New York: Academic Press.

Agarwala, U., Rees, C. E. & Thode, H. G. (1965). *Can. J. Chem.* **43**, 2802.

Ahlström, L., von Euler, H., Gernow, I. & Hägglund, B. (1944). *Ark. Kemi. Miner. Geol.* 18 A, No. 20.

Akagi, J. M. (1964). *J. Bact.* **88**, 813.

Akagi, J. M. (1965). *Biochem. biophys. Res. Commun.* **21**, 72.

Akagi, J. M. (1967). *J. biol. Chem.* **242**, 2478.

Akagi, J. M. & Campbell, L. L. (1961). *J. Bact.* **82**, 927.

Akagi, J. M. & Campbell, L. L. (1962*a*). *Bact. Proc.* p. 102.

Akagi, J. M. & Campbell, L. L. (1962*b*). *J. Bact.* **84**, 1194.

Akagi, J. M. & Campbell, L. L. (1963). *J. Bact.* **86**, 563.

Aleem, M. I. H. (1965). *J. Bact.* **90**, 95.

Aleem, M. I. H. (1966*a*). *Biochim. biophys. Acta* **128**, 1.

Aleem, M. I. H. (1966*b*). *J. Bact.* **91**, 729.

Aleem, M. I. H. & Huang, E. (1965). *Biochem. biophys. Res. Commun.* **20**, 515.

Aleem, M. I. H., Ross, A. J. & Schoenhoff, R. L. (1968). *Bact. Proc.* p. 140.

Allen, E. & Roy, A. B. (1968). *Biochim. biophys. Acta* **168**, 243.

Alexander, M. (1961). *Introduction to Soil Microbiology.* New York: John Wiley & Sons Ltd.

Alexander, P. (1960). In *Radiation Protection and Recovery*, p. 3. Ed. by Hollaender, A. New York: Pergamon Press.

Alico, R. K. & Liegey, F. W. (1966). *J. Bact.* **91**, 1112.

Ames, B. N. & Garry, B. (1959). *Proc. natn. Acad. Sci. U.S.A.* **45**, 1453.

Ammon, R. & Keutel, G. (1960). *Annls. Univ. Sarav.* **8**, 37.

Ammon, R. & Ney, K. H. (1957). *Archs. Biochem. Biophys.* **69**, 178.

Ammon, R. & Ney, K. H. (1959). *Hoppe-Seyler's Z.* **314**, 240.

Anastasi, A., Bertaccini, G. & Erspamer, V. (1966). *Br.J.pharmac. Chemotherap.* **27**, 479.

Anastasi, A., Erspamer, V. & Endean, R. (1967). *Experientia* **23**, 699.

Andersen, S. O. (1959*a*). *Acta chem. scand.* **13**, 120.

Andersen, S. O. (1959*b*). *Acta chem. scand.* **13**, 884.

Andersen, S. O. (1959*c*). *Acta chem. scand.* **13**, 1671.

Andrews, L. W. (1903). *J. Am. chem. Soc.* **25**, 756.

Arcos, M. & Lieberman, S. (1967). *Biochemistry* **6**, 2032.

Arnon, D. I. (1961). In *Biological Structure and Function*, Vol. II, p. 339. Ed. by Goodwin, T. W. & Lindberg, O. London: Academic Press.

Arnon, D. I., Losada, M., Nozaki, M. & Tagawa, K. (1961). *Nature, Lond.* **190**, 601.

Arrigoni, O. (1959). *Ital. J. Biochem.* **8**, 181.

Arrigoni, O. & Rossi, G. (1961). *G. Biochim.* **10**, 463.

Artman, M. (1956). *J. gen. Microbiol.* **14**, 315.

Asada, K. (1967). *J. biol. Chem.* **242**, 3646.

Asada, K. & Bandurski, R. S. (1967). Quoted by Thompson (1967).

Asada, K., Tamura, G. & Bandurski, R. S. (1966). *Biochem. biophys. Res. Commun.* **25**, 529.

Asada, K., Tamura, G. & Bandurski, R. S. (1968). *Biochem. biophys. Res. Commun.* **30**, 554.

Asahi, T. (1963). *Agric. biol. Chem.* **27**, 734.

Asahi, T. (1964). *Biochim. biophys. Acta* **82**, 58.

Asahi, T., Bandurski, R. S. & Wilson, L. G. (1961). *J. biol. Chem.* **236**, 1830.

Asahi, T. & Minamikawa, T. (1960). *J. Biochem., Tokyo* **48**, 548.

Aubert, J-P. (1956). In *Colloque sur la Biochimie du Soufre*, p. 81. Paris: Editions CNRS.

Aubert, J-P., Milhaud, G. & Millet, J. (1956). *C. r. hebd. Seanc. Acad. Sci., Paris* **242**, 2059.

Aubert, J-P., Milhaud, G. & Millet, J. (1957). *Annls. Inst. Pasteur, Paris* **92**, 515.

Aubert, J-P., Milhaud, G., Moncel, C. & Millet, J. (1958). *C. r. hebd. Seanc. Acad. Sci., Paris* **246**, 1616.

Ault, W. U. (1959). In *Researches in Geochemistry*, p. 241. Ed. by Abelson, P. H. New York: John Wiley & Sons.

Ault, W. U. & Kulp, J. L. (1960). *Econ. Geol.* **55**, 73.

Austin, J. H. (1965). In *Medical Aspects of Mental Retardation*, p. 768. Ed. by Carter, C. H. Springfield: C. C. Thomas.

Austin, J., Armstrong, D. & Shearer, L. (1965). *Archs. Neurol., Chicago* **13**, 593.

## BIBLIOGRAPHY

Austin, J., Armstrong, D., Shearer, L. & McAfee, D. (1966). *Archs. Neurol., Chicago* **14**, 259.

Austin, J. H., Balasubramanian, A. S., Pattabiraman, T. N., Saraswathi, S., Basu, D. K. & Bachhawat, B. K. (1963). *J. Neurochem.* **10**, 805.

Austin, J. H. & Bischel, M. (1961). *Blood* **17**, 216.

Austin, J., McAfee, D. & Shearer, L. (1965). *Archs. Neurol., Chicago* **12**, 447.

Austin, J., McAfee, D., Armstrong, D., O'Rourke, M., Shearer, L. & Bachhawat, B. (1964). *Biochem. J.* **93**, 15c.

Avers, C. J. (1961). *Am. J. Bot.* **48**, 137.

Awapara, J. (1953). *J. biol. Chem.* **203**, 183.

Awapara, J. & Doctor, V. M. (1955). *Archs. Biochem. Biophys.* **58**, 506.

Awapara, J. & Wingo, W. J. (1953). *J. biol. Chem.* **203**, 189.

Awtrey, A. D. & Connick, R. E. (1951). *J. Am. chem. Soc.* **73**, 1842.

Azoulay, E., Puig, J. & Pichinoty, F. (1967). *Biochem. biophys. Res. Commun.* **27**, 270.

Baalsrud, K. (1954). In *Autotrophic Micro-organisms*, p. 54. Ed. by Fry, B. A. & Peel, J. L. London: Cambridge University Press.

Baalsrud, K. & Baalsrud, K. S. (1952). In *Phosphorus Metabolism*, Vol. 2, p. 544. Ed. by McElroy, W. D. & Glass, B. Baltimore: Johns Hopkins Press.

Baalsrud, K. & Baalsrud, K. S. (1954). *Arch. Mikrobiol.* **20**, 34.

Baars, E. K. (1930). *Over Sulfaatreductie door Bacteriën.* Dissertation W. D. Meinema, N. V. Delft. Holland.

Baas Becking, L. G. M. & Moore, D. (1961). *Econ. Geol.* **56**, 259.

Bachhawat, B. K., Austin, J. & Armstrong, D. (1967). *Biochem. J.* **104**, 15c.

Baddiley, J., Buchanan, J. G. & Letters, R. (1957). *J. chem. Soc.* p. 1067.

Baddiley, J., Buchanan, J. G. & Letters, R. (1958). *J. chem. Soc.* p. 1000.

Baddiley, J., Buchanan, J. G., Letters, R. & Sanderson, A. R. (1959). *J. chem. Soc.* p. 1731.

Baddiley, J. & Sanderson, A. R. (1963). *Biochem. Prep.* **10**, 3.

Baer, J. E. & Carmack, M. (1949). *J. Am. chem. Soc.* **71**, 1215.

Baker, F. D., Papiska, H. R. & Campbell, L. L. (1962). *J. Bact.* **84**, 973.

Balasubramanian, A. S. & Bachhawat, B. K. (1961). *J. scient. ind. Res.* **20C**, 202.

Balasubramanian, A. S. & Bachhawat, B. K. (1962). *Biochim. biophys. Acta* **59**, 389.

Balasubramanian, A. S. & Bachhawat, B. K. (1963). *J. Neurochem.* **10**, 201.

Balasubramanian, A. S. & Bachhawat, B. K. (1964). *J. Neurochem.* **11**, 877.

Balasubramanian, A. S. & Bachhawat, B. K. (1965a). *Biochim. biophys. Acta* **106**, 218.

Balasubramanian, A. S. & Bachhawat, B. K. (1965b). *Indian J. Biochem.* **2**, 212.

Balasubramanian, A. S., Spolter, L., Rice, L. I., Sharon, J. B. & Marx, W. (1967). *Analyt. Biochem.* **21**, 22.

Balharry, G. J. E. & Nicholas, D. J. D. (1968). *Proc. Aust. biochem. Soc.* p. 60.

Baliga, B. S., Vartak, H. G. & Jagannathan, V. (1961). *J. scient. ind. Res.* **20C**, 33.

Bandurski, R. S. (1965). In *Plant Biochemistry*, p. 467. Ed. by Bonner, J. & Varner, J. E. New York: Academic Press.

Bandurski, R. S., Wilson, L. G. & Asahi, T. (1960). *J. Am. chem. Soc.* **82**, 3218.

Banerjee, R. K. & Roy, A. B. (1966). *Molec. Pharm.* **2**, 56.
Banerjee, R. K. & Roy, A. B. (1967*a*). *Proceedings of the Second International Congress on Hormonal Steroids*, p. 397. Excerpta Medica International Congress Series, No. 132.
Banerjee, R. K. & Roy, A. B. (1967*b*). *Biochim. biophys. Acta* **137**, 211.
Banerjee, R. K. & Roy, A. B. (1968). *Biochim. biophys. Acta* **151**, 573.
Barber, M., Brooksbank, B. W. L. & Kluper, S. W. A. (1951). *J. path. Bact.* **63**, 57.
Barker, B. T. P. (1929). *Long Ashton Agr. Hort. Res. Sta. Ann. Rept.* p. 130.
Barker, B. T. P. (1930). *Rev. appl. Mycol.* **9**, 732.
Barker, H. A. & Kornberg, A. (1954). *J. Bact.* **68**, 655.
Barker, S. A., Cruickshank, C. N. D. & Webb, T. (1965). *Carb. Res.* **1**, 62.
Barra, H. S. & Caputto, R. (1965). *Biochim. biophys. Acta* **101**, 367.
Bartlett, J. K. & Skoog, D. A. (1954). *Analyt. Chem.* **26**, 1008.
Bartlett, P. D., Colter, A. K., Davis, R. E. & Roderick, W. R. (1961). *J. Am. chem. Soc.* **83**, 109.
Bartlett, P. D., Cox, E. F. & Davis, R. E. (1961). *J. Am. chem. Soc.* **83**, 103.
Bartlett, P. D. & Davis, R. E. (1958). *J. Am. chem. Soc.* **80**, 2513.
Barton, A. D. & Young, L. (1943). *J. Am. chem. Soc.* **65**, 294.
Bastin, E. S. (1926). *J. Geol.* **34**, 773.
Batts, B. D. (1966*a*). *J. chem. Soc.* (B), p. 547.
Batts, B. D. (1966*b*). *J. chem. Soc.* (B), p. 551.
Bauchop, T. & Elsden, S. R. (1960). *J. gen. Microbiol.* **23**, 457.
Baulieu, E. E. (1962). *J. clin. Endocr. Metab.* **22**, 501.
Baulieu, E. E., Corpéchot, C., Dray, F., Emiliozzi, R., Lebeau, M. C., Mauvais-Jarvis, P. & Robel, P. (1965). *Recent Prog. Horm. Res.* **21**, 411.
Baulieu, E. E., Fabre-Jung, I. & Huis in't Veld, L. G. (1967). *Endocrinology* **81**, 34.
Baum, H. & Dodgson, K. S. (1957). *Nature, Lond.* **179**, 312.
Baum, H. & Dodgson, K. S. (1958). *Biochem. J.* **69**, 573.
Baum, H., Dodgson, K. S. & Spencer, B. (1958). *Biochem. J.* **69**, 567.
Baum, H., Dodgson, K. S. & Spencer, B. (1959). *Clinica chim. Acta* **4**, 453.
Bauman, E. (1876). *Pfluger's Arch. ges. Physiol.* **12**, 63, 69; **13**, 285.
Baumgarten, P. (1926). *Ber. dt. chem. Ges.* **59**, 1166.
Baumgarten, P. (1930). *Ber. dt. chem. Ges.* **63**, 1330.
Baxter, C. F. & van Reen, R. (1958). *Biochim. biophys. Acta* **28**, 567, 573.
Baxter, C. F., van Reen, R. & Pearson, P. B. (1956). *Fedn. Proc.* **15**, 215.
Baxter, C. F., van Reen, R., Pearson, P. B. & Rosenberg, C. (1958). *Biochim. biophys. Acta* **27**, 584.
Beck, J. V. (1960). *J. Bact.* **79**, 502.
Beck, J. V. & Elsden, S. R. (1958). *J. gen. Microbiol.* **19**, i.
Beck, J. V. & Shafia, F. M. (1964). *J. Bact.* **88**, 850.
Beijerinck, M. W. (1895). *Zentbl. Bakt. ParasitKde.* Abt II, **1**, 1.
Beijerinck, M. W. (1900). *Zentbl. Bakt. ParasitKde.* Abt II, **6**, 193.
Beijerinck, M. W. (1901). *Archs. neerl. Sci. Ser.* 2, **4**, 1.
Beijerinck, M. W. (1904). *Zentbl. Bakt. ParasitKde.* Abt II, **11**, 593.
Bénard, H., Gajdos, A. & Gajdos-Török, M. (1948*a*). *C. r. Seanc. Soc. Biol.* **142**, 150.
Bénard, H., Gajdos, A. & Gajdos-Török, M. (1948*b*). *Presse med.* **56**, 269.
Bénard, H., Gajdos, A. & Gajdos-Török, M. (1949). *Revue Path. comp. Hyg. gen.* **49**, 72.

Bénard, H., Gajdos-Török, M. & Gajdos, A. (1947). *C. r. Seanc. Soc. Biol.* **141**, 702.

Benesch, R. E. & Benesch, R. (1958). *J. Am. chem. Soc.* **80**, 1666.

Benkovic, S. J. & Benkovic, P. A. (1966). *J. Am. chem. Soc.* **88**, 5504.

Bennet, M. A. (1937). *Biochem. J.* **31**, 962.

Benson, A. A. (1963). *Adv. Lipid Res.* **1**, 387.

Benson, A. A. & Atkinson, M. R. (1967). *Fedn. Proc.* **26**, 394.

Bergeret, B. & Chatagner, F. (1954). *Biochim. biophys. Acta* **14**, 297.

Berglund, F., Helander, C.-G. & Howe, R. B. (1960). *Am. J. Physiol.* **198**, 586

Berglund, F. & Sörbo, B. (1960). *Scand. J. clin. Lab. Invest.* **12**, 147.

Bergmann, W. (1962). In *Comparative Biochemistry*, Vol. 3A, p. 103. Ed. by Florkin, M. & Mason, H. S. New York: Academic Press.

Berlow, S. (1967). *Adv. clin. Chem.* **9**, 165.

Bernstein, S., Dusza, J. P. & Joseph, J. P. (1968). *Physical Properties of Steroid Conjugates.* New York: Springer Verlag.

Bernstein, S., Dusza, J. P. & Joseph, J. P. (1969). In *Chemical and Biological Aspects of Steroid Conjugation*, Ed. Bernstein, S. & Solomon, S. New York: Springer Verlag.

Bernstein, S. & McGilvery, R. W. (1952*a*). *J. biol. Chem.* **198**, 195.

Bernstein, S. & McGilvery, R. W. (1952*b*). *J. biol. Chem.* **199**, 745.

Bethge, P. O. (1956). *Analyt. Chem.* **28**, 119.

Bianchi, M. (1955). *Pathologica*, **47**, 39, 43.

Bielig, H.-J., Pfleger, K., Rummel, W. & Seifen, E. (1961). *Hoppe-Seyler's Z.* **327**, 35.

Bighi, C. & Trabanelli, G. (1955*a*). *Boll. scient. Fac. Chim. ind. Univ. Bologna* **13**, 100.

Bighi, C. & Trabanelli, G. (1955*b*). *Annali Chim.* **45**, 1186.

Bighi, C., Trabanelli, G. & Pancaldi, G. (1958). *Boll. scient. Fac. Chim. ind. Univ. Bologna* **16**, 92.

Binkley, F. & Okeson, D. (1950). *J. biol. Chem.* **182**, 273.

Bisset, K. A. & Grace, J. B. (1954). In *Autotrophic Micro-organisms*, p. 28. Ed. by Fry, B. A. & Peel, J. L. London: Cambridge University Press.

Błeszyński, W. (1967). *Enzymologia* **32**, 169.

Błeszyński, W. & Działoszyński, L. M. (1965). *Biochem. J.* **97**, 360.

Błeszyński, W. & Leźnicki, A. (1967). *Enzymologia*, **33**, 373.

Block, R. J. & Stekol, J. A. (1950). *Proc. Soc. exp. Biol. Med.* **73**, 391.

Block, R. J., Stekol, J. A. & Loosli, J. K. (1951). *Archs. Biochem. Biophys.* **33**, 353.

Booth, G. H. (1964). *J. appl. Bact.* **27**, 174.

Booth, G. H. & Mercer, S. J. (1963). *Nature, Lond.* **199**, 622.

Booth, G. H., Miller, J. D. A., Paisley, H. M. & Saleh, A. M. (1966). *J. gen. Microbiol.* **44**, 83.

Borichewski, R. M. (1965). *Bact. Proc.* p. 77.

Borichewski, R. M. & Umbreit, W. W. (1966). *Archs. Biochem. Biophys.* **116**, 97.

Boström, H. & Åqvist, S. (1952). *Acta chem. scand.* **6**, 1557.

Boström, H., Berntsen, K. & Whitehouse, M. W. (1964). *Biochem. Pharmac.* **13**, 413.

Boström, H., Franksson, C. & Wengle, B. (1964). *Acta Endocr., Copenh.* **47**, 633.

Boström, H. & Odeblad, E. (1953). *Ark. Kemi* **6**, 39.

Boström, H. & Wengle, B. (1964). *Acta Soc. Med. Upsal.* **69**, 41.
Boström, H. & Wengle, B. (1967). *Acta Endocr., Copenh.* **56**, 691.
Bowden, D. H., Fraser, D., Jackson, S. H. & Ford Walker, N. (1956). *Medicine, Baltimore* **35**, 335.
Bowen, T. J., Butler, P. J. & Happold, F. C. (1965*a*). *Biochem. J.* **95**, 5P.
Bowen, T. J., Butler, P. J. & Happold, F. C. (1965*b*). *Biochem. J.* **97**, 651.
Bowen, T. J. & Cook, W. K. T. (1966). *J. Chromatog.* **22**, 488.
Bowen, T. J., Happold, F. C. & Taylor, B. F. (1966). *Biochim. biophys. Acta* **118**, 566.
Boxer, G. E. & Rickards, J. C. (1952). *Archs. Biochem. Biophys.* **39**, 7.
Boyland, E., Manson, D. & Orr, S. F. D. (1957). *Biochem. J.* **65**, 417.
Boyland, E., Manson, D. & Sims, P. (1953). *J. chem. Soc.* p. 3623.
Boyland, E., Manson, D., Sims, P. & Williams, D. C. (1956). *Biochem. J.* **62**, 68.
Boyland, E. & Sims, P. (1958). *J. chem. Soc.* p. 4198.
Boyland, E., Sims, P. & Williams, D. C. (1956). *Biochem. J.* **62**, 546.
Boyland, E., Wallace, D. M. & Williams, D. C. (1955). *Br. J. Cancer* **9**, 62.
Bradlow, H. L. (1969). In *Chemical and Biological Aspects of Steroid Conjugation.* Ed. by Bernstein, S. & Solomon, S. New York: Springer-Verlag.
Brauer, E. & Staude, H. (1953). *Z. wiss Photogr.* **48**, 16.
Braun, H. & Silberstein, W. (1942). *İnstab. Üniv. Fen. Fak. Mecm.* B7, 1.
Bray, H. G., Humphris, B. G., Thorpe, W. V., White, K. & Wood, P. B. (1952). *Biochem. J.* **52**, 419.
Bregoff, H. M. & Kamen, M. D. (1952). *J. Bact.* **63**, 147.
Bridgwater, R. J. & Ryan, D. A. (1957). *Biochem. J.* **65**, 24P.
Brodskii, A. I. (1954). *Usp. Khim.* **23**, 614.
Brodskii, A. I. & Eremenko, R. K. (1954). *Zh. obshch. Khim.* **24**, 1142.
Broekhuysen, J. (1958). *Analytica chim. Acta* **19**, 542.
Bruff, B. S. & Johnson, E. J. (1966). *Bact. Proc.* p. 94.
Brüggemann, J., Schlossmann, K., Merkenschlager, M. & Waldschmidt, M. (1962). *Biochem. Z.* **335**, 392.
Brüggemann, J. & Waldschmidt, M. (1962). *Biochem. Z.* **335**, 408.
Brunngraber, E. G. (1958). *J. biol. Chem.* **233**, 472.
Buch Andersen, A. (1936). *Z. phys. Chem.* B32, 237.
Buchanan, R. E. & Fulmer, E. I. (1930). *Physiology and Biochemistry of Bacteria*, Vol. III, p. 201. Baltimore: Williams & Wilkins Co.
Bunker, H. J. (1936). *A Review of the Physiology and Biochemistry of the Sulphur Bacteria.* Dept. Sci. Ind. Res. Spec. Rep. No. 3. London: H.M.S.O.
Bunton, C. A. & Hendy, B. N. (1963). *J. chem. Soc.* p. 3130.
Burke, J. J. & Jones, G. E. (1968). *Bact. Proc.* p. 57.
Burkhardt, G. N., Evans, A. G. & Warhurst, E. (1936). *J. chem. Soc.* p. 25.
Burkhardt, G. N., Ford, W. G. K. & Singleton, E. (1936). *J. chem. Soc.* p. 17.
Burkhardt, G. N., Horrex, C. & Jenkins, D. I. (1936). *J. chem. Soc.* pp. 1649 and 1654.
Burkhardt, G. N. & Lapworth, A. (1926). *J. chem. Soc.* p. 684.
Burkhardt, G. N. & Wood, H. (1929). *J. chem. Soc.* p. 141.
Burns, G. R. (1967). *Oxidation of Sulphur in Soils.* The Sulphur Institute. Tech. Bull. No. 13.
Burstein, S. (1962). *Biochim. biophys. Acta* **62**, 576.
Burstein, S. (1967). *Biochim. biophys. Acta* **146**, 529.
Burstein, S. & Dorfman, R. I. (1963). *J. biol. Chem.* **238**, 1656.

Burstein, S. & Lieberman, S. (1958a). *J. Am. chem. Soc.* **80**, 5235.
Burstein, S. & Lieberman, S. (1958b). *J. biol. Chem.* **233**, 331.
Burstein, S. & Westort, C. (1967). *Endocrinology* **80**, 1120.
Burton, S. D. & Morita, R. Y. (1964). *J. Bact.* **88**, 1755.
Burwell, R. L. (1952). *J. Am. chem. Soc.* **74**, 1462.
Butenandt, A., Biekert, E., Koga, N. & Traub, P. (1960). *Hoppe-Seyler's Z.* **321**, 258.
Butenandt, A. & Hofstetter, H. (1939). *Hoppe-Seyler's Z.* **259**, 222.
Butler, R. G. & Umbreit, W. W. (1965). *Bact. Proc.* p. 84.
Butler, R. G. & Umbreit, W. W. (1966). *J. Bact.* **91**, 661.
Butlin, K. R. (1949). *Proc. Soc. appl. Bact.* **2**, 39.
Butlin, K. R. (1953). *Research, Lond.* **6**, 184.
Butlin, K. R. & Adams, M. E. (1947). *Nature, Lond.* **160**, 154.
Butlin, K. R., Adams, M. E. & Thomas, M. (1949). *J. gen. Microbiol.* **3**, 46.
Butlin, K. R. & Postgate, J. R. (1954). In *Autotrophic Micro-organisms*, p. 271. Ed. by Fry, B. A. & Peel, J. L. London: Cambridge University Press.
Calvin, H. I. & Lieberman, S. (1966). *J. clin. Endocr. Metab.* **26**, 402.
Campbell, L. L., Frank, H. A. & Hall, E. R. (1957). *J. Bact.* **73**, 516.
Campbell, L. L., Kasprzycki, M. A. & Postgate, J. R. (1966). *J. Bact.* **92**, 1122.
Campbell, L. L. & Postgate, J. R. (1965). *Bact. Rev.* **29**, 359.
Carroll, J. & Spencer, B. (1965a). *Biochem. J.* **94**, 20P.
Carroll, J. & Spencer, B. (1965b). *Biochem. J.* **96**, 79P.
Castella Bertran, E. (1954). *10° Congr. int Industr. Agric. aliment. Madrid.* p. 769.
Castor, L. N. & Chance, B. (1959). *J. biol. Chem.* **234**, 1587.
Cataldi, M. S. (1940). *Rev. Inst. bact. Dep. nac. Hig. B. Aires* **9**, 393.
Cavallaro, L., Bighi, C., Pancaldi, G. & Trabanelli, G. (1958). *Ann. Chim.* **48**, 466.
Cavallini, D., De Marco, C. & Mondovi, B. (1958). *J. biol. Chem.* **230**, 25.
Cavallini, D., De Marco, C., Mondovi, B. & Mori, B. G. (1960). *Enzymologia* **22**, 161.
Cavallini, D., Mondovi, B., De Marco, C. & Scioscia-Santoro, A. (1962a). *Archs. Biochem. Biophys.* **96**, 456.
Cavallini, D., Mondovi, B., De Marco, C. & Scioscia-Santoro, A. (1962b). *Enzymologia* **24**, 253.
Cavallini, D., Scandurra, R. & De Marco, C. (1963). *J. biol. Chem.* **238**, 2999.
Cecil, R. (1950). *Biochem. J.* **47**, 572.
Cecil, R. & McPhee, J. R. (1955). *Biochem. J.* **60**, 496.
Chadwick, B. T. & Wilkinson, J. H. (1960). *Biochem. J.* **76**, 102.
Challenger, F. (1955). *Q. Rev. chem. Soc.* **9**, 255.
Challenger, F. (1959). *Aspects of the Organic Chemistry of Sulphur.* London: Butterworths.
Chapeville, F. & Fromageot, P. (1955). *Biochim. biophys. Acta* **17**, 275.
Chapeville, F. & Fromageot, P. (1957). *Biochim. biophys. Acta* **26**, 538.
Chapeville, F. & Fromageot, P. (1958). *Bull. Soc. Chim. biol.* **40**, 1965.
Chapeville, F. & Fromageot, P. (1960). *Bull. Soc. Chim. biol.* **42**, 877.
Chapeville, F. & Fromageot, P. (1961). *Biochim. biophys. Acta* **49**, 328.
Chapeville, F., Fromageot, P., Brigelhuber, A. & Henry, M. (1956). *Biochim. biophys. Acta* **20**, 351.
Charles, A. M. & Suzuki, I. (1965). *Biochem. biophys. Res. Commun.* **19**, 686.

Charles, A. M. & Suzuki, I. (1966a). *Biochim. biophys. Acta* **128**, 510.
Charles, A. M. & Suzuki, I. (1966b). *Biochim. biophys. Acta* **128**, 522.
Chaste, J. & Pierfitte, M. (1965). *Bull. Soc. Pharm. Nancy* **64**, 13.
Chatagner, F., Bergeret, B., Séjourné, T. & Fromageot, C. (1952). *Biochim. biophys. Acta* **9**, 340.
Chauncey, H. H., Lionetti, F., Winer, R. A. & Lisanti, V. F. (1954). *J. dent. Res.* **33**, 321.
Cheney, E. S. & Jensen, M. L. (1962). *Econ. Geol.* **57**, 624.
Cherayil, J. D. & Van Kley, H. (1961). *Fedn. Proc.* **20**, 235.
Cherayil, J. D. & Van Kley, H. (1962). *Fedn. Proc.* **21**, 230.
Cherayil, J. D. & Van Kley, H. (1963). *Fedn. Proc.* **22**, 241.
Cherniak, R. & Davidson, E. A. (1964). *J. biol. Chem.* **239**, 2986.
Chowrenko, M. A. (1912). *Z. physiol. Chem.* **80**, 253.
Christiansen, J. A. & Drost-Hansen, W. (1949). *Nature, Lond.* **164**, 759.
Christiansen, J. A., Drost-Hansen, W. & Nielsen, A. E. (1952). *Acta chem. scand.* **6**, 333.
Chu, T. M. & Slaunwhite, W. R. (1968). *Steroids* **12**, 309.
Cilento, G. & Tedeschi, P. (1961). *J. biol. Chem.* **236**, 907.
Clarke, H. T. (1932). *J. biol. Chem.* **97**, 235.
Clarke, P. H. (1953). *J. gen. Microbiol.* **8**, 397.
Clausen, J. & Asboe-Hansen, G. (1967). *Clinica chim. Acta* **16**, 131.
Cleland, W. W. (1963). *Biochim. biophys. Acta* **67**, 104.
Clemedson, C. J., Sörbo, B. & Ullberg, S. (1960). *Acta physiol. scand.* **48**, 382.
Cloud, P. E. (1968). *Science, N.Y.* **160**, 729.
Clowes, R. C. (1958). *J. gen. Microbiol.* **18**, 140.
Cohen, H. & Bates, R. W. (1949). *Endocrinology* **44**, 317.
Cohn, G. L. (1965). *Nature, Lond.* **208**, 80.
Colefax, A. (1908). *J. chem. Soc.* **93**, 798.
Coleman, G. S. (1960). *J. gen. Microbiol.* **22**, 423.
Colmer, A. R., Temple, K. L. & Hinkle, M. E. (1950). *J. Bact.* **59**, 317.
Coltorti, M. & Giusti, G. (1956). *Boll. Soc. ital. Biol. sper.* **32**, 1094.
Connor, R. (1943). In *Organic Chemistry*, 2nd ed., Vol. 1, p. 835. Ed. by Gilman, H. New York: John Wiley.
Cook, T. M. (1964). *J. Bact.* **88**, 620.
Cook, T. M. & Umbreit, W. W. (1963). *Biochemistry* **2**, 194.
Cook, W. K. T. (1967). *Biochem. J.* **102**, 5P.
Corner, E. D. S., Leon, Y. A. & Bulbrook, R. D. (1960). *J. mar. biol. Ass. U.K.* **39**, 51.
Cosby, E. L. & Sumner, J. B. (1945). *Archs. Biochem.* **7**, 457.
Coval, M. L., Horio, T. & Kamen, M. D. (1961). *Biochim. biophys. Acta* **51**, 246.
Cowie, D. B., Bolton, E. T. & Sands, M. K. (1950). *J. Bact.* **60**, 233.
Cowie, D. B., Bolton, E. T. & Sands, M. K. (1951). *J. Bact.* **62**, 63.
Cramer, F., Kenner, G. W., Hughes, N. A. & Todd, A. (1957). *J. chem. Soc.* p. 3297.
Crawford, I. P. & Ito, J. (1964). *Proc. natn. Acad. Sci. U.S.A.* **51**, 390.
Creange, J. E. & Szego, C. M. (1967). *Biochem. J.* **102**, 898.
Crépy, O. & Judas, O. (1960). *Revue fr. Etud. clin. biol.* **5**, 284.
Crépy, O., Judas, O. & Lachese, B. (1964). *J. Chromat.* **16**, 340.
Crum, E. H. & Siehr, D. J. (1967). *J. Bact.* **94**, 2069.

# BIBLIOGRAPHY

Cugini, G. (1876). *Just's bot. Jber.* **4**, 113.

D'Abramo, F. & Lipmann, F. (1957). *Biochim. biophys. Acta* **25**, 211.

Darby, F. J., Heenan, M. P. & Smith, J. N. (1966). *Life Sci.* **5**, 1499.

Daughaday, W. H. & Parker, M. L. (1965). *A. Rev. Med.* **16**, 47.

Davidson, B. & Westley, J. (1965). *J. biol. Chem.* **240**, 4463.

Davidson, C. F. (1962). *Econ. Geol.* **57**, 265.

Davidson, E. A. & Riley, J. G. (1960). *J. biol. Chem.* **235**, 3367.

Davies, W. H., Mercer, E. I. & Goodwin, T. W. (1966). *Biochem. J.* **98**, 369.

Davis, E. A. & Johnson, E. J. (1967). *Can. J. Microbiol.* **13**, 873.

Davis, J. B. (1967). *Petroleum Microbiology.* Amsterdam: Elsevier.

Davis, R. E. (1958). *J. Am. chem. Soc.* **80**, 3565.

Davis, R. E. (1962). *Symp. Anti-radiation Drugs, Am. chem. Soc. Meeting, Spring,* Washington D.C.

Davis, R. E. (1964). *Surv. Prog. Chem.* **2**, 189.

Day, F. H. (1963). *The Chemical Elements in Nature,* p. 284. London: George C. Harrap & Co. Ltd.

Deevey, E. S., Nakai, N. & Stuiver, M. (1963). *Science N. Y.* **139**, 407.

de Kruyff, C. D., van der Walt, J. P. & Schwartz, H. M. (1957). *Antonie van Leeuwenhoek* **23**, 305.

De Ley, J. & van Poucke, M. (1961). *Biochim. biophys. Acta* **50**, 371.

De Marco, C. & Coletta, M. (1961a). *Biochim. biophys. Acta* **47**, 257.

De Marco, C. & Coletta, M. (1961b). *Biochim. biophys. Acta* **47**, 262.

De Marco, C., Coletta, M. & Cavallini, D. (1961). *Archs. Biochem. Biophys.* **93**, 178.

De Marco, C., Coletta, M., Mondovi, B. & Cavallini, D. (1960). *G. Biochim.* **9**, 1.

De Marco, C., Mondovi, B., Scandurra, R. & Cavallini, D. (1962). *Enzymologia* **25**, 94.

De Meio, R. H. (1952). *Acta physiol. latinoam.* **2**, 251.

De Meio, R. H., Wizerkaniuk, M. & Fabiani, E. (1953). *J. biol. Chem.* **203**, 257.

De Meio, R. H., Wizerkaniuk, M. & Schreibman, I. (1955). *J. biol. Chem.* **213**, 439.

Demolon, A. (1921). *C. r. hebd. Séanc. Acad. Sci., Paris* **173**, 1408.

De Renzo, E. C. (1962). In *Mineral Metabolism,* Vol. 2B, p. 483. Ed. by Comar, C. L. & Bronner, F. New York: Academic Press.

de Rey-Pailhade, J. (1888a). *C. r. hebd. Séanc. Acad. Sci., Paris* **106**, 1683.

de Rey-Pailhade, J. (1888b). *C. r. hebd. Séanc. Acad. Sci., Paris* **107**, 43.

de Rey-Pailhade, J. (1898). *C. r. Séanc. Soc. Biol.* **50**, 372.

De Ritis, F., Coltorti, M. & Giusti, G. (1954). *Biochim. appl.* **1**, 57.

De Ritis, F., Coltorti, M. & Giusti, G. (1956). *G. Clin. med.* **37**, 285.

Dessau, G., Jensen, M. L. & Nakai, N. (1962). *Econ. Geol.* **57**, 410.

De Vito, P. C. & Dreyfuss, J. (1964). *J. Bact.* **88**, 1341.

Deyrup, I. J. (1956). *J. gen. Physiol.* **39**, 893.

Deyrup, I. J. (1963). *Fedn. Proc.* **22**, 332.

Deyrup, I. J. (1964). *Am. J. Physiol.* **207**, 84.

Dicken, C. H. & Decker, R. H. (1966). *J. invest. Derm.* **47**, 426.

Dinegar, R. H., Smellie, R. H. & La Mer, V. K. (1951). *J. Am. chem. Soc.* **73**, 2050.

Dingle, J. T. (1961). *Biochem. J.* **79**, 509.

Dodd, G. & Griffith, R. O. (1949). *Trans. Faraday Soc.* **45**, 546.

Dodgson, K. S. (1956). In *Colloque sur la Biochimie du Soufre*, p. 123. Paris: Editions CNRS.
Dodgson, K. S. (1959*a*). *Enzymologia* **20**, 301.
Dodgson, K. S. (1959*b*). *Biochim. biophys. Acta* **35**, 532.
Dodgson, K. S. (1961*a*). *Biochem. J.* **78**, 312.
Dodgson, K. S. (1961*b*). *Biochem. J.* **78**, 324.
Dodgson, K. S., Gatehouse, P. W., Lloyd, A. G. & Powell, G. M. (1965). *Biochem. J.* **95**, 18 P.
Dodgson, K. S., Lewis, J. I. M. & Spencer, B. (1953). *Biochem. J.* **55**, 253.
Dodgson, K. S. & Lloyd, A. G. (1957). *Biochem. J.* **66**, 532.
Dodgson, K. S. & Lloyd, A. G. (1958). *Biochem. J.* **68**, 88.
Dodgson, K. S. & Lloyd, A. G. (1961). *Biochem. J.* **78**, 319.
Dodgson, K. S., Lloyd, A. G. & Spencer, B. (1957). *Biochem. J.* **65**, 131.
Dodgson, K. S., Lloyd, A. G. & Tudball, N. (1961). *Biochem. J.* **79**, 111.
Dodgson, K. S. & Powell, G. M. (1959). *Biochem. J.* **73**, 666, 672.
Dodgson, K. S. & Price, R. G. (1962). *Biochem. J.* **84**, 106.
Dodgson, K. S. & Rose, F. A. (1966). *Nutr. abstr. Rev.* **36**, 327.
Dodgson, K. S., Rose, F. A. & Spencer, B. (1955). *Biochem. J.* **60**, 346.
Dodgson, K. S., Rose, F. A. & Spencer, B. (1957). *Biochem. J.* **66**, 357.
Dodgson, K. S., Rose, F. A. & Tudball, N. (1959). *Biochem. J.* **71**, 10.
Dodgson, K. S. & Spencer, B. (1953). *Biochem. J.* **55**, 436.
Dodgson, K. S. & Spencer, B. (1954). *Biochem. J.* **57**, 310.
Dodgson, K. S. & Spencer, B. (1956*a*). *Clinica chim. Acta* **1**, 478.
Dodgson, K. S. & Spencer, B. (1956*b*). *Rep. Prog. Chem.* **53**, 318.
Dodgson, K. S. & Spencer, B. (1957*a*). *Meth. biochem. Anal.* **4**, 211.
Dodgson, K. S. & Spencer, B. (1957*b*). *Biochem. J.* **65**, 668.
Dodgson, K. S., Spencer, B. & Thomas, J. (1953). *Biochem. J.* **53**, 452.
Dodgson, K. S., Spencer, B. & Thomas, J. (1954). *Biochem. J.* **56**, 177.
Dodgson, K. S., Spencer, B. & Williams, K. (1955). *Biochem. J.* **61**, 374.
Dodgson, K. S., Spencer, B. & Williams, K. (1956*a*). *Biochem. J.* **64**, 216.
Dodgson, K. S., Spencer, B. & Williams, K. (1956*b*). *Nature, Lond.* **177**, 432.
Dodgson, K. S., Spencer, B. & Wynn, C. H., (1956). *Biochem. J.* **62**, 500.
Dodgson, K. S. & Tudball, N. (1961). *Biochem. J.* **81**, 68.
Dodgson, K. S. & Wynn, C. H. (1958). *Biochem. J.* **68**, 387.
Dohlman, C. H. (1956). *Acta physiol. scand.* **37**, 220.
Dohlman, C. H. (1957). *Acta ophthal.* **35**, 115.
Donohue, J. (1961). In *Organic Sulfur Compounds*, Vol. 1, p. 1. Ed. by Kharasch, N. New York: Pergamon Press.
Donohue, J. (1965). In *Elemental Sulfur*, p. 13. Ed. by Meyer, B. New York: Interscience.
Doolittle, R. F. & Blombäck, B. (1964). *Nature, Lond.* **202**, 147.
Dornow, A. (1939). *Ber. dt. chem. Ges.* **72**, 568.
Dostálek, M. (1961). *Fol. microbiol., Praha* **6**, 10.
Dostálek, M. & Spurný, M. (1957). *Čslká Mikrobiol.* **2**, 300.
Dostálek, M. & Spurný, M. (1958). *Fol. biol. Praha*, **4**, 166.
Drayer, N. M. & Giroud, C. J. P. (1965). *Steroids* **5**, 289.
Dreyfuss, J. (1964). *J. biol. Chem.* **239**, 2292.
Dreyfuss, J. & Monty, K. J. (1963*a*). *J. biol. Chem.* **238**, 1019.
Dreyfuss, J. & Monty, K. J. (1963*b*). *J. biol. Chem.* **238**, 3781.
Dreyfuss, J. & Pardee, A. B. (1965). *Biochim. biophys. Acta* **104**, 308.
Dreyfuss, J. & Pardee, A. B. (1966). *J. Bact.* **91**, 2275.

**BIBLIOGRAPHY**

Dumas, M. (1874). *Annls. Chim. Phys. Ser. 5*, **3**, 57.

Durrant, P. J. & Durrant, B. (1962). *Introduction to Advanced Inorganic Chemistry*, p. 776. London: Longmans.

Działoszyński, L. M. (1951). *Bull. Soc. Amis Sci. Lett. Poznan. Ser. B*, **11**, 58, 80, 87.

Działoszyński, L. M. (1957). *Clinica chim. Acta* **2**, 542.

Działoszyński, L. M., Błeszyński, W. & Lewosz, J. (1966). *Zesz. nauk. Uniw. Mikolaja Kopernika Torun* **9**, 15.

Działoszyński, L. M. & Gniot-Szulżycka, J. (1967). *Clinica chim. Acta* **15**, 381.

Działoszyński, L. M., Gniot-Szulżycka, J. & Barancewicz, Z. (1966). *Zesz. nauk. Uniw. Mikolaja Kopernika Torun* **9**, 75.

Działoszyński, L. M., Kroll, J. L. & Fröhlich, A. (1966). *Clinica chim. Acta* **14**, 450.

Działoszyński, L. M., Kuik, K. & Leźnicki, A. (1966). *Zesz. nauk. Uniw. Mikolaja Kopernika Torun* **9**, 87.

Działoszyński, L. M. & Zawielak, I. J. (1955). *Acta biochim. pol.* **2**, 429.

Dziewiatkowski, D. D. (1945). *J. biol. Chem.* **161**, 723.

Dziewiatkowski, D. D. (1946). *J. biol. Chem.* **164**, 165.

Dziewiatkowski, D. D. (1956). *J. biol. Chem.* **223**, 239.

Dziewiatkowski, D. D. & Di Ferrante, N. (1957). *J. biol. Chem.* **227**, 347.

Eagle, H., Washington, C. & Friedman, S. M. (1966). *Proc. natn. Acad. Sci. U.S.A.* **56**, 156.

Egami, F., Ishimoto, M. & Taniguchi, S. (1961). In *Haematin Enzymes*, p. 392. Ed. by Falk, J. E., Lemberg, R. & Morton, R. K. London: Pergamon Press.

Egami, F. & Itahashi, M. (1951). Quoted in *Chem. Abstr.* (1951), **45**, 10278c.

Egami, F. & Takahashi, N. (1962). In *Biochemistry and Medicine of Mucopolysaccharides*, p. 53. Ed. by Egami, F. & Oshima, Y. Tokyo: Maruzen Co. Ltd.

Ehrlich, H. L. & Fox, S. I. (1967). *Biotech. Bioeng.* **9**, 471.

Eisenman, R. A., Balasubramanian, A. S. & Marx, W. (1967). *Archs. Biochem. Biophys.* **119**, 387.

Ekert, H. & Denett, X. (1966). *Australas. Ann. Med.* **15**, 152.

Eldjarn, L. (1954). *Scand. J. clin. lab. Invest.* **6**, Suppl. 13.

Eldjarn, L. (1965). *Scand. J. clin. lab. Invest.* **17**, Suppl. 86, 7.

Ellis, R. J. (1964a). *Biochim. biophys. Acta* **85**, 335.

Ellis, R. J. (1964b). *Biochem. J.* **93**, 19 p.

Ellis, R. J. (1966). *Nature, Lond.* **211**, 1266.

Ellis, R. J., Humphries, S. K. & Pasternak, C. A. (1964). *Biochem. J.* **92**, 167.

Ellis, R. J. & Pasternak, C. A. (1962). *Biochem. J.* **84**, 97 p.

Emiliozzi, R. (1960). *Bull. Soc. chim. Fr.* p. 911.

Emoto, Y. (1933). *Bot. Mag., Tokyo* **47**, 405 (see p. 569).

Eremenko, R. K. & Brodskii, A. I. (1955). *J. gen. Chem. U.S.S.R.* **25**, 1189.

Ericsson, J. L. E. & Helminen, H. J. (1967). *Histochemie* **9**, 170.

Eriksson, B. & Sörbo, B. (1967). *Acta chem. scand.* **21**, 958.

Ettlinger, M. G., Dateo, G. P., Harrison, B. W., Mabry, T. J. & Thompson, C. P. (1961). *Proc. natn. Acad. Sci. U.S.A.* **47**, 1875.

Ettlinger, M. G. & Lundeen, A. J. (1956). *J. Am. chem. Soc.* **78**, 4172.

Ettlinger, M. G. & Lundeen, A. J. (1957). *J. Am. chem. Soc.* **79**, 1764.

Evans, R. J. & St John, J. L. (1944). *Ind. Engng. Chem. analyt. Edn.* **16**, 630.

Ewetz, L. & Sörbo, B. (1966). *Biochim. biophys. Acta* **128**, 296.

Eymers, J. G. & Wassink, E. C. (1937). *Enzymologia* **2**, 258.

Fanshier, D. W. & Kun, E. (1962). *Biochim. biophys. Acta* **58**, 266.

342

Faust, L. & Wolfe, R. S. (1961). *J. Bact.* **81**, 99.

Fava, A. (1953). *Gazz. chim. Ital.* **83**, 87.

Fava, A. & Bresadola, S. (1955). *J. Am. chem. Soc.* **77**, 5792.

Fava, A. & Divo, D. (1952). *Gazz. chim. Ital.* **82**, 558.

Fava, A. & Iliceto, A. (1958). *J. Am. chem. Soc.* **80**, 3478.

Fava, A. & Pajaro, G. (1954). *J. chim. phys.* **51**, 594.

Fava, A. & Pajaro, G. (1956). *J. Am. chem. Soc.* **78**, 5203.

Fedorov, V. D. & Maksimov, V. N. (1965). *Dokl. Akad. Nauk. S.S.S.R.* (Engl. Trans.) **162**, 100.

Feely, H. W. & Kulp, J. L. (1957). *Bull. Am. Assoc. Petrol. Geol.* **41**, 1802.

Fehér, F. (1963). In *Handbook of Preparative Inorganic Chemistry*, p. 341. Ed by Brauer, G. New York: Academic Press.

Fell, H. B. & Dingle, J. T. (1963). *Biochem. J.* **87**, 403.

Fell, H. B., Mellanby, E. & Pelc, S. R. (1956). *J. Physiol.* **134**, 179.

Fendler, E. J. & Fendler, J. H. (1968). *J. org. Chem.* **33**, 3852.

Fiedler, H. & Wood, J. L. (1956). *J. biol. Chem.* **222**, 387.

Flavin, M. (1962). *J. biol. Chem.* **237**, 768.

Fletcher, J. C. & Robson, A. (1962). *Biochem. J.* **84**, 439.

Fletcher, J. C. & Robson, A. (1963). *Biochem. J.* **87**, 553.

Flynn, T. G., Dodgson, K. S., Powell, G. M. & Rose, F. A. (1967). *Biochem. J.* **105**, 1003.

Foerster, F. & Centner, K. (1926). *Z. anorg. Allg. Chem.* **157**, 45.

Foerster, F. & Hornig, A. (1923). *Z. anorg. Allg. Chem.* **125**, 86.

Fogarty, L. M. & Rees, W. R. (1962). *Nature, Lond.* **193**, 1180.

Footner, H. B. & Smiles, S. (1925). *J. chem. Soc.* p. 2887.

Ford, E. A. & Ruoff, P. M. (1965). *Chem. Commun.* p. 630.

Foss, O. (1945). Studies on Polythionates and related compounds. *Kgl. norske Videnskab. Selskabs. Skrifter* NR 2.

Foss, O. (1947). *Acta chem. scand.* **1**, 307.

Foss, O. (1950). *Acta chem. scand.* **4**, 404.

Foss, O. (1960). *Adv. inorg. Chem. Radiochem.* **2**, 237.

Foss, O. (1961*a*). In *Organic Sulfur Compounds*, Vol. 1, p. 75. Ed. by Kharasch, N. New York: Pergamon Press.

Foss, O. (1961*b*). In *Organic Sulfur Compounds*, Vol. 1, p. 83. Ed. by Kharasch, N. Oxford: Pergamon Press.

Foss, O. (1961*c*). *Acta chem. Scand.* **15**, 1610.

Foss, O. & Kringlebotn, I. (1961). *Acta chem. Scand.* **15**, 1608.

Foster, A. B. & Huggard, A. J. (1955). *Adv. Carbohyd. Chem.* **10**, 335.

Fowler, L. R. & Rammler, D. H. (1964). *Biochemistry* **3**, 230.

Franklin, T. J. (1962). *Biochem. J.* **82**, 118.

French, A. P. & Warren, J. C. (1965). *Steroids* **6**, 865.

French, A. P. & Warren, J. C. (1966). *Steroids* **8**, 79.

French, A. P. & Warren, J. C. (1967). *Biochem. J.* **105**, 233.

Frendo, J., Koj, A. & Górniak, A. (1963). *Przegl. Lek.* **2**, 141.

Freney, J. R. (1958). *Nature, Lond.* **182**, 1318.

Freney, J. R. (1960). *Aust. J. biol. Sci.* **13**, 387.

Freney, J. R. (1967*a*). In *Soil Biochemistry*, p. 229. Ed. by McClaren, A. D. & Peterson, G. H. New York: Marcel Dekker, Inc.

Freney, J. R. (1967*b*). *Mineralium Deposita* **2**, 181.

Freney, J. R. & Stevenson, F. J. (1966). *Soil Sci.* **101**, 307.

Fridovich, I. & Handler, P. (1956*a*). *J. biol. Chem.* **221**, 323.

343

## BIBLIOGRAPHY

Fridovich, I. & Handler, P. (1956*b*). *J. biol. Chem.* **223**, 321.

Fridovich, I. & Handler, P. (1957). *J. biol. Chem.* **228**, 67.

Fridovich, I. & Handler, P. (1958). *J. biol. Chem.* **233**, 1578, 1581.

Fridovich, I. & Handler, P. (1961). *J. biol. Chem.* **236**, 1836.

Friess, S. L., Durant, R. C., Chanley, J. D. & Fash, F. J. (1967). *Biochem. Pharmac.* **16**, 1617.

Fromageot, C. (1938). *Ergebn. Enzymforsch.* **7**, 50.

Fromageot, C. (1947). *Adv. Enzymol.* **7**, 369.

Fromageot, C. (1950). In *The Enzymes*, Vol. 1, p. 517. Ed. by Sumner, J. B. & Myrback, K. New York: Academic Press.

Fromageot, C., Chatagner, F. & Bergeret, B. (1948). *Biochim. biophys. Acta* **2**, 294.

Fromageot, C., Wookey, E. & Chaix, P. (1939). *C. r. hebd. Séanc. Acad. Sci., Paris* **209**, 1019.

Fromageot, P. & Chapeville, F. (1961). *Biochim. biophys. Acta* **50**, 325, 334.

Fromageot, P. & Perez-Milan, H. (1956). *C. r. hebd. Séanc. Acad. Sci., Paris*, **243**, 1061.

Fromageot, P. & Perez-Milan, H. (1959). *Biochim. biophys. Acta* **32**, 457.

Fromageot, P., Roderick, U. R. & Pouzat, J. (1963). *Biochim. biophys. Acta* **78**, 126.

Fujimoto, D. & Ishimoto, M. (1961). *J. Biochem., Tokyo* **50**, 533.

Fujino, Y. & Negishi, T. (1957). *Bull. agric. chem. Soc. Japan* **21**, 225.

Fuller, R. (1963). *Chem. Revs.* **63**, 21.

Fuller, E. C. & Crist, R. H. (1941). *J. Am. chem. Soc.* **63**, 1644.

Funaki, H., Shibata, K., Yamoaka, S. & Watanabe, H. (1958). *Kyoto Furitsu Ikadaigaku Zasshi* **63**, 565.

Furness, W. (1950). *J. Soc. Dyers Colour.* **66**, 270.

Furness, W. & Davies, W. C. (1952). *Analyst* **77**, 697.

Furusaka, C. (1961). *Nature, Lond.* **192**, 427.

Gaines, R. D. & Goering, K. J. (1960). *Biochem. biophys. Res. Commun.* **2**, 207.

Gaines, R. D. & Goering, K. J. (1962). *Archs. Biochem. Biophys.* **96**, 13.

Gale, N. L. & Beck, J. V. (1966). *Bact. Proc.* p. 94.

Gale, N. L. & Beck, J. V. (1967). *J. Bact.* **94**, 1052.

Garnier, Y. & Duval, C. (1959). *J. Chromatog.* **2**, 72.

Geison, R. L., Rogers, W. E. & Johnson, B. C. (1968). *Biochim. biophys. Acta* **165**, 448.

Gemeinhardt, K. (1938). *Ber. dt. bot. Ges.* **56**, 275.

Gemeinhardt, K. (1939). *Süddeut. Apoth.-Ztg.* **79**, 256.

Gerber, M. I. & Shusharina, A. D. (1950). *Zh. analit. Khim.* **5**, 262.

Gerlach, U. (1963). *Klin. Wschr.* **41**, 873.

Gest, H., San Pietro, A. & Vernon, L. P. (1963). Eds. *Bacterial Photosynthesis.* Yellow Springs, Ohio: Antioch Press.

Gianetto, R. & Viala, R. (1955). *Science, N.Y.* **121**, 801.

Gibian, H. & Bratfisch, G. (1956). *Hoppe-Seyler's Z.* **305**, 265.

Gilbert, E. E. (1962). *Chem. Rev.* **62**, 549.

Gilman, A., Philips, F. S., Koelle, E. S., Allen, R. P. & St John, E. (1946). *Am. J. Physiol.* **147**, 115.

Gleen, H. & Quastel, J. H. (1953). *Appl. Microbiol.* **1**, 70.

Gmelin (1960). Gmelins Handbuch der Anorganischen Chemie, 8th ed. System 9, Schwefel, Teil B, Lief 2. Weinheim: Verlag Chemie.

344

Gmelin (1963). Gmelins Handbook of Inorganic Chemistry, 8th ed. System 9, Sulfur. Part B, section 3. New York: Walter J. Johnson.

Gniot-Szulżycka, J. & Działoszyński, L. M. (1966). *Acta biochim. pol.* **13**, 171.

Goehring, M. (1952). *Forts. Chem. Forsch.* **2**, 444.

Goehring, M. & Feldmann, U. (1948). *Z. anorg. allg. Chem.* **257**, 223.

Goehring, M., Feldmann, U. & Helbing, W. (1949). *Z. analyt. Chem.* **129**, 346.

Goehring, M., Helbing, W. & Appel, I. (1947). *Z. anorg. allg. Chem.* **254**, 185.

Goldberg, I. H. (1960). *Fedn. Proc.* **19**, 220.

Goldberg, I. H. & Delbruck, A. (1959). *Fedn. Proc.* **18**, 235.

Goldfischer, S. (1965). *J. Histochem. Cytochem.* **13**, 520.

Goldstein, F. & Rieders, F. (1953). *Am. J. Physiol.* **173**, 287.

Gordon, R. S. & Sizer, I. W. (1955). *Science, N.Y.* **122**, 1270.

Gorska-Brylass, A. (1965). *Acta Soc. Bot. Pol.* **34**, 589.

Gottfried, H., Dorfman, R. I. & Wall, P. E. (1967). *Nature, Lond.* **215**, 409.

Gottfried, H. & Lusis, O. (1966). *Nature, Lond.* **212**, 1488.

Gracie, D. S., Rizk, M., Moukhtar, A. & Moustafa, A. H. I. (1934). *Egypt. Min. Agric. Tech. Bull. No. 148.*

Grant, W. M. (1947). *Ind. Engng. Chem. analyt. Edn.* **19**, 345.

Grassini, G. & Lederer, M. (1959). *J. Chromatog.* **2**, 326.

Green, J. R. & Westley, J. (1961). *J. biol. Chem.* **236**, 3047.

Gregory, H., Hardy, P. M., Jones, D. S., Kenner, G. W. & Sheppard, R. C. (1964). *Nature, Lond.* **204**, 931.

Gregory, J. D. (1955). *J. Am. chem. Soc.* **77**, 3922.

Gregory, J. D. & Lipmann, F. (1957). *J. biol. Chem.* **229**, 1081.

Griffith, R. O. & Irving, R. (1949). *Trans. Faraday Soc.* **45**, 563.

Grimes, A. J. (1959). *Biochem. J.* **73**, 723.

Gross, D. (1957). *Chemy Ind.* p. 1597.

Grossman, J. P. & Postgate, J. R. (1953). *Nature, Lond.* **171**, 600.

Grunberg-Manago, M., Campillo-Campbell, A., Dondon, L. & Michelson, A. M (1966). *Biochim. biophys. Acta* **123**, 1.

Guarraia, L. J., Laishley, E. J., Forget, N. & Peck, H. D. (1968). *Bact. Proc.* p. 133.

Guarraia, L. J. & Peck, H. D. (1967). *Bact. Proc.* p 117.

Gubin, V. & Tzechomskaja, V. (1930). *Zentbl. Bakt. ParasitKde*, Abt II, **81**, 396.

Guha, A. & Roels, O. A. (1965). *Biochim. biophys. Acta* **111**, 364.

Guiseley, K. B. & Ruoff, P. M. (1961). *J. org. Chem.* **26**, 1248.

Guittonneau, G. (1925a). *C. r. hebd. Séanc. Acad. Sci., Paris* **180** 1142.

Guittonneau, G. (1925b). *C. r. hebd. Séanc. Acad. Sci., Paris* **181**, 261.

Guittonneau, G. (1926). *C. r. hebd. Séanc. Acad. Sci., Paris* **182**, 661.

Guittonneau, G. (1927). *C. r. hebd. Séanc. Acad. Sci., Paris* **184**, 45.

Guittonneau, G. & Keilling, J. (1927). *C. r. hebd. Séanc. Acad. Sci., Paris* **184**, 898.

Guittonneau, G. & Keilling, J. (1932a). *C. r. hebd. Séanc. Acad. Sci., Paris* **195**, 679.

Guittonneau, G. & Keilling, J. (1932b). *Annls. agron. N.S.* **2**, 690.

Gutiérrez, R. & Ruiz-Herrera, J. (1968). *Revta lat.-am. Microbiol. Parasitol.* **10**, 181.

Hall, M. E. (1950). *Analyt. Chem.* **22**, 1137.

Hall, M. E. (1953). *Analyt. Chem.* **25**, 556.

345

**BIBLIOGRAPHY**

Hall, M. O. & Straatsma, B. R. (1966). *Biochim. biophys. Acta* **124**, 246.
Hall, M. R. & Berk, R. S. (1968). *Can. J. Microbiol.* **14**, 515.
Hansen, C. J. (1933). *Ber. dt. chem. Ges.* **66**, 817.
Happold, F. C., Johnstone, K. I., Rogers, H. J. & Youatt, J. B. (1954). *J. gen. Microbiol.* **10**, 261.
Happold, F. C., Jones, G. L. & Pratt, D. B. (1958). *Nature, Lond.* **182**, 266.
Harada, T. (1952). *J. agric. chem. Soc. Japan* **26**, 95.
Harada, T. (1957). *Bull. agric. chem. Soc. Japan* **21**, 267.
Harada, T. (1959). *Bull. agric. chem. Soc. Japan* **23**, 222.
Harada, T. (1963). *Mem. Inst. scient. ind. Res. Osaka Univ.* **20**, 111.
Harada, T. (1964). *Biochim. biophys. Acta* **81**, 193.
Harada, T. & Hattori, F. (1956). *Bull. agric. chem. Soc. Japan* **20**, 110.
Harada, T. & Kono, K. (1954). *J. agric. chem. Soc. Japan* **28**, 608.
Harada, T., Kono, K. & Yagi, K. (1954). *Mem. Inst. scient. ind. Res. Osaka Univ.* **11**, 193.
Harada, T., Shimizu, S., Nakanishi, Y. & Suzuki, S. (1967). *J. biol. Chem.* **242**, 2288.
Harada, T. & Spencer, B. (1960). *J. gen. Microbiol.* **22**, 520.
Harada, T. & Spencer, B. (1962). *Biochem. J.* **82**, 148.
Harada, T. & Spencer, B. (1964). *Biochem. J.* **93**, 373.
Hare, R., Wildy, P., Billet, F. S. & Twort, D. N. (1952). *J. Hyg., Camb.* **50**, 295.
Harrison, A. G. & Thode, H. G. (1957). *Trans. Faraday Soc.* **53**, 1648.
Harrison, A. G. & Thode, H. G. (1958). *Trans. Faraday Soc.* **54**, 84.
Haschke, R. & Campbell, L. L. (1967). *Bact. Proc.* p. 118.
Haschke, R. & Campbell, L. L. (1968). *Fedn. Proc.* **27**, 390.
Hasegawa, F., Delbruck, A. & Lipmann, F. (1961). *Fedn. Proc.* **20**, 86.
Haslewood, G. A. D. (1964). *Biochem. J.* **90**, 309.
Haslewood, G. A. D. (1966). *Biochem. J.* **100**, 233.
Haslewood, G. A. D. (1967). *Bile Salts*, p. 19. London: Methuen.
Havinga, E., de Jongh, R. O. & Dorst, W. (1956). *Recl. Trav. chim. Pays-Bas Belg.* **75**, 378.
Hawkins, J. B. & Young, L. (1954). *Biochem. J.* **56**, 166.
Hayward, H. R. & Stadtman, T. C. (1959). *J. Bact.* **78**, 557.
Hearse, D. J., Powell, G. M., Olavesen, A. H. & Dodgson, K. S. (1967). *Biochem. J.* **105**, 33P.
Hearse, D. J., Powell, G. M., Olavesen, A. H. & Dodgson, K. S. (1968). *Biochem. J.* **107**, 24P.
Heffter, A. (1908). *Med-naturw. Arch.* **1**, 81.
Heimberg, M., Fridovich, I. & Handler, P. (1953). *J. biol. Chem.* **204**, 913.
Held, E. & Buddecke, E. (1967). *Hoppe-Seyler's Z.* **348**, 1047.
Hempelmann, L. H., Lisco, H. & Hoffman, J. G. (1952). *Ann. intern. Med.* **36**, 279.
Hempfling, W. P. (1964). *Comparative Aspects of Obligate Autotrophy.* Ph.D. Thesis, Yale.
Hempfling, W. P., Trudinger, P. A. & Vishniac, W. (1967). *Arch. Mikrobiol.* **59**, 149.
Hempfling, W. P. & Vishniac, W. (1965). *Biochem. Z.* **342**, 272.
Henderson, R. J. & Loughlin, R. E. (1968). *Biochim. biophys. Acta* **156**, 195.
Herrmann, I. & Repke, K. (1964). *Naunyn-Schmiedebergs Arch. exp. Path. Pharmak.* **248**, 370.

346

Hidaka, H., Nagatsu, T. & Yaka, K. (1967). *Analyt. Biochem.* **19**, 388.
Higaki, M., Takahashi, M., Suzuki, T. & Sahashi, Y. (1965). *J. Vitam.* **11**, 261, 266.
Hilz, H. & Kittler, M. (1958). *Biochim. biophys. Acta* **30**, 650.
Hilz, H. & Kittler, M. (1960). *Biochem. biophys. Res. Commun.* **3**, 140.
Hilz, H., Kittler, M. & Knape, G. (1959). *Biochem. Z.* **332**, 151.
Hilz, H. & Lipmann, F. (1955). *Proc. natn. Acad. Sci. U.S.A.* **41**, 880.
Himwich, W. A. & Saunders, J. P. (1948). *Am. J. Physiol.* **153**, 348.
Hitchcock, M. & Smith, J. N. (1964). *Biochem. J.* **93**, 392.
Hockenhull, D. J. D. (1948). *Biochem. J.* **43**, 498.
Hockenhull, D. J. D. (1949). *Biochim. biophys. Acta* **3**, 326.
Hodgkin, J. H., Craigie, J. S. & McInnes, A. G. (1966). *Can. J. Chem.* **44**, 74.
Hodson, R. C., Schiff, J. A. & Scarsella, A. J. (1968). *Pl. Physiol.* **43**, 570.
Hodson, R. C., Schiff, J. A., Scarsella, A. J. & Levinthal, M. (1968). *Pl. Physiol.* **43**, 563.
Hofman-Bang, N. (1949). *Acta chem. scand.* **3**, 872.
Hofman-Bang, N. (1950). *Acta chem. scand.* **4**, 456.
Hofman-Bang, N. (1951). *Acta chem. scand.* **5**, 1375.
Holcenberg, J. S. & Rosen, S. W. (1965). *Archs. Biochem. Biophys.* **110**, 551.
Holt, S. J. (1959). *Expl Cell Res.* Suppl. 7, 1.
Hope, D. B. (1955). *Biochem. J.* **59**, 497.
Hopsu, V. K., Arstila, A. & Glenner, G. G. (1965). *Annls. Med. exp. Biol. Fenn.* **43**, 114.
Hopsu-Havu, V. K., Arstila, A. U., Helminen, H. J., Kalimo, H. O. & Glenner, G. G. (1967). *Histochemie* **8**, 54.
Horhammer, L. & Hansel, R. (1953). *Arch. Pharm., Berl.* **286**, 153.
Horio, T. & Kamen, M. D. (1961). *Biochim. biophys. Acta* **48**, 266.
Horowitz, N. H. (1950). *Adv. Genet.* **3**, 33.
Horowitz, N. H. (1955). In *Amino Acid Metabolism*, p. 631. Ed. by McElroy, W. D. & Glass, B. Baltimore: The Johns Hopkins Press.
Horsfall, J. G. (1956). *Principles of Fungicidal Action*, p. 159. Waltham, Mass.: Chronica Botanica Co.
Horwitz, J. P., Chua, J., Noel, M., Donatti, J. T. & Freisler, J. (1966). *J. mednl. pharm. Chem.* **9**, 447.
Howell, L. G. & Fridovich, I. (1968). *J. biol. Chem.* **243**, 5941.
Hsu, L. & Tappel, A. L. (1965). *Biochim. biophys. Acta* **101**, 113.
Hsu, Y.-C. (1963). *Nature, Lond.* **200**, 1091.
Hsu, Y.-C. (1965). *Nature, Lond.* **207**, 385.
Huggins, C. & Smith, D. R. (1947). *J. biol. Chem.* **170**, 391.
Hunt, S. & Jevons, F. R. (1966). *Biochem. J.* **98**, 522.
Hunt, W. F. (1915). *Econ. Geol.* **10**, 543.
Hunter, F. E. & Ford, L. (1954). *Fedn. Proc.* **13**, 234.
Huovinen, J. A. & Gustafsson, B. E. (1967). *Biochim. biophys. Acta* **136**, 441.
Hurley, K. E. & Williams, R. J. (1955). *Archs. Biochem. Biophys.* **54**, 384.
Hussey, C., Orsi, B. A., Scott, J. & Spencer, B. (1965). *Nature, Lond.* **207**, 632.
Hussey, C. & Spencer, B. (1967). *Biochem. J.* **103**, 56 P.
Hutchinson, M., Johnstone, K. I. & White, D. (1965). *J. gen. Microbiol.* **41**, 357.
Hutchinson, M., Johnstone, K. I. & White, D. (1966). *J. gen. Microbiol.* **44**, 373.
Hutchinson, M., Johnstone, K. I. & White, D. (1967). *J. gen. Microbiol.* **47**, 17.

**BIBLIOGRAPHY**

Huzisige, H. & Haga, K. (1944). *Acta phytochim., Tokyo* **14**, 141.
Hvid-Hansen, N. (1951). *Acta path. microbiol. scand.* **29**, 314.
Hylin, J. W., Fiedler, H. & Wood, J. L. (1959). *Proc. Soc. exptl Biol. Med.* **100**, 165.
Hylin, J. W. & Wood, J. L. (1959). *J. biol. Chem.* **234**, 2141.
Ibanez, M. L. & Lindstrom, E. S. (1962). *J. Bact.* **84**, 451.
Ichihara, A. (1959). *Mem. Inst. Protein Res. Osaka Univ.* **1**, 177.
Iguchi, A. (1958*a*). *Bull. chem. Soc. Japan* **31**, 600.
Iguchi, A. (1958*b*). *Bull. chem. Soc. Japan* **31**, 597.
Imai, K., Okuzumi, M. & Katagiri, H. (1962). *Koso Kagaku Shimpoziumu* **17**, 132; *Chem. Abstr.* **59**, 6746*f* (1963).
Irreverre, F., Mudd, S. H., Heizer, W. D. & Laster, L. (1967). *Biochem. Med.* **1**, 187.
Ishimoto, M. (1959). *J. Biochem. Tokyo* **46**, 105.
Ishimoto, M. & Fujimoto, D. (1959). *Proc. Japan Acad.* **35**, 243.
Ishimoto, M. & Fujimoto, D. (1961). *J. Biochem., Tokyo* **50**, 299.
Ishimoto, M., Kondo, Y., Kameyama, T., Yagi, T. & Shiraki, M. (1958). *Int. Symp. Enzyme Chem.* Tokyo and Kyoto, p. 229.
Ishimoto, M. & Koyama, J. (1957). *J. Biochem., Tokyo* **44**, 233.
Ishimoto, M., Koyama, J. & Nagai, Y. (1954). *J. Biochem., Tokyo* **41**, 763.
Ishimoto, M., Koyama, J. & Nagai, Y. (1955). *J. Biochem., Tokyo* **42**, 41.
Ishimoto, M., Koyama, J., Omura, T. & Nagai, Y. (1954). *J. Biochem., Tokyo* **41**, 537.
Ishimoto, M., Koyama, J., Yagi, T. & Shiraki, M. (1957). *J. Biochem., Tokyo* **44**, 413.
Ishimoto, M. & Yagi, T. (1961). *J. Biochem., Tokyo* **49**, 103.
Ishimoto, M., Yagi, T. & Shiraki, M. (1957). *J. Biochem., Tokyo* **44**, 707.
Ishimoto, M. & Yamashina, I. (1949). *Symp. Enzyme Chem.* **2**, 36.
Itahashi, M. (1961). *J. Biochem., Tokyo* **50**, 52.
Ivanov, M. V. (1957). *Mikrobiologiya* **26**, 338.
Ivanov, M. V. (1960). *Microbiology* **29**, 180.
Ivanov, M. V. (1968). *Microbiological Processes in the Formation of Sulfur Deposits.* Springfield Va.: U.S. Dept. Commerce. Clearing House for Fed. Sci. Tech. Inf.
Ivanov, V. I. & Lyalikova, N. N. (1962). *Microbiology* **31**, 382.
Iverson, W. P. (1966). *Appl. Microbiol.* **14**, 529.
Iverson, W. P. (1967). *Science, N.Y.* **156**, 1112.
Iverson, W. P. (1968). *Nature, Lond.* **217**, 1265.
Iwatsuka, H., Kuno, M. & Maruyama, M. (1962). *Pl. Cell Physiol., Tokyo* **3**, 157.
Iwatsuka, H. & Mori, T. (1960). *Pl. Cell Physiol., Tokyo* **1**, 163.
Jackson, C. & Black, R. E. (1967). *Biol. Bull. mar. biol. Lab. Woods Hole*, **132**, 1.
Jackson, J. F., Moriarty, D. J. W. & Nicholas, D. J. D. (1968). *J. gen. Microbiol.* **53**, 53.
Jacobsen, J. G. & Smith, L. H. (1963). *Nature, Lond.* **200**, 575.
Jacobsen, J. G. & Smith, L. H. (1968). *Physiol. Rev.* **48**, 424.
Jamieson, G. S. (1915). *Am. J. Sci.* **39**, 639.
Jankowski, G. J. & ZoBell, C. E. (1944). *J. Bact.* **47**, 447.
Jarrige, P. (1962). *Purification et Proprietes des Sulfatase du suc Digestif D'Helix Pomatia.* Paris: R. Foulon & Cie.

348

Jarrige, P. (1963). *Bull. Soc. Chim. biol.* **45**, 761.
Jarrige, P., Yon, J. & Jayle, M. F. (1963). *Bull. Soc. Chim. biol.* **45**, 783.
Jatzkewitz, H. (1960). *Hoppe-Seyler's Z.* **318**, 265.
Jay, R. R. (1953). *Analyt. Chem.* **25**, 288.
Jensen, M. L. (1962). In *Biogeochemistry of Sulfur Isotopes*, p. 1. Ed. by Jensen, M. L. Dept. of Geology, Yale University, New Haven, U.S.A.
Jerfy, A. & Roy, A. B. (1969). *Biochim. biophys. Acta* **175**, 355.
Jevons, F. R. (1963). *Biochem. J.* **89**, 621.
John, D. W. & Miller, L. L. (1965). *Lab. Invest.* **14**, 1402.
John, R. A., Rose, F. A., Wusteman, F. S. & Dodgson, K. S. (1966). *Biochem. J.* **100**, 278.
Johnson, C. M. & Nishita, H. (1952). *Analyt. Chem.* **24**, 736.
Johnson, E. J. & Peck, H. D. (1965). *J. Bact.* **89**, 1041.
Johnstone, K. I., Townshend, M. & White, D. (1961). *J. gen. Microbiol.* **24**, 201.
Jones, A. S. & Letham, D. S. (1954). *Chemy. Ind.* p. 662.
Jones, G. E. & Benson, A. A. (1965). *J. Bact.* **89**, 260.
Jones, G. E. & Starkey, R. L. (1957). *Appl. Microbiol.* **5**, 111.
Jones, G. E. & Starkey, R. L. (1961). *J. Bact.* **82**, 788.
Jones, G. E. & Starkey, R. L. (1962). In *Biogeochemistry of Sulfur Isotopes*, p. 61. Ed. by Jensen, M. L. Dept. of Geology, Yale University, New Haven, U.S.A.
Jones, G. E., Starkey, R. L., Feely, H. W. & Kulp, J. L. (1956). *Science, N.Y.* **123**, 1124.
Jones, G. L. & Happold, F. C. (1961). *J. gen. Microbiol.* **26**, 361.
Jones, J. G. & Dodgson, K. S. (1965). *Biochem. J.* **94**, 331.
Jones, J. G., Scotland, S. M. & Dodgson, K. S. (1966). *Biochem. J.* **98**, 138.
Jones, W. G. M. & Peat, S. (1942). *J. chem. Soc.* p. 225.
Jones-Mortimer, M. C. (1968). *Biochem. J.* **106**, 33 P.
Jones-Mortimer, M. C., Wheldrake, J. F. & Pasternak, C. A. (1968). *Biochem. J.* **107**, 51.
Joseph, J. P., Dusza, J. P. & Bernstein, S. (1966). *Steroids* **7**, 577.
Kaiser, E. T., Panar, M. & Westheimer, F. H. (1963). *J. Am. chem. Soc.* **85**, 602.
Kaji, A. & Gregory, J. D. (1959). *J. biol. Chem.* **234**, 3007.
Kaji, A. & McElroy, W. D. (1958). *Biochim. biophys. Acta* **30**, 190.
Kaji, A. & McElroy, W. D. (1959). *J. Bact.* **77**, 630.
Kamin, H., Masters, B. S. S., Siegel, L. M., Vorhaben, J. E. & Gibson, Q. H. (1968). *Abstr. 7th Intern. Congr. Biochem., Tokyo*, p. 187.
Kanner, L. C. & Kozloff, L. M. (1964). *Biochemistry* **3**, 215.
Kaplan, I. R. & Rittenberg, S. C. (1962 *a*). In *Biogeochemistry of Sulfur Isotopes*, p. 80. Ed. by Jensen, M. L., Dept of Geology, Yale University, New Haven, U.S.A.
Kaplan, I. R. & Rittenberg, S. C. (1962 *b*). *Nature, Lond.* **194**, 1098.
Kaplan, I. R. & Rittenberg, S. C. (1964). *J. gen. Microbiol.* **34**, 195.
Kates, M., Palameta, B., Perry, M. P. & Adams, G. A. (1967). *Biochim. biophys. Acta* **137**, 213.
Kato, A., Ogura, M., Kimura, H., Kawai, T. & Suda, M. (1966). *J. Biochem., Tokyo* **59**, 34.
Kawakami, M., Iizuka, T. & Mitsuhashi, S. (1957). *Jap. J. exp. Med.* **27**, 317.
Kawashima, N. & Asahi, T. (1961). *J. Biochem., Tokyo* **49**, 52.

## BIBLIOGRAPHY

Kawiak, J. (1964). *Acta biochim. pol.* **11**, 91.

Kawiak, J., Sawicki, W. & Miks, B. (1964). *Acta Histochem.* **19**, 184.

Kay, R. E., Harris, D. C. & Entenman, C. (1956). *Am. J. Physiol.* **186**, 175.

Kearney, E. B. & Singer, T. P. (1953). *Biochim. biophys. Acta* **11**, 276.

Keller, H. & Blennemann, H. (1961*a*). *Hoppe-Seyler's Z.* **324**, 125.

Keller, H. & Blennemann, H. (1961*b*). *Hoppe-Seyler's Z.* **324**, 138.

Kelly, D. P. (1965). *J. gen. Microbiol.* **41**, v.

Kelly, D. P. (1966). *Biochem. J.* **100**, 9 P.

Kelly, D. P. (1967*a*). *Sci. Prog. Oxf.* **55**, 35.

Kelly, D. P. (1967*b*). *Arch. Mikrobiol.* **56**, 91.

Kelly, D. P. (1967*c*). *Arch. Mikrobiol.* **58**, 99.

Kelly, D. P. (1968). *Aust. J. Sci.* **31**, 165.

Kelly, D. P., Chambers, L. A. & Trudinger, P. A. (1969). *Analyt. Chem.* **41**, 898.

Kelly, D. P. & Syrett, P. J. (1963). *Nature, Lond.* **197**, 1087.

Kelly, D. P. & Syrett, P. J. (1964*a*). *J. gen. Microbiol.* **34**, 307.

Kelly, D. P. & Syrett, P. J. (1964*b*). *Nature, Lond.* **202**, 597.

Kelly, D. P. & Syrett, P. J. (1966*a*). *Biochem. J.* **98**, 537.

Kelly, D. P. & Syrett, P. J. (1966*b*). *J. gen. Microbiol.* **43**, 109.

Kemp, A. L. W. & Thode, H. G. (1968). *Geochim. cosmochim. Acta* **32**, 71.

Kemp, J. D., Atkinson, D. E., Ehret, A. & Lazzarini, R. A. (1963). *J. biol. Chem.* **238**, 3466.

Kent, P. W. & Whitehouse, M. W. (1955). *Analyst, Lond.* **80**, 630.

Kessler, E. (1957). *Planta* **49**, 435.

Kharasch, N. (1961). *Organic Sulfur Compounds*, Vol. 1. London: Pergamon.

Kharasch, N. & Meyers, C. Y. (1966). *The Chemistry of Organic Sulfur Compounds*, Vol. 2. London: Pergamon.

Kice, J. L. (1968). *Accts. chem. Res.* **1**, 58.

Kice, J. L. & Anderson, J. M. (1966). *J. Am. chem. Soc.* **88**, 5242.

Kice, J. L., Anderson, J. M. & Pawlowski, N. E. (1966). *J. Am. chem. Soc.* **88**, 5245.

Kikal, T. & Smith, J. N. (1959). *Biochem. J.* **71**, 48.

King, H. F. (1967). *Mineralium Deposita* **2**, 142.

King, N. E. & Winfield, M. E. (1955). *Biochim. biophys. Acta* **18**, 431.

Kinsel, N. A. (1960). *J. Bact.* **80**, 628.

Kishimoto, Y., Takahashi, N. & Egami, F. (1961). *J. Biochem., Tokyo* **49**, 436.

Kjaer, A. (1960). *Fortschr. Chem. org. NatStoffe*, **18**, 122.

Klimek, R., Skarżyński, B. & Szczepkowski, T. W. (1956). *Acta biochim. pol.* **3**, 261.

Kline, B. C. & Schoenhard, D. E. (1968). *Bact. Proc.* p. 141.

Kluyver, A. J. (1953). In *Microbial Metabolism*, p. 71. Symp. 1st. Sup. di Sanita. Rome, Italy.

Kluyver, A. J. & van Niel, C. B. (1936). *Zentbl. Bakt. ParasitKde.* Abt II, **94**, 369.

Knaak, J. B., Kozbelt, S. J. & Sullivan, L. J. (1966). *Toxic. appl. Pharmac.* **8**, 369.

Knaysi, G. (1943). *J. Bact.* **46**, 451.

Knaysi, G. (1951). In *Bacterial Physiology*, p. 28. Ed. by Werkman, C. H. & Wilson, P. W. New York: Academic Press.

Knight, E. & Hardy, R. W. F. (1966). *J. biol. Chem.* **241**, 2752.

Knobloch, K. (1966*a*). *Planta* **70**, 73.

Knobloch, K. (1966*b*). *Planta* **70**, 172.
Knox, R. (1945). *Brit. J. exp. Path.* **26**, 146.
Knox, R., Gell, P. G. H. & Pollock, M. R. (1943). *J. Hyg.* **43**, 147.
Knox, R. & Pollock, M. R. (1944). *Biochem. J.* **38**, 299.
Kodama, Y. (1961). Quoted in *Chem. Abstr.* (1964), **60**, 9751 f.
Kodicek, E. & Loewi, G. (1955). *Proc. R. Soc., B.* **144**, 100.
Köhler, W., Ghatak, S. N., Rische, H. & Ziesché, K. (1966). *Zentbl. Bakt. ParasitKde.* Abt. 1, **201**, 482.
Koh, T. (1965). *Bull. chem. Soc. Japan* **38**, 1510.
Koh, T. & Iwasaki, I. (1965). *Bull. chem. Soc. Japan* **38**, 2135.
Koh, T. & Iwasaki, I. (1966*a*). *Bull. chem. Soc. Japan* **39**, 352.
Koh, T. & Iwasaki, I. (1966*b*). *Bull. chem. Soc. Japan* **39**, 703.
Koj, A. (1968). *Acta biochim. pol.* **15**, 161.
Koj, A. & Frendo, J. (1962). *Acta biochim. pol.* **9**, 373.
Koj, A., Frendo, J. & Borysiewicz, J. (1964). *Acta med. pol.* **5**, 109.
Koj, A., Frendo, J. & Janik, Z. (1967). *Biochem. J.* **103**, 791.
Kolthoff, I. M. & Lingane, J. J. (1952). *Polarography*, Vol. 2, Chap. 33. New York: Interscience Pub.
Kolthoff, I. M. & Miller, C. S. (1941). *J. Am. chem. Soc.* **63**, 1405.
Kondo, Y., Kameyama, T. & Tamiya, N. (1956). *J. Biochem., Tokyo* **43**, 749.
Kondrat'eva, E. N. & Mekhtieva, V. L. (1966). *Microbiology* **35**, 481.
Korn, E. D. (1959*a*). *J. biol. Chem.* **234**, 1321.
Korn, E. D. (1959*b*). *J. biol. Chem.* **234**, 1647.
Korn, E. D. & Payza, A. N. (1956). *J. biol. Chem.* **223**, 859.
Kosower, E. M. (1956). *J. Am. chem. Soc.* **78**, 3497.
Kosower, E. M. (1962). *Molecular Biochemistry*, p. 263. New York: McGraw-Hill.
Koyama, J., Tamiya, N., Ishimoto, M. & Nagai, H. (1954). *Seikagaku* **26**, 304.
Kozik, M. & Wenclewski, A. (1965). *Acta Histochem.* **21**, 135.
Krebs, H. (1957). *Z. Naturforsch.* B **12**, 795.
Krebs, H. A. (1929). *Biochem. Z.* **204**, 343.
Kredich, N. M. & Tomkins, G. M. (1966). *J. biol. Chem.* **241**, 4955.
Kredich, N. M. & Tomkins, G. M. (1967). In *Organisational Biosynthesis*, p. 189. Ed. by Vogel, H. J., Lampen, J. O. & Bryson, V. New York: Academic Press.
Krouse, H. R., McCready, R. G. L., Husain, S. A. & Campbell, J. N. (1967). *Can. J. Microbiol.* **13**, 21.
Kubica, G. P. & Vestal, A. L. (1961). *Am. Rev. resp. Dis.* **83**, 728, 733, 737.
Kuczynski, J., Pydzik, T. & Wenclewski, A. (1964). Quoted in *Chem. Abstr.* (1964), **61**, 13687 e.
Kun, E. (1961). In *Metabolic Pathways*, Vol. 2, p. 237. Ed. Greenberg, D. New York: Academic Press.
Kun, E. & Fanshier, D. W. (1958). *Biochim. biophys. Acta* **27**, 659.
Kun, E. & Fanshier, D. W. (1959*a*). *Biochim. biophys. Acta* **33**, 26.
Kun, E. & Fanshier, D. W. (1959*b*). *Biochim. biophys. Acta* **32**, 338.
Kun, E. & Fanshier, D. W. (1961). *Biochim. biophys. Acta* **48**, 187.
Kurtenacker, A. (1938). *Analytische Chemie der Sauerstoffsäuren des Schwefels.* Stuttgart: F. Enke.
Kurtenacker, A. & Goldbach, E. (1927). *Z. anorg. allg. Chem.* **166**, 177.
Kurtenacker, A. & Kaufmann, M. (1925). *Z. anorg. allg. Chem.* **148**, 256.
Kurtenacker, A., Mutschin, A. & Stastny, F. (1935). *Z. anorg. allg. Chem.* **224**, 399.

## BIBLIOGRAPHY

Kuznetsov, S. I., Ivanov, M. V. & Lyalikova, N. N. (1963). *Introduction to Geological Microbiology*. New York: McGraw-Hill.

Lampen, J. O., Roepke, R. R. & Jones, M. J. (1947). *Archs. Biochem. Biophys.* **13**, 55.

Landesman, J., Duncan, D. W. & Walden, C. C. (1966). *Can. J. Microbiol.* **12**, 957.

Lang, K. (1933). *Biochem. Z.* **259**, 243.

Lang, K. (1949). *Z. Vitam. Horm. u Ferment-forsch.* **2**, 288.

Lange-Posdeeva, I. P. (1930). *Archs. Sci. biol. St Petersb.* (*Arkh. biol. Nauk*) **30**, 189.

Laporte, G., Laurent, M. & Kaiser, P. (1965). *Annls. Inst. Pasteur, Paris* **109**, 563.

La Rivière, J. W. M. (1955). *Antonie van Leeuwenhoek* **21**, 1, 9.

La Rivière, J. W. M. (1963). In *Symp. on Marine Microbiology*, p. 61. Ed. by Oppenheimer, C. H. Springfield, Ill.: Ch. C. Thomas.

Larsen, H. (1952). *J. Bact.* **64**, 187.

Larsen, H. (1953). *K. norske Vidensk. Selsk. Skr.* NR1.

Lash, J. W. & Whitehouse, M. W. (1960). *Biochem. J.* **74**, 351.

Lash, J. W. & Whitehouse, M. W. (1961). *Lab. Invest.* **10**, 388.

Lasnitzki, I., Dingle, J. T. & Adams, S. (1966). *Expl Cell. Res.* **43**, 120.

Laughland, D. H. & Young, L. (1944). *J. Am. chem. Soc.* **66**, 657.

Law, G. L., Mansfield, G. P., Muggleton, D. F. & Parnell, E. W. (1963). *Nature, Lond.* **197**, 1024.

Lawrinson, W. & Gross, S. (1964). *Lab. Invest.* **13**, 1612.

Layton, L. L. (1951*a*). *Proc. Soc. exp. Biol. Med.* **76**, 596.

Layton, L. L. (1951*b*). *Archs. Biochem. Biophys.* **32**, 224.

Leathen, W. W., Kinsel, N. A. & Braley, S. A. (1956). *J. Bact.* **72**, 700.

Leban, M., Edwards, V. H. & Wilke, C. R. (1966). *J. ferm. Technol.* **44**, 334.

Lecher, H. Z. & Hardy, E. M. (1955). *J. org. Chem.* **20**, 475.

Lederer, M. (1957). *Analytica chim. Acta* **17**, 606.

Lees, H. (1955). *Biochemistry of Autotrophic Bacteria*. London: Butterworth Sci. Publs.

Lees, H. (1960). *A. Rev. Microbiol.* **14**, 83.

Le Gall, J. (1963). *J. Bact.* **86**, 1120.

Le Gall, J. (1968). *Annls. Inst. Pasteur, Paris* **114**, 109.

Le Gall, J. & Dragoni, N. (1966). *Biochem. biophys. Res. Commun.* **23**, 145.

Le Gall, J. & Hatchikian, E. C. (1967). *C. r. hebd. Séanc. Acad. Sci., Paris* **264**, 2580.

Le Gall, J., Mazza, G. & Dragoni, N. (1963). *Biochim. biophys. Acta* **56**, 666.

Leininger, K. R. & Westley, J. (1967). *Fedn. Proc.* **26**, 390.

Leininger, K. R. & Westley, J. (1968). *J. biol. Chem.* **243**, 1892.

Leinweber, F-J. & Monty, K. J. (1961). *Biochem. biophys. res. Commun.* **6**, 355.

Leinweber, F-J. & Monty, K. J. (1962). *Biochim. biophys. Acta* **63**, 171.

Leinweber, F-J. & Monty, K. J. (1963). *J. biol. Chem.* **238**, 3775.

Leinweber, F-J. & Monty, K. J. (1965). *J. biol. Chem.* **240**, 782.

Leinweber, F-J., Siegel, L. M. & Monty, K. J. (1965). *J. biol. Chem.* **240**, 2699.

Lé John, H. B., Van Caeseele, L. & Lees, H. (1967). *J. Bact.* **94**, 1484.

Lemann, J., Relman, A. S. & Connors, H. P. (1959). *J. clin. Invest.* **38**, 2215.

Le Minor, L. & Pichinoty, F. (1963). *Annls. Inst. Pasteur, Paris* **104**, 384.

Leon, Y. A., Bulbrook, R. D. & Corner, E. D. S. (1960). *Biochem. J.* **75**, 612.

Le Page, G. A. & Umbreit, W. W. (1943). *J. biol. Chem.* **148**, 255.

Levenson, G. I. P. (1954). *J. appl. Chem. Lond.* **4**, 13.

Levi, A. S., Geller, S., Root, D. M. & Wolf, G. (1968). *Biochem. J.* **109**, 69.

Levi, A. S. & Wolf, G. (1969). *Biochim. biophys Acta* **178**, 262.

Levin, R. E., Ng, H., Nagel, C. W. & Vaughn, R. H. (1959). *Bact. Proc.* p. 7.

Levin, R. E. & Vaughn, R. H. (1966). *Fd. Res.* **31**, 768.

Levinthal, M. (1967). *Diss. Abstr.* **27**B, 4281.

Levinthal, M. & Schiff, J. A. (1965). *Pl. Physiol.* **40**, Suppl. xvi.

Levinthal, M. & Schiff, J. A. (1968). *Pl. Physiol.* **43**, 555.

Levitz, M. (1963). *Steroids* **1**, 117.

Levitz, M. (1966). *J. clin. Endocr. Metab.* **26**, 773.

Levitz, M., Condon, G. P., Money, W. L. & Dancis, J. (1960). *J. biol. Chem.* **235**, 973.

Levitz, M., Katz, J. & Twombly, G. H. (1966). *Steroids* **6**, 553.

Lewbart, M. L. & Schneider, J. J. (1955). *Nature, Lond.* **176**, 1175.

Lewis, D. (1954). *Biochem. J.* **56**, 391.

Ley, H. & König, E. (1938). *Z. phys. Chem.* B **41**, 365.

Lieberman, S. & Roberts, K. (1969). In *Chemical and Biological Aspects of Steroid Conjugation.* Ed. by Bernstein, S. & Solomon, S. New York: Springer-Verlag.

Lieske, R. (1912). *Ber. dt. bot. Ges.* **30**, 12.

Linke, W. F. (1965). *Solubilities of Inorganic and Metal Organic Compounds,* Vol. II, p. 1400. Washington DC: Am. Chem. Soc.

Linker, A., Hoffman, P., Meyer, K., Sampson, P. & Korn, E. D. (1960). *J. biol. Chem.* **235**, 3061.

Lipman, C. B. & McLees, E. (1940). *Soil Sci.* **50**, 429.

Littlewood, D. & Postgate, J. R. (1956). *Biochim. biophys. Acta* **20**, 399.

Littlewood, D. & Postgate, J. R. (1957). *J. gen. Microbiol.* **16**, 596.

Ljunggren, P. (1960). *Econ. Geol.* **55**, 531.

Lloyd, A. G. (1959). *Biochem. J.* **72**, 133.

Lloyd, A. G. (1960). *Biochem. J.* **75**, 478.

Lloyd, A. G. (1961). *Biochem. J.* **80**, 572.

Lloyd, A. G. (1962*a*). *Biochem. J.* **83**, 455.

Lloyd, A. G. (1962*b*). *Biochim. biophys. Acta* **58**, 1.

Lloyd, A. G. (1964). *Biochem. J.* **91**, 4P.

Lloyd, A. G. (1966*a*). *Meth. Enzym.* **8**, 670.

Lloyd, A. G. (1966*b*). *Meth. Enzym.* **8**, 663.

Lloyd, A. G. & Dodgson, K. S. (1961). *Biochim. biophys. Acta* **46**, 116.

Lloyd, A. G., Dodgson, K. S., Price, R. G. & Rose, F. A. (1961). *Biochim. biophys. Acta* **46**, 108.

Lloyd, A. G., Embery, G., Powell, G. M., Curtis, C. G. & Dodgson, K. S. (1966). *Biochem. J.* **98**, 33P.

Lloyd, A. G., Embery, G., Wusteman, F. S. & Dodgson, K. S. (1966). *Biochem. J.* **98**, 33P.

Lloyd, A. G., Large, P. J., Davies, M., Olavesen, A. H. & Dodgson, K. S. (1968). *Biochem. J.* **108**, 393.

Lloyd, A. G., Olavesen, A. H., Woolley, P. A. & Embery, G. (1967). *Biochem. J.* **102**, 37P.

Lloyd, A. G., Tudball, N. & Dodgson, K. S. (1961). *Biochim. biophys. Acta* **52**, 413.

## BIBLIOGRAPHY

Lloyd, A. G., Wusteman, F. S., Tudball, N. & Dodgson, K. S. (1964). *Biochem. J.* **92**, 68.

Lloyd, P. F. & Fielder, R. J. (1967). *Biochem. J.* **105**, 33 P.

Lloyd, P. F. & Fielder, R. J. (1968). *Biochem. J.* **109**, 14 P.

Lloyd, P. F. & Lloyd, K. O. (1963). *Nature, Lond.* **199**, 287.

Lloyd, P. F., Lloyd, K. O. & Owen, O. (1962). *Biochem. J.* **85**, 193.

Loiselet, J. & Chatagner, F. (1964). *Biochim. biophys. Acta* **89**, 330.

London, J. (1963 a). *Arch. Mikrobiol.* **46**, 329.

London, J. (1963 b). *Science, N.Y.* **140**, 409.

London, J. P. (1964). *Path of Sulfur in Sulfide and Thiosulfate Oxidation by Members of the Genus Thiobacillus.* Ph.D. Thesis. Univ. Southern California.

London, J. & Rittenberg, S. C. (1964). *Proc. natn. Acad. Sci. U.S.A.* **52**, 1183.

London, J. & Rittenberg, S. C. (1966). *J. Bact.* **91**, 1062.

London, J. & Rittenberg, S. C. (1967). *Arch. Mikrobiol.* **59**, 218.

Longenecker, J. B. & Snell, E. E. (1957). *J. biol. Chem.* **225**, 409.

Lorenz, L. & Samuel, R. (1931). *Z. phys. Chem.* B **14**, 219.

Losada, M., Nozaki, M. & Arnon, D. I. (1961). In *Light and Life*, p. 570. Ed. by McElroy, W. D. & Glass, B. Baltimore: Johns Hopkins Press.

Lovatt Evans, C. (1967). *Q. Jl exp. Physiol.* **52**, 231.

Lovenberg, W. & Sobel, B. E. (1965). *Proc. natn. Acad. Sci. U.S.A.* **54**, 193.

Lowe, I. P. & Roberts, E. (1955). *J. biol. Chem.* **212**, 477.

Lucy, J. A., Dingle, J. T. & Fell, H. B. (1961). *Biochem. J.* **79**, 500.

Ludewig, S. & Chanutin, A. (1950). *Archs. Biochem.* **29**, 441.

Lugg, J. W. H. (1932). *Biochem. J.* **26**, 2144.

Luke, C. L. (1943). *Ind. Engng. Chem. analyt. Edn.* **15**, 602.

Luke, C. L. (1949). *Analyt. Chem.* **21**, 1369.

McCallan, S. E. A. (1964). *Agrochimica* **9**, 15.

McCallan, S. E. A. & Wilcoxon, F. (1931). *Contr. Boyce Thompson Inst. Pl. Res.* **3**, 13.

McChesney, C. A. (1958). *Nature, Lond.* **181**, 347.

McElroy, W. D. & Glass, B. (1961). Eds. *Light & Life.* Baltimore: Johns Hopkins Press.

Machlin, L. J., Pearson, P. B. & Denton, C. A. (1955). *J. biol. Chem.* **212**, 469.

Machlin, L. J., Pearson, P. B., Denton, C. A. & Bird, H. R. (1953). *J. biol. Chem.* **205**, 213.

McKenna, J. & Norymberski, J. K. (1957 a). *J. chem. Soc.* p. 3889.

McKenna, J. & Norymberski, J. K. (1957 b). *J. chem. Soc.* p. 3893.

McKenna, J. & Norymberski, J. K. (1960). *Biochem. J.* **76**, 60 P.

McKhann, G. M., Levy, R. & Ho, W. (1965). *Biochem. biophys. Res. Commun.* **20**, 109.

MacLeod, R. M., Farkas, W., Fridovich, I. & Handler, P. (1961). *J. biol. Chem.* **236**, 1841.

MacLeod, R. M., Fridovich, I. & Handler, P. (1961). *J. biol. Chem.* **236**, 1847.

Mager, J. (1960). *Biochim. biophys. Acta* **41**, 553.

Mahoney, R. P. & Edwards, M. R. (1966). *J. Bact.* **92**, 487.

Maickel, R. P., Jondorf, W. R. & Brodie, B. B. (1958). *Fedn. Proc.* **17**, 390.

Maliszewski, T. F. & Bass, D. E. (1955). *J. appl. Physiol.* **8**, 289.

Maloof, F. & Soodak, M. (1964). *J. biol. Chem.* **239**, 1995.

Marchlewitz, B. & Schwartz, W. (1961). *Z. allg. Mikrobiol.* **1**, 100.

Margalith, P., Silver, M. & Lundgren, D. G. (1966). *J. Bact.* **92**, 1706.

Margolis, D. & Mandl, R. M. (1958). *Contr. Boyce Thompson Inst. Pl. Res.* **19**, 509.

Marroudos, A. (1968). *Biophys. J.* **8**, 575.

Martelli, H. L. & Benson, A. A. (1964). *Biochim. biophys. Acta* **93**, 169.

Martin, H. (1964). *The Scientific Principles of Crop Protection*, 5th ed., p. 110. London: Edward Arnold Ltd.

Marunouchi, T. & Mori, T. (1968). *Bot. Mag., Tokyo* **81**, 179.

Mason, V. C., Hansen, J. H. & Jakobsen, P. E. (1965). *Int. atom. Energy Ag. Symp., Prague* 1964, p. 421.

Mathews, M. B. (1958). *Nature, Lond.* **181**, 421.

Mathewson, J. H., Burger, L. J. & Millstone, H. G. (1968). *Fedn. Proc.* **27**, 774.

Mayers, D. F. & Kaiser, E. T. (1968). *J. Am. chem. Soc.* **90**, 6192.

Mayers, G. L. & Haines, T. H. (1967). *Biochemistry* **6**, 1665.

Mayeux, J. V. & Johnson, E. J. (1966). *Bact. Proc.* p. 94.

Mead, J. A. R., Smith, J. N. & Williams, R. T. (1955). *Biochem. J.* **61**, 569.

Mechalas, B. J. & Rittenberg, S. C. (1960). *J. Bact.* **80**, 501.

Mechsner, Kl. (1957). *Arch. Mikrobiol.* **26**, 32.

Medes, G. (1937). *Biochem. J.* **31**, 1330.

Meezan, E. & Davidson, E. A. (1967*a*). *J. biol. Chem.* **242**, 1685.

Meezan, E. & Davidson, E. A. (1967*b*). *J. biol. Chem.* **242**, 4956.

Meezan, E., Olavesen, A. H. & Davidson, E. A. (1964). *Biochim. biophys. Acta* **83**, 256.

Mehl, E. & Jatzkewitz, H. (1963). *Hoppe-Seyler's Z.* **331**, 292.

Mehl, E. & Jatzkewitz, H. (1964). *Hoppe-Seyler's Z.* **339**, 260.

Mehl, E. & Jatzkewitz, H. (1965). *Biochem. biophys. Res. Commun.* **19**, 407.

Mehl, E. & Jatzkewitz, H. (1968). *Biochim. biophys. Acta* **151**, 619.

Meister, A. (1953). *Fedn. Proc.* **12**, 245.

Meister, A., Fraser, P. E. & Tice, S. V. (1954). *J. biol. Chem.* **206**, 561.

Mekhtieva, V. L. (1964). *Geokhimiya*, No 1, p. 61.

Mekhtieva, V. L. & Kondrat'eva, E. N. (1966). *Dokl. Akad. Nauk, S.S.S.R.* **166**, 465.

Mellor, J. W. (1930). *A Comprehensive Treatise on Inorganic and Theoretical Chemistry*, Vol. 10. London: Longmans, Green & Co.

Metz, L. (1929). *Z. anal. Chem.* **76**, 347.

Metzenberg, R. L. & Parson, J. W. (1966). *Proc. natn. Acad. Sci. U.S.A.* **55**, 629.

Metzler, D. E., Ikawa, M. & Snell, E. E. (1954). *J. Am. chem. Soc.* **76**, 648.

Metzler, D. E. & Snell, E. E. (1952). *J. biol. Chem.* **198**, 353.

Meyer, B. (1964). *Chem. Revs.* **64**, 429.

Michels, F. G. & Smith, J. T. (1965). *J. Nutr.* **87**, 217.

Michelson, A. M. & Wold, F. (1962). *Biochemistry* **1**, 1171.

Mietkiewski, K. & Kozik, M. (1966). *Acta histochem.* **25**, 205.

Milazzo, F. H. & Fitzgerald, J. W. (1966). *Can. J. Microbiol.* **12**, 735.

Milazzo, F. H. & Lougheed, G. J. (1967). *Can. J. Bot.* **45**, 532.

Milhaud, G., Aubert, J.-P. & Millet, J. (1956). *C. r. hebd. Séanc. Acad. Sci., Paris* **243**, 102.

Milhaud, G., Aubert, J.-P. & Millet, J. (1957). *C. r. hebd. Séanc. Acad. Sci., Paris* **244**, 1289.

Milhaud, G., Aubert, J.-P. & Millet, J. (1958). *C. r. hebd. Séanc. Acad. Sci., Paris* **246**, 1766.

Miller, J. D. A., Hughes, J. E., Saunders, G. F. & Campbell, L. L. (1968). *J. gen. Microbiol.* **52**, 173.

Miller, J. D. A. & Saleh, A. M. (1964). *J. gen. Microbiol.* **37**, 419.

Miller, J. D. A. & Wakerley, D. S. (1966). *J. gen. Microbiol.* **43**, 101.

Miller, L. P. (1949). *Contr. Boyce Thompson Inst. Pl. Res.* **15**, 467.

Miller, L. P. (1950). *Contr. Boyce Thompson Inst. Pl. Res.* **16**, 85.

Miller, L. P., McCallan, S. E. A. & Weed, R. M. (1953). *Contr. Boyce Thompson Inst. Pl. Res.* **17**, 151.

Millet, J. (1955). *C. r. hebd. Séanc. Acad. Sci., Paris* **240**, 253.

Millet, J. (1956). In *Colloque sur la Biochimie du Soufre*, pp. 77, 79. Paris: Editions CNRS.

Milligan, B. & Swan, J. M. (1962). *Rev. pure appl. Chem.* **12**, 72.

Minatoya, S., Aoe, I. & Nagai, I. (1935). *Ind. Engng. Chem. analyt. Edn.* **7**, 414.

Mintel, R. & Westley, J. (1966a). *J. biol. Chem.* **241**, 3381.

Mintel, R. & Westley, J. (1966b). *J. biol. Chem.* **241**, 3386.

Miquel, P. (1879). *Bull. Soc. chim. Paris* **32**, Part 2, 127.

Miraglia, R. J., Martin, W. G., Spaeth, D. G. & Patrick, H. (1966). *Proc. Soc. exp. Biol. Med.* **123**, 725.

Mitsuhashi, S. & Matsuo, Y. (1950). *Jap. J. exp. Med.* **20**, 729.

Mitsuhashi, S. & Matsuo, Y. (1953). *Jap. J. exp. Med.* **23**, 1.

Mizobuchi, K., Demerec, M. & Gillespie, D. H. (1962). *Genetics* **47**, 1617.

Moffatt, J. G. & Khorana, H. G. (1961). *J. Am. chem. Soc.* **83**, 663.

Moffitt, W. (1950). *Proc. Roy. Soc.* A **200**, 409.

Mori, T., Kodama, A. & Marunouchi, T. (1967). *Abstr. 7th Int. Congr. Biochem.* p. 953.

Moriarty, D. J. W. & Nicholas, D. J. D. (1968). *Proc. Aust. biochem. Soc.* p. 97.

Morris, H. E., Lacombe, R. E. & Lane, W. H. (1948). *Analyt. Chem.* **20**, 1037.

Morton, R. K. (1958). *Rev. pure appl. Chem.* **8**, 161.

Moser, U. S. & Olson, R. V. (1953). *Soil Sci.* **76**, 251.

Mross, G. A. & Doolittle, R. F. (1967). *Archs. Biochem. Biophys.* **122**, 674.

Mudd, S. H., Finkelstein, J. D., Irreverre, F. & Laster, L. (1965). *J. biol. Chem.* **240**, 4382.

Mudd, S. H., Irreverre, F. & Laster, L. (1967). *Science, N.Y.* **156**, 1599.

Muir, H. & Jacobs, S. (1967). *Biochem. J.* **103**, 367.

Mukherji, B. & Bachhawat, B. K. (1966). *Indian J. Biochem.* **3**, 1, 4.

Mukherji, B. & Bachhawat, B. K. (1967). *Biochem. J.* **104**, 318.

Mumma, R. O. (1966). *Lipids* **1**, 221.

Mumma, R. O. (1968). *Biochim. biophys. Acta* **165**, 571.

Nagashima, Z. & Uchiyama, M. (1959a). Quoted in *Chem. Abstr.* (1962) **57**, 3787h.

Nagashima, Z. & Uchiyama, M. (1959b). Quoted in *Chem. Abstr.* (1962), **57**, 15498f.

Nagashima, Z. & Uchiyama, M. (1959c). Quoted in *Chem. Abstr.* (1963), **58**, 4769a.

Naiki, N. (1964). *Pl. Cell Physiol.* **5**, 71.

Naiki, N. (1965). *Pl. Cell Physiol.* **6**, 179.

Nakai, N. & Jensen, M. L. (1964). *Geochim. cosmochim. Acta* **28**, 1893.

Nakamura, T. (1962). *J. gen. Microbiol.* **27**, 221.

Nakamura, T. & Sato, R. (1960). *Nature, Lond.* **185**, 163.

Nakamura, T. & Sato, R. (1962). *Nature, Lond.* **193**, 481.
Nakamura, T. & Sato, R. (1963*a*). *Biochem. J.* **86**, 328.
Nakamura, T. & Sato, R. (1963*b*). *Nature, Lond.* **198**, 1198.
Nathansohn, A. (1902). *Mitt. zool. Stn. Neapel.* **15**, 655.
Neuberg, C. & Hofmann, E. (1931). *Biochem. Z.* **234**, 345.
Neuberg, C. & Schonebeck, O. (1933). *Biochem. Z.* **265**, 223.
Neuberg, C. & Wagner, J. (1927). *Z. ges. exp. Med.* **56**, 334.
Neuberg, C. & Welde, E. (1914). *Biochem. Z.* **67**, 111.
Newburgh, R. W. (1954). *J. Bact.* **68**, 93.
Newton, W. A. & Snell, E. E. (1964). *Proc. natn. Acad. Sci. U.S.A.* **51**, 382.
Ney, K. H. & Ammon, R. (1959). *Hoppe-Seyler's Z.* **315**, 145.
Nichol, L. W. & Roy, A. B. (1964). *J. Biochem. Tokyo* **55**, 643.
Nichol, L. W. & Roy, A. B. (1965). *Biochemistry* **4**, 386.
Nichol, L. W. & Roy, A. B. (1966). *Biochemistry* **5**, 1379.
Nietzel, O. A. & DeSesa, M. A. (1955). *Analyt. Chem.* **27**, 1839.
Nightingale, G. T., Schermerhorn, L. G. & Robbins, W. R. (1932). *Pl. Physiol.* **7**, 565.
Nimgade, N. M. (1968). Trans. 9th Int. Congr. Soil Sci. Adelaide, Vol. 2, p. 765.
Nissen, P. & Benson, A. A. (1961). *Science, N.Y.* **134**, 1759.
Nissen, P. & Benson, A. A. (1964). *Biochim. biophys. Acta* **82**, 400.
Nonami, Y. (1959). Quoted in *Chem. Abstr.* (1962) **57**, 17229f.
Nose, Y. & Lipmann, F. (1958). *J. biol. Chem.* **233**, 1348.
Nyiri, W. (1923). *Biochem. Z.* **141**, 160.
Oertel, G. W. (1961*a*). *Biochem. Z.* **334**, 431.
Oertel, G. W. (1961*b*). *Naturwissenschaften* **48**, 621.
Oertel, G. W. (1963*a*). *Biochem. Z.* **339**, 125.
Oertel, G. W. (1963*b*). *Biochem. Z.* **339**, 135.
Oertel, G. W. & Groot, K. (1965). *Hoppe-Seyler's Z.* **341**, 1.
Oertel, G. W., Tornero, M. C. & Groot, K. (1964). *J. Chromat.* **14**, 509.
Oertel, G. W., Treiber, L. & Rindt, W. (1967). *Experientia* **23**, 97.
Ogata, G. & Bower, C. A. (1965). *Soil Sci. Soc. Am. Proc.* **29**, 23.
Ogston, A. G. (1966). *Fedn. Proc.* **25**, 986.
Ohara, M. & Kurata, J. (1950). Quoted in *Chem. Abstr.* (1951) **45**, 1187g.
Ohmura, H. & Yasoda, T. (1960). Quoted in *Biol. Abstr.* (1962), **39**, 837.
Oike, S. (1958). *Kyoto Furitsu Ikadaigaku Zasshi*, **63**, 285.
Okuda, S. & Uemura, T. (1965). *Biochim. biophys. Acta* **97**, 154.
Okuzumi, M. (1965). *Agric. biol. Chem.* **29**, 1069.
Okuzumi, M. (1966*a*). *Agric. biol. Chem.* **30**, 313.
Okuzumi, M. (1966*b*). *Agric. biol. Chem.* **30**, 713.
Okuzumi, M. & Imai, K. (1965). *J. Ferment. Technol., Osaka* **43**, 10.
Okuzumi, M. & Kita, T. (1965). *Agric. biol. Chem.* **29**, 1063.
Olitzki, A. L. (1954). *J. gen. Microbiol.* **11**, 160.
Olszewska, M. J. & Gabara, B. (1964). *Protoplasma* **59**, 163.
Ormerod, J. G. & Gest, H. (1962). *Bact. Rev.* **26**, 51.
Orsi, B. A. & Spencer, B. (1964). *J. Biochem., Tokyo* **56**, 81.
Orzel, R. (1966). *Can. J. Zool.* **45**, 134.
Ostrowski, W. & Krawczyk, A. (1957). *Acta biochim. pol.* **4**, 249.
Osuntokun, B. O. (1968). *Brain* **91**, 215.
Osuntokun, B. O., Durowoju, J. E., McFarlane, H. & Wilson, J. (1968). *Br. med. J.* **3**, 647.

Palmer, R. H. (1967). *Proc. natn. Acad. Sci., U.S.A.* **58**, 1047.

Pankhurst, E. S. (1964). *J. gen. Microbiol.* **34**, 427.

Pankhurst, E. S. (1968). *J. appl. Bact.* **31**, 179.

Panikkar, K. R. & Bachhawat, B. K. (1968). *Biochim. biophys. Acta* **151**, 725.

Pardee, A. B. (1966). *J. biol. Chem.* **241**, 5886.

Pardee, A. B. (1967). *Science* **156**, 1627.

Pardee, A. B. & Prestidge, L. S. (1966). *Proc. natn. Acad. Sci., U.S.A.* **55**, 189.

Pardee, A. B., Prestidge, L. S., Whipple, M. B. & Dreyfuss, J. (1966). *J. biol. Chem.* **241**, 3962.

Pardee, A. B. & Watanabe, K. (1968). *J. Bact.* **96**, 1049.

Parker, A. J. & Kharasch, N. (1959). *Chem. Rev.* **59**, 583.

Parker, C. D. (1945). *Aust. J. exp. Biol. med. Sci.* **23**, 81.

Parker, C. D. (1947). *Nature, Lond.* **159**, 439.

Parker, C. D. & Prisk, J. (1953). *J. gen. Microbiol.* **8**, 344.

Pasternak, C. A. (1961). *Biochem. J.* **81**, 2P.

Pasternak, C. A. (1962). *Biochem. J.* **85**, 44.

Pasternak, C. A., Ellis, R. J., Jones-Mortimer, M. C. & Crichton, C. E. (1965). *Biochem. J.* **96**, 270.

Pasternak, C. A., Humphries, S. K. & Pirie, A. (1963). *Biochem. J.* **86**, 382.

Pasqualini, J. R., Cedard, L., Nguyen, B. L. & Alsatt, E. (1967). *Biochim. biophys. Acta* **139**, 177.

Pasqualini, J. R., Zelnik, R. & Jayle, M. F. (1962). *Bull. Soc. chim. Fr.* p. 1171,

Patchornik, A., Lawson, W. B. & Witkop, B. (1958). *J. Am. chem. Soc.* **80**, 4747.

Paterson, J. Y. F. & Klyne, W. (1948). *Biochem. J.* **43**, 614.

Patrick, A. D. (1962). *Biochem. J.* **83**, 248.

Pauling, L. (1949). *Proc. natn. Acad. Sci., U.S.A.* **35**, 495.

Payne, W. J., Williams, J. P. & Mayberry, W. R. (1965). *Appl. Microbiol.* **13**, 698.

Payne, W. J., Williams, J. P. & Mayberry, W. R. (1967). *Nature, Lond.* **214**, 623.

Peat, S. & Rees, D. A. (1961). *Biochem. J.* **79**, 7.

Peat, S., Turvey, J. R., Clancy, M. J. & Williams, T. P. (1960). *J. chem. Soc.* p. 4761.

Peche, K. (1913). *Ber. dt. bot. Ges.* **31**, 458.

Peck, H. D. (1959). *Proc. natn. Acad. Sci., U.S.A.* **45**, 701.

Peck, H. D. (1960). *Proc. natn. Acad. Sci., U.S.A.* **46**, 1053.

Peck, H. D. (1961*a*). *J. Bact.* **82**, 933.

Peck, H. D. (1961*b*). *Biochim. biophys. Acta* **49**, 621.

Peck, H. D. (1962*a*). *Bact. Rev.* **26**, 67.

Peck, H. D. (1962*b*). *J. biol. Chem.* **237**, 198.

Peck, H. D. (1966). *Biochem. biophys. Res. Commun.* **22**, 112.

Peck, H. D. (1966–7). *Some Evolutionary Aspects of Inorganic Sulfur Metabolism. Lect. theor. appl. Aspects. mod. microbiol.* Univ. Maryland, Maryland, U.S.A.

Peck, H. D. & Davidson, J. T. (1967). *Bact. Proc.* p. 118.

Peck, H. D., Deacon, T. E. & Davidson, J. T. (1965). *Biochim. biophys. Acta* **96**, 429.

Peck, H. D. & Fisher, E. (1962). *J. biol. Chem.* **237**, 190.

Peck, H. D. & Stulberg, M. P. (1962). *J. biol. Chem.* **237**, 1648.

Pepler, W. J., Loubser, E. & Kooij, R. (1958). *Dermatologica* **117**, 468.

Percival, E. G. V. (1949). *Q. Rev. chem. Soc.* **3**, 369.

Petrova, E. A. (1959). *Microbiology* **28**, 758.

Pfansteil, R. (1946). *Inorg. Synth.* **2**, 170.

Pichinoty, F. (1963). *Enzymologia* **26**, 176.

Pichinoty, F. (1965). In *Mécanismes de Régulation des Activités Cellulaires chez les Micro-organisms*, p. 507. Paris: Editions CNRS.

Pichinoty, F. & Bigliardi-Rouvier, J. (1962). *Antonie van Leeuwenhoek* **28**, 134.

Pichinoty, F. & Bigliardi-Rouvier, J. (1963). *Biochim. biophys. Acta* **67**, 366.

Pichinoty, F. & Senez, J. C. (1958). *C. r. hebd. Séanc. Acad. Sci., Paris* **247**, 361.

Picou, D. & Waterlow, J. C. (1963). *Nature, Lond.* **197**, 1103.

Piéchaud, M., Puig, J., Pichinoty, F., Azoulay, E. & Le Minor, L. (1967). *Annls. Inst. Pasteur, Paris* **112**, 24.

Pihl, A. & Lange, R. (1962). *J. biol. Chem.* **237**, 1356.

Pirie, N. W. (1934a). *Biochem. J.* **28**, 305.

Pirie, N. W. (1934b). *Biochem. J.* **28**, 1063.

Pirie, N. W. & Hele, T. S. (1933). *Biochem. J.* **27**, 1716.

Plager, J. E. (1965). *J. clin. Invest.* **44**, 1234.

Pochon, J., Coppier, O. & Tchan, Y. T. (1951). *Chim. Ind.* **65**, 496.

Poczekaj, J., Hejduk, J. & Wenclewski, A. (1963). Quoted in *Chem. Abstr.* (1964), **61**, 1067f.

Poczekaj, J. & Wenclewski, A. (1963). Quoted in *Chem. Abstr.* (1964), **61**, 1047d.

Pollacci, E. (1875). *Gazz. chim. ital.* **5**, 451.

Pollard, F. H. (1954). *Brit. med. Bull.* **10**, 187.

Pollard, F. H. & Jones, D. J. (1958). In *Recent Work on the Inorganic Chemistry of Sulphur*. The Chemical Society, London. Spec. Pub. No. 12, p. 363.

Pollard, F. H., McOmie, J. F. W. & Jones, D. J. (1955). *J. chem. Soc.* p. 4337.

Pollard, F. H., McOmie, J. F. W. & Stevens, H. M. (1951). *J. chem. Soc.* p. 771.

Pollard, F. H., Nickless, G. & Burton, K. W. C. (1962). *J. Chromatog.* **8**, 507.

Pollard, F. H., Nickless, G. & Glover, R. B. (1964a). *J. Chromatog.* **15**, 518.

Pollard, F. H., Nickless, G. & Glover, R. B. (1964b). *J. Chromatog.* **15**, 533.

Pollock, M. R. & Knox, R. (1943). *Biochem. J.* **37**, 476.

Pollock, M. R., Knox, R. & Gell, P. G. H. (1942). *Nature, Lond.* **150**, 94.

Postgate, J. R. (1949). *Nature, Lond.* **164**, 670.

Postgate, J. R. (1951a). *J. gen. Microbiol.* **5**, 714.

Postgate, J. R. (1951b). *J. gen. Microbiol.* **5**, 725.

Postgate, J. R. (1952a). *Research, Lond.* **5**, 189.

Postgate, J. R. (1952b). *J. gen. Microbiol.* **6**, 128.

Postgate, J. R. (1954). *Biochem. J.* **56**, xi.

Postgate, J. R. (1956a). *J. gen. Microbiol.* **14**, 545.

Postgate, J. R. (1956b). *J. gen. Microbiol.* **15**, viii.

Postgate, J. R. (1956c). *J. gen. Microbiol.* **15**, 186.

Postgate, J. R. (1959a). *A Rev. Microbiol.* **13**, 505.

Postgate, J. R. (1959b). *J. Sci. Fd. Agric.* **10**, 669.

Postgate, J. R. (1959c). *Nature, Lond.* **183**, 481.

Postgate, J. R. (1960a). *Z. allg. Mikrobiol.* **1**, 53.

Postgate, J. R. (1960b). *Prog. Ind. Microbiol.* **2**, 49.

Postgate, J. R. (1961). In *Haematin Enzymes*, p. 407. Ed. by Falk, J. E., Lemberg, R. & Morton, R. K. London: Pergamon Press.

Postgate, J. R. (1963a). *Appl. Microbiol.* **11**, 265.

Postgate, J. R. (1963b). *J. Bact.* **85**, 1450.

Postgate, J. R. (1963c). *Arch. Mikrobiol.* **46**, 287.

**BIBLIOGRAPHY**

Postgate, J. R. (1963d). *J. gen. Microbiol.* **30**, 481.
Postgate, J. R. (1965a). *Bact. Rev.* **29**, 425.
Postgate, J. R. (1965b). *Anreicherungskultur u Mutantenauslese*, p. 190. Ed. by Schlegel, H. G. & Kroger, E. Stuttgart: Fischer.
Postgate, J. R. (1966). *Lab. Pract.* **15**, 1239.
Postgate, J. R. (1967). Quoted by W. S. Silver, *Science* **157**, 101.
Postgate, J. R. & Campbell, L. L. (1963). *J. Bact.* **86**, 274.
Postgate, J. R. & Campbell, L. L. (1966). *Bact. Rev.* **30**, 732.
Poulton, F. C. J. & Tarrant, L. (1951). *J. appl. Chem.* **1**, 29.
Poux, N. (1966). *J. Histochem. Cytochem.* **14**, 932.
Powell, G. M., Curtis, C. G. & Dodgson, K. S. (1967). *Biochem. Pharmac.* **16**, 1997.
Prabhakararao, K. & Nicholas, D. J. D. (1968). *Proc. Aust. biochem. Soc.* p. 32.
Pratt, D. P. (1958). *Nature, Lond.* **181**, 1075.
Pryor, W. A. (1962). *Mechanisms of Sulfur Reactions.* New York: McGraw Hill.
Puig, J., Azoulay, E. & Pichinoty, F. (1967). *C. r. hebd. Séanc. Acad. Sci., Paris* **264**, 1507.
Pulkkinen, M. O. (1961). *Acta physiol. Scand.* **52**, suppl. 180.
Pulkkinen, M. O. & Hakkarainen, H. (1965). *Acta endocr., Copenh.* **48**, 313.
Pulkkinen, M. O. & Paunio, I. (1963). *Annls. Med. exp. Biol. Fenn.* **41**, 283.
Quentin, K.-E. & Pachmayr, F. (1961). *Vom. Wass* **28**, 79.
Ragland, J. B. (1959). *Archs. Biochem. Biophys.* **84**, 541.
Ragland, J. B. & Liverman, J. L. (1958). *Archs. Biochem. Biophys.* **76**, 496.
Rambaut, P. C. & Miller, S. A. (1965). *Fedn. Proc.* **24**, 373.
Rammler, D. H., Grado, C. & Fowler, L. R. (1964). *Biochemistry* **3**, 224.
Ramseyer, J., Williams, J. S. & Hirschman, H. (1967). *Steroids* **9**, 347.
Rankine, B. C. (1963). *J. Sci. Fd. Agric.* **14**, 79.
Rankine, B. C. (1964). *J. Sci. Fd. Agric.* **15**, 872.
Rao, G. S. & Gorin, G. (1959). *J. org. Chem.* **24**, 749.
Raschig, F. (1908). *Chemikerzeitung* **32**, 1203.
Raschig, F. (1915). *Ber. dt. chem. Ges.* **48**, 2088.
Raschig, F. (1920). *Z. angew. Chem.* **33**, 260.
Rasmussen, H., Sallis, J., Fang, M., DeLuca, H. F. & Young, R. (1964). *Endocrinology* **74**, 388.
Ratsisalovanina, O., Chapeville, F. & Fromageot, P. (1961). *Biochim. biophys. Acta* **49**, 322.
Razzell, W. E. & Trussell, P. C. (1963). *J. Bact.* **85**, 595.
Rees, D. A. (1961a). *Biochem. J.* **78**, 25 P.
Rees, D. A. (1961b). *Biochem. J.* **80**, 449.
Rees, D. A. (1963). *Biochem. J.* **88**, 343.
Rees, D. A. (1966). *Rep. Prog. Chem.* p. 469.
Reese, E. T., Clapp, R. C. & Mandels, M. (1958). *Archs. Biochem. Biophys.* **75**, 228.
Reichard, P. & Ringertz, N. R. (1959). *J. Am. chem. Soc.* **81**, 878.
Revenda, J. (1934). *Colln. Czech. chem. Commun. Engl. Edn.* **6**, 453.
Rice, L. I., Spolter, L., Tokes, Z., Eisenman, R. & Marx, W. (1967). *Archs. Biochem. Biophys.* **118**, 374.
Robbins, P. W. (1962). *Meth. Enzym.* **5**, 964.
Robbins, P. W. (1963). *Meth. Enzym.* **6**, 770.
Robbins, P. W. & Lipmann, F. (1956a). *J. Am. chem. Soc.* **78**, 2652.

Robbins, P. W. & Lipmann, F. (1956*b*). *J. Am. chem. Soc.* **78**, 6409.
Robbins, P. W. & Lipmann, F. (1957). *J. biol. Chem.* **229**, 837.
Robbins, P. W. & Lipmann, F. (1958*a*). *J. biol. Chem.* **233**, 681.
Robbins, P. W. & Lipmann, F. (1958*b*). *J. biol. Chem.* **233**, 686.
Roberts, K. D., Bandi, L., Calvin, H. I., Drucker, W. D. & Lieberman, S. (1964). *Biochemistry* **3**, 1983.
Roberts, K. D., Bandi, L. & Lieberman, S. (1967). *Biochem. biophys. Res. Commun.* **29**, 741.
Roberts, K. D., VandeWiele, R. L. & Lieberman, S. (1961). *J. biol. Chem.* **236**, 2213.
Roberts, R. B., Abelson, P. H., Cowie, D. B., Bolton, E. T. & Britten, R. J. (1955). *Carnegie Inst. Wash. Publ.* **607**, 318.
Roberts, W. M. B. (1967). *Mineralium Deposita* **2**, 188.
Robinson, D., Smith, J. N., Spencer, B. & Williams, R. T. (1952). *Biochem. J.* **51**, 202.
Robinson, D., Smith, J. N. & Williams, R. T. (1953). *Biochem. J.* **53**, 125.
Robinson, H. C. (1965). *Biochem. J.* **94**, 687.
Robinson, H. C. & Pasternak, C. A. (1964). *Biochem. J.* **93**, 487.
Roche, J., Michel, R., Closon, J. & Michel, O. (1959). *Biochim. biophys. Acta* **33**, 461.
Roe, D. A., Miller, P. S. & Lutwak, L. (1966). *Analyt. Biochem.* **15**, 313.
Roels, O. A. (1967). In *The Vitamins*, 2nd ed., Vol. 1, p. 234. Ed. by Sebrell, W. H. & Harris, R. S. New York: Academic Press.
Rolls, J. P. & Lindstrom, E. S. (1967*a*). *J. Bact.* **94**, 784.
Rolls, J. P. & Lindstrom, E. S. (1967*b*). *J. Bact.* **94**, 860.
Rose, F. A., Flanagan, T. H. & John, R. A. (1966). *Biochem. J.* **98**, 168.
Rosenfeld, L. O. & Ruchelman, A. A. (1940). *Ukr. biokhem. Zh.* **16**, 53.
Ross, A. J., Schoenhoff, R. L. & Aleem, M. I. H. (1968). *Biochem. biophys. Res. Commun.* **32**, 301.
Roth, H. (1951). *Mikrochemie mikrochem. Acta* **36**, 379.
Rowden, G. (1967). *Nature, Lond.* **215**, 1283.
Roy, A. B. (1953*a*). *Biochem. J.* **53**, 12.
Roy, A. B. (1953*b*). *Biochem. J.* **55**, 653.
Roy, A. B. (1954*a*). *Biochem. J.* **57**, 465.
Roy, A. B. (1954*b*). *Biochim. biophys. Acta* **14**, 149.
Roy, A. B. (1955). *Biochem. J.* **59**, 8.
Roy, A. B. (1956*a*). *Biochem. J.* **62**, 41.
Roy, A. B. (1956*b*). *Biochem. J.* **63**, 294.
Roy, A. B. (1956*c*). *Biochem. J.* **64**, 651.
Roy, A. B. (1957*a*). *Experientia* **13**, 32.
Roy, A. B. (1957*b*). *Biochem. J.* **66**, 700.
Roy, A. B. (1958). *Biochem. J.* **68**, 519.
Roy, A. B. (1960*a*). *Biochem. J.* **74**, 49.
Roy, A. B. (1960*b*). *Biochem. J.* **77**, 380.
Roy, A. B. (1960*c*). *Adv. Enzymol.* **22**, 205.
Roy, A. B. (1961). *Biochem. J.* **79**, 253.
Roy, A. B. (1962*a*). *Biochem. J.* **82**, 66.
Roy, A. B. (1962*b*). *J. Histochem. Cytochem.* **10**, 106.
Roy, A. B. (1963*a*). *Hoppe-Seyler's Z.* **333**, 166.
Roy, A. B. (1963*b*). *Aust. J. exp. Biol. med. Sci.* **41**, 331.
Roy, A. B. (1964). *J. molec. Biol.* **10**, 176.

## BIBLIOGRAPHY

Roy, A. B. & Kerr, L. M. H. (1956). *Nature, Lond.* **178**, 376.
Ruelius, H. W. & Gauhe, A. (1950). *Justus Liebigs Annln. Chem.* **570**, 121.
Rutenburg, A. M., Cohen, R. B. & Seligman, A. M. (1952). *Science, N.Y.* **116**, 539.
Rutenburg, A. M. & Seligman, A. M. (1956). *Archs. Biochem. Biophys.* **60**, 198.
Rzymkowski, J. (1925). *Z. Elektrochem.* **31**, 371.
Sadana, J. C. & Jagannathan, V. (1954). *Biochim. biophys. Acta* **14**, 287.
Sachs, G. & Braun-Falco, O. (1960). *J. invest. Derm.* **34**, 439.
Sadler, W. R. & Stanier, R. Y. (1960). *Proc. natn. Acad. Sci., U.S.A.* **46**, 1328.
Sadler, W. R. & Trudinger, P. A. (1967). *Mineralium Deposita* **2**, 158.
Sahashi, Y., Suzuki, T., Higaki, M. & Asano, T. (1967). *J. Vitam.* **13**, 33.
Sahashi, Y., Suzuki, T., Nishikawa, T., Tanaka, T., Takahashi, M., Inaba, S. & Miyazawa, E. (1962). *J. Vitam.* **8**, 121.
Saito, H., Yamagata, T. & Suzuki, S. (1968). *J. biol. Chem.* **243**, 1536.
Sakai, S. (1960). *Kyoto Furitsu Ikadaigaku Zasshi* **67**, 759.
Saleh, A. M. (1964). *J. gen. Microbiol.* **37**, 113.
Salmon, W. D. & Daughaday, W. H. (1957). *J. Lab. clin. Med.* **49**, 825.
Salmon, W. D., Von Hagen, M. J. & Thompson, E. Y. (1967). *Endocrinology* **80**, 999.
Sandved, K. & Holte, J. B. (1940). *K. norske Vidensk. Selsk. Forh.* **13**, 9.
Santer, M. (1959). *Biochem. biophys. Res. Commun.* **1**, 9.
Santer, M., Boyer, J. & Santer, U. (1959). *J. Bact.* **78**, 197.
Santer, M., Margulies, M., Klinman, N. & Kaback, R. (1960). *J. Bact.* **79**, 313.
Santer, M. & Vishniac, W. (1955). *Biochim. biophys. Acta* **18**, 157.
Sargeant, K., Buck, P. W., Ford, J. W. S. & Yeo, R. G. (1966). *Appl. Microbiol.* **14**, 998.
Sasaki, S. (1967). *Clin. chim. Acta* **17**, 215.
Sasaki, T. & Otsuka, I. (1912). *Biochem. Z.* **39**, 208.
Saslavsky, A. S. (1927). *Zentbl. Bakt. ParasitKde.* Abt II, **72**, 236.
Satake, T. (1960). Quoted in *Chem. Abstr.* (1962), **57**, 15736d.
Sato, T. & Hayashi, T. (1952). *J. Jap. biochem. Soc.* **24**, 123.
Saunders, G. F. & Campbell, L. L. (1966). *J. Bact.* **92**, 515.
Saunders, G. F., Campbell, L. L. & Postgate, J. R. (1964). *J. Bact.* **87**, 1073.
Saunders, J. P. & Himwich, W. A. (1950). *Am. J. Physiol.* **163**, 404.
Savard, K., Bagnoli, E. & Dorfman, R. I. (1954). *Fedn. Proc.* **13**, 289.
Sciarini, L. J. & Nord, F. F. (1943). *Archs. Biochem.* **3**, 261.
Schaeffer, W. I., Holbert, P. E. & Umbreit, W. W. (1963). *J. Bact.* **85**, 137.
Schaeffer, W. I. & Umbreit, W. W. (1963). *J. Bact.* **85**, 492.
Schanderl, H. (1952). *Microbiologia Esp.* **5**, 17.
Schellenburg, M. & Schwarzenbach, G. (1962). *Proc. 7th Intern. Conf. Coordination Chem.* Stockholm & Uppsala, Sweden, p. 157.
Schiff, J. A. (1959). *Pl. Physiol.* **34**, 73.
Schiff, J. A. & Levinthal, M. (1968). *Pl. Physiol.* **43**, 547.
Schlossmann, K., Brüggemann, J. & Lynen, F. (1962). *Biochem. Z.* **336**, 258.
Schlossmann, K. & Lynen, F. (1957). *Biochem. Z.* **328**, 591.
Schmidt, M. (1957*a*). *Z. anorg. allg. Chem.* **289**, 175.
Schmidt, M. (1957*b*). *Z. anorg. allg. Chem.* **289**, 193.
Schmidt, M. (1963). *Ost. Chem. Ztg.* **64**, 236.
Schmidt, M. (1965). In *Elemental Sulfur*, p. 301. Ed. by Meyer, B. New York: Interscience.
Schmidt, M. & Sand, T. (1964*a*). *J. inorg. nucl. Chem.* **26**, 1173.

Schmidt, M. & Sand, T. (1964*b*). *J. inorg. nucl. Chem.* **26**, 1179.
Schmidt, M. & Sand, T. (1964*c*). *J. inorg. nucl. Chem.* **26**, 1165.
Schmidt, M. & Talsky, G. (1959). *Z. anal. Chem.* **166**, 274.
Schmidt-Nielsen, S. & Holmsen, J. (1921). *Archs. int. Physiol.* **18**, 128.
Schneegans, D. (1935). *Congr. intern. mines mét. géol. appl.* 7e session Paris. Géol. **1**, 351.
Schneider, J. F. & Westley, J. (1963). *J. biol. Chem.* **238**, PC3516.
Schneider, J. J. & Lewbart, M. L. (1956). *J. biol. Chem.* **222**, 787.
Schöön, N.-H. (1959). *Acta chem. Scand.* **13**, 525.
Schroeter, L. C. (1963). *J. pharm. Sci.* **52**, 559.
Schroeter, L. C. (1966). *Sulfur Dioxide*, Chap. 2. Oxford: Pergamon Press.
Schwab, G. M. & Strohmeyer, M. (1956). *Z. phys. Chem.* (Frankfurt) (NF), **7**, 132.
Schwartz, W. (1958). In *Encyclopedia of Plant Physiology*, Vol. IX, p. 89. Ed. by Ruhland, W. Berlin: Springer-Verlag.
Schwimmer, S. (1960). *Acta chem. scand.* **14**, 1439.
Schwimmer, S. (1961). *Acta chem. scand.* **15**, 535.
Scoffone, E. & Carini, E. (1955). *Ricerca scient.* **25**, 2109.
Scott, F. L. & Spillane, W. J. (1967). *Chemy. Ind.* p. 1999.
Scott, J. M. & Spencer, B. (1965). *Biochem. J.* **96**, 78 P.
Scott, J. M. & Spencer, B. (1968). *Biochem. J.* **106**, 471.
Scotten, H. L. & Stokes, J. L. (1962). *Arch. Mikrobiol.* **42**, 353.
Segal, H. L. & Mologne, L. A. (1959). *J. biol. Chem.* **234**, 909.
Segel, I. H. & Johnson, M. J. (1961). *J. Bact.* **81**, 91.
Segel, I. H. & Johnson, M. J. (1963*a*). *Biochim. biophys. Acta* **69**, 433.
Segel, I. H. & Johnson, M. J. (1963*b*). *Archs. Biochem. Biophys.* **103**, 216.
Seligman, A. M., Nachlas, M. M., Manheimer, L. M., Friedman, O. M. & Wolf, G. (1949). *Ann. Surg.* **130**, 333.
Selmi, F. (1876). *Just's bot. Jber.* **4**, 116.
Semadeni, E. G. (1967). *Planta* **72**, 91.
Senez, J. C. (1954). *J. gen. Microbiol.* **11**, vi.
Senez, J. C. (1955). *Bull. Soc. Chim. biol.* **37**, 1135.
Senez, J. C., Geoffray, C. & Pichinoty, F. (1956). *Gaz. France*, M 102, 1.
Senez, J. C. & Pascal, M. (1961). *Z. allg. Mikrobiol.* **1**, 142.
Senez, J. C. & Pichinoty, F. (1958*a*). *Biochim. biophys. Acta* **27**, 569.
Senez, J. C. & Pichinoty, F. (1958*b*). *Biochim. biophys. Acta* **28**, 355.
Senez, J. C., Pichinoty, F. & Konavaltchikoff-Mazoyer, M. (1956). *C. r. hebd. Séanc. Acad. Sci., Paris* **242**, 570.
Senez, J. C. & Volcani, B. E. (1951). *C. r. hebd. Séanc. Acad. Sci., Paris* **232**, 1035.
Sentenac, A., Chapeville, F. & Fromageot, P. (1963). *Biochim. biophys. Acta* **67**, 672.
Sentenac, A. & Fromageot, P. (1964). *Biochim. biophys. Acta* **81**, 289.
Servigne, Y. & Duval, C. (1957). *C. r. hebd. Séanc. Acid. Sci., Paris* **245**, 1803.
Shaposhnikov, V. N., Kondrat'eva, E. N. & Fedorov, V. D. (1958). *Microbiology* **27**, 521.
Shaulis, L. G. (1968). *Bact. Proc.* p. 140.
Shepherd, C. J. (1956). *J. gen. Microbiol.* **15**, 29.
Sherman, W. R. & Stanfield, E. F. (1967). *Biochem. J.* **102**, 905.
Shibko, S. & Tappel, A. L. (1965). *Biochem. J.* **95**, 731.
Shimodo, K. (1957). Quoted in *Chem. Abstr.* (1962), **56**, 6570a.

Shively, J. M. & Benson, A. A. (1967). *J. Bact.* **94**, 1679.

Siebenthal, C. E. (1915). *Origin of the Lead and Zinc Deposits of the Joplin Region. Bull. 606*, U.S. Geological Survey.

Siegel, L. M., Click, E. M. & Monty, K. J. (1964). *Biochem. biophys. Res. Commun.* **17**, 125.

Siegel, L. M., Leinweber, F.-J. & Monty, K. J. (1965). *J. biol. Chem.* **240**, 2705.

Siegel, L. M. & Monty, K. J. (1964). *Biochem. biophys. Res. Commun.* **17**, 201.

Sigal, N., Senez, J. C., Le Gall, J. & Sebald, M. (1963). *J. Bact.* **85**, 1315.

Sijderius, R. (1946). *Heterotrophe Bacterien, die Thiosulfaat Oxydeeren.* Thesis. Amsterdam, Netherlands.

Silver, M. & Lundgren, D. G. (1968). *Can. J. Biochem.* **46**, 457.

Silverman, M. P. & Ehrlich, H. L. (1964). *Adv. appl. Microbiol.* **6**, 153.

Silverman, M. P. & Lundgren, D. G. (1959). *J. Bact.* **78**, 326.

Singer, T. P. & Kearney, E. B. (1954). *Biochim. biophys. Acta* **14**, 570.

Sinha, D. B. & Walden, C. C. (1966). *Can. J. Microbiol.* **12**, 1041.

Sisler, F. D. & ZoBell, C. E. (1950). *J. Bact.* **60**, 747.

Sisler, F. D. & ZoBell, C. E. (1951). *J. Bact.* **62**, 117.

Skarżyński, B., Klimek, R. & Szczepkowski, T. W. (1956). *Bull. Acad. pol. Sci. Cl. II*, **4**, 299.

Skarżyński, B., Ostrowski, W. & Krawczyk, A. (1957). *Bull. Acad. pol. Sci. Cl. II*, **5**, 159.

Skarżyński, B. & Szczepkowski, T. W. (1959). *Nature, Lond.* **183**, 1413.

Skarżyński, B., Szczepkowski, T. W. & Weber, M. (1960). *Acta biochim. pol.* **7**, 105.

Skerman, V. B. D., Dementjeva, G. & Carey, B. J. (1957). *J. Bact.* **73**, 504.

Skerman, V. B. D., Dementjeva, G. & Skyring, G. W. (1957). *Nature, Lond.* **179**, 742.

Smith, A. D. M., Duckett, S. & Waters, A. H. (1963). *Nature, Lond.* **200**, 179.

Smith, A. J. (1964). *J. gen. Microbiol.* **34**, ix.

Smith, A. J. (1965). *Biochem. J.* **94**, 27 P.

Smith, A. J. (1966). *J. gen. Microbiol.* **42**, 371.

Smith, A. J. & Lascelles, J. (1966). *J. gen. Microbiol.* **42**, 357.

Smith, A. J., London, J. & Stanier, R. Y. (1967). *J. Bact.* **94**, 972.

Smith, H. W. (1951). *The Kidney.* New York: Oxford University Press.

Smith, J. N. (1951). *J. chem. Soc.* p. 2861.

Smith, J. N. (1964). In *Comparative Biochemisty*, Vol. 6, p. 403. Ed. by Florkin, M. & Mason, H. S. New York: Academic Press.

Smith, J. N. (1968). *Adv. comp. Physiol. Biochem.* **3**, 173.

Smith, P. A. S. (1948). *J. Am. chem. Soc.* **70**, 323.

Smythe, C. V. (1945). *Adv. Enzymol.* **5**, 237.

Sneddon, A. & Marrian, G. F. (1963). *Biochem. J.* **86**, 385.

Sobel, A. E., Drekter, I. J. & Natelson, S. (1936). *J. biol. Chem.* **115**, 381.

Sobel, A. E. & Spoerri, P. E. (1941). *J. Am. chem. Soc.* **63**, 1259.

Soda, K., Novogrodsky, A. & Meister, A. (1964). *Biochemistry* **3**, 1450.

Soda, T. (1936). *J. Fac. Sci. Tokyo Univ. 1*, **3**, 149.

Soda, T. & Egami, F. (1938). *J. chem. Soc. Japan* **59**, 1202.

Soda, T. & Hattori, C. (1931). *Proc. imp. Acad. Japan* **7**, 267.

Sokolova, G. A. (1962). *Microbiology* **31**, 264.

Sörbo, B. H. (1951*a*). *Acta chem. scand.* **5**, 724.

Sörbo, B. H. (1951*b*). *Acta chem. scand.* **5**, 1218.
Sörbo, B. H. (1953*a*). *Acta chem. scand.* **7**, 1129.
Sörbo, B. H. (1953*b*). *Acta chem. scand.* **7**, 238.
Sörbo, B. H. (1953*c*). *Acta chem. scand.* **7**, 1137.
Sörbo, B. H. (1953*d*). *Acta chem. scand.* **7**, 32.
Sörbo, B. (1954). *Acta chem. scand.* **8**, 694.
Sörbo, B. (1955). *Acta chem. scand.* **9**, 1656.
Sörbo, B. (1956). *Biochim. biophys. Acta* **21**, 393.
Sörbo, B. (1957*a*). *Biochim. biophys. Acta* **23**, 412.
Sörbo, B. (1957*b*). *Biochim. biophys. Acta* **24**, 324.
Sörbo, B. (1957*c*). *Acta chem. scand.* **11**, 628.
Sörbo, B. (1958*a*). *Acta chem. scand.* **12**, 1990.
Sörbo, B. (1958*b*). *Biochim. biophys. Acta* **27**, 324.
Sörbo, B. (1960). *Biochim. biophys. Acta* **38**, 349.
Sörbo, B. (1962*a*). *Acta chem. scand.* **16**, 243.
Sörbo, B. (1962*b*). *Acta chem. scand.* **16**, 2455.
Sörbo, B. (1962*c*). *Acta radiol.* **58**, 186.
Sörbo, B. (1963*a*). *Acta chem. scand.* **17**, S107.
Sörbo, B. (1963*b*). *Acta chem. scand.* **17**, 2205.
Sörbo, B. (1964). *Acta chem. scand.* **18**, 821.
Sörbo, B. (1965). *Scand. J. clin. Lab. Invest.* **17**, Suppl. 86, 21.
Sörbo, B. & Ewetz, L. (1965). *Biochem. biophys. Res. Commun.* **18**, 359.
Sörbo, B. & Heyman, T. (1957). *Biochim. biophys. Acta* **23**, 624.
Sörbo, B. & Ljunggren, J. G. (1958). *Acta chem. scand.* **12**, 470.
Sorokin, Y. I. (1954*a*). *Dokl. Akad. Nauk. S.S.S.R.* **95**, 661.
Sorokin, Y. I. (1954*b*). *Trudy Inst. Mikrobiol. Mosk.* **3**, 21.
Sorokin, Y. I. (1960). *Dokl. Akad. Nauk. S.S.S.R.* **132**, 464.
Spencer, B. (1958). *Biochem. J.* **69**, 155.
Spencer, B. (1960*a*). *Biochem. J.* **75**, 435.
Spencer, B. (1960*b*). *Biochem. J.* **77**, 294.
Spencer, B. & Harada, T. (1960). *Biochem. J.* **77**, 305.
Spencer, B., Hussey, E. C., Orsi, B. A. & Scott, J. M. (1968). *Biochem. J.* **106**, 461.
Spencer, B. & Raftery, J. (1966). *Biochem. J.* **99**, 35 P.
Spillane, W. J. & Scott, F. L. (1967). *Tetrahedron Lett.* p. 1251.
Spolter, L., Rice, L. I., Yamada, R. & Marx, W. (1967). *Biochem. Pharmac.* **16**, 229.
Stamm, H. & Goehring, M. (1942). *Z. anorg. allg. Chem.* **250**, 226.
Stamm, H., Seipold, O. & Goehring, M. (1941). *Z. anorg. allg. Chem.* **247**, 277.
Starka, J. (1951). *Biol. Listy* **32**, 108.
Starka, L., Sulcova, J. & Silink, K. (1962). *Clin. chim. Acta* **7**, 309.
Starkey, R. L. (1925). *J. Bact.* **10**, 135.
Starkey, R. L. (1934*a*). *J. Bact.* **28**, 365.
Starkey, R. L. (1934*b*). *J. Bact.* **28**, 387.
Starkey, R. L. (1935*a*). *J. gen. Physiol.* **18**, 325.
Starkey, R. L. (1935*b*). *Soil Sci.* **39**, 197.
Starkey, R. L. (1937). *J. Bact.* **33**, 545.
Starkey, R. L. (1950). *Soil Sci.* **70**, 55.
Starkey, R. L. (1966), *Soil Sci.* **101**, 297.
Starkey, R. L., Jones, G. E. & Frederick, L. R. (1956). *J. gen. Microbiol.* **15**, 329.

Starkey, R. L. & Wight, K. M. (1945). *Anaerobic Corrosion of Iron in Soil.* New York: Am. Gas Assoc.

Stearns, R. N. (1953). *J. cell. comp. Physiol.* **41**, 163.

Steigmann, A. (1942). *J. Soc. chem. Ind. Lond.* **61**, 18.

Steigmann, A. (1945). *J. Soc. chem. Ind. Lond.* **64**, 119.

Stekol, J. A. (1936). *J. biol. Chem.* **113**, 675.

Stelmaszyńska, T. & Szczepkowski, T. W. (1961). *Roczniki. Chem.* **35**, 571.

Stephenson, M. & Stickland, L. H. (1931). *Biochem. J.* **25**, 215.

Stickland, R. G. (1961). *Nature, Lond.* **190**, 648.

Stone, R. W. & ZoBell, C. E. (1952). *Ind. engng. Chem. Int. Edn.* **44**, 2564.

Stranks, D. R. & Wilkins, R. G. (1957). *Chem. Rev.* **57**, 743.

Stricks, W. & Kolthoff, I. M. (1951). *J. Am. chem. Soc.* **73**, 4569.

Stricks, W., Kolthoff, I. M. & Kapoor, R. C. (1955). *J. Am. chem. Soc.* **77**, 2057.

Stuart, C. H. (1966). Personal communication.

Subba Rao, K. & Ganguly, J. (1964). *Biochem. J.* **90**, 104.

Subba Rao, K. & Ganguly, J. (1966). *Biochem. J.* **98**, 693.

Subba Rao, K., Sastry, P. S. & Ganguly, J. (1963). *Biochem. J.* **87**, 312.

Subrahmanya, R. S. (1955). *Proc. Indian Acad. Sci.* **42**A, 267.

Suckow, R. & Schwartz, W. (1963). In *Marine Microbiology*, p. 187. Ed. by Oppenheimer, C. H. Springfield, Ill.: Charles C. Thomas.

Sugimoto, H. & Aoshima, M. (1964). *Chem. pharm. Bull., Tokyo* **12**, 362.

Suh, B. & Akagi, J. M. (1966). *J. Bact.* **91**, 2281.

Suh, B., Nakatsukasa, W. & Akagi, J. M. (1968). *Bact. Proc.* p. 133.

Sumizu, K. (1961). *Biochim. biophys. Acta* **53**, 435.

Sundaresan, P. R. (1966). *Biochim. biophys. Acta* **113**, 95.

Suter, C. M. (1944). *The Organic Chemistry of Sulfur.* New York: John Wiley.

Suzuki, I. (1965*a*). *Biochim. biophys. Acta* **104**, 359.

Suzuki, I. (1965*b*). *Biochim. biophys. Acta* **110**, 97.

Suzuki, I. & Silver, M. (1966). *Biochim. biophys. Acta* **122**, 22.

Suzuki, I. & Werkman, C. H. (1958). *Iowa St Coll. J. Sci.* **32**, 475.

Suzuki, I. & Werkman, C. H. (1959). *Proc. natn. Acad. Sci. U.S.A.* **45**, 239.

Suzuki, I. & Werkman, C. H. (1960). *Biochem. J.* **74**, 359.

Suzuki, S., Saito, H., Yamagata, T., Anno, K., Seno, N., Kawai, Y. & Furuhashi, T. (1968). *J. biol. Chem.* **243**, 1543.

Suzuki, S. & Strominger, J. L. (1960). *J. biol. Chem.* **235**, 257, 267, 274.

Suzuki, S., Takahashi, N. & Egami, F. (1957). *Biochim. biophys. Acta* **24**, 444.

Suzuki, S., Takahashi, N. & Egami, F. (1959). *J. Biochem., Tokyo* **46**, 1.

Suzuki, S., Trenn, R. H. & Strominger, J. L. (1961). *Biochim. biophys. Acta* **50**, 169.

Sweetman, B. J. (1959). *Nature, Lond.* **183**, 744.

Szafran, Z. & Szafran, H. (1964). *Acta biochim. pol.* **11**, 227.

Szczepkowski, T. W. (1958). *Nature, Lond.* **182**, 934.

Szczepkowski, T. W. (1961*a*). *Acta biochim. pol.* **8**, 265.

Szczepkowski, T. W. (1961*b*). *Acta biochim. pol.* **8**, 251.

Szczepkowski, T. W. & Skarżyński, B. (1952). *Acta microbiol pol.* **1**, 93.

Szczepkowski, T. W. & Wood, J. L. (1967). *Biochim. biophys. Acta* **139**, 469.

Tagawa, K. & Arnon, D. I. (1962). *Nature, Lond.* **195**, 537.

Tager, J. M. & Rautanen, N. (1955). *Biochim. biophys. Acta* **18**, 111.

Tager, J. M. & Rautanen, N. (1956). *Physiologia Pl.* **9**, 665.

Takahashi, N. (1960*a*). *J. Biochem., Tokyo* **47**, 230.

Takahashi, N. (1960*b*). *J. Biochem., Tokyo* 48, 508, 691.

Takahashi, N., & Egami, F. (1960). *Biochim. biophys. Acta* **38**, 375.

Takahashi, N. & Egami, F. (1961*a*). *J. Biochem., Tokyo* **49**, 358.

Takahashi, N. & Egami, F. (1961*b*). *Biochem. J.* **80**, 384.

Takahashi, K., Titani, K. & Minakami, S. (1959). *J. Biochem., Tokyo* **46**, 1323.

Takebe, I. (1961). *J. Biochem., Tokyo* **50**, 245.

Tamura, G. (1964). *Bot. Mag., Tokyo* **77**, 239.

Tamura, G. (1965). *J. Biochem., Tokyo* **57**, 207.

Tamiya, H., Haga, K. & Huzisige, H. (1941). *Acta phytochim., Tokyo* **12**, 173.

Tanaka, K. R., Valentine, W. N. & Fredricks, R. E. (1962). *Br. J. Haemat.* **8**, 86.

Tanaka, T., Takahashi, M., Miyazawa, E., Higaki, M., Inaba, S., Nishikawa, T., Suzuki, T. & Sahashi, Y. (1963). *J. Vitam.* **9**, 62.

Tanner, F. W. (1917). *J. Bact.* **2**, 585.

Tanner, F. W. (1918). *J. Am. chem. Soc.* **40**, 663.

Tano, T., Asano, H. & Imai, K. (1968). *Agric. biol. Chem.* **32**, 140.

Tano, T. & Imai, K. (1968*a*). *Agric. biol. Chem.* **32**, 51.

Tano, T. & Imai, K. (1968*b*). *Agric. biol. Chem.* **32**, 284.

Tano, T. & Imai, K. (1968*c*). *Agric. biol. Chem.* **32**, 401.

Tano, T., Kagawa, H. & Imai, K. (1968). *Agric. biol. Chem.* **32**, 279.

Tappel, A. L., Zalkin, H., Caldwell, K. A., Desai, I. D. & Shibko, S. (1962). *Archs. Biochem. Biophys.* **96**, 340.

Tarr, H. L. A. (1933). *Biochem. J.* **27**, 1869.

Tarr, H. L. A. (1934). *Biochem. J.* **28**, 192.

Tarshis, M. S. (1963). *Am. Rev. resp. Dis.* **88**, 847, 852, 854.

Tarshis, M. S. (1965). *Acta tuberc. scand.* **46**, 81.

Taylor, C. B. & Hutchinson, G. H. (1947). *J. Soc. chem. Ind., Lond.* **66**, 54.

Telser, A., Robinson, H. C. & Dorfman, A. (1965). *Proc. natn. Acad. Sci., U.S.A.* **54**, 912.

Temple, K. L. & Colmer, A. R. (1951). *J. Bact.* **62**, 605.

Temple, K. L. & Delchamps, E. W. (1953). *Appl. Microbiol.* **1**, 255.

Temple, K. L. & Le Roux, N. W. (1964). *Econ. Geol.* **59**, 271.

Thaysen, A. C., Bunker, H. J. & Adams, M. E. (1945). *Nature, Lond.* **155**, 322.

Thiele, H. H. (1968). *Antonie van Leeuwenhoek* **34**, 350.

Thode, H. G., Wanless, R. K. & Wallouch, R. (1954). *Geochim. Cosmochim. Acta* **5**, 286.

Thomas, J. H. & Tudball, N. (1967). *Biochem. J.* **105**, 467.

Thompson, J. F. (1967). *A Rev. Pl. Physiol.* **18**, 59.

Thomson, A. E. R. & O'Connor, T. W. E. (1966). *Biochem. J.* **101**, 28 P.

Tigert, N. J. & Smith, J. T. (1965). *Fedn. Proc.* **24**, 169.

Toennies, G. & Lavine, T. F. (1936). *J. biol. Chem.* **113**, 571, 583.

Torii, K. & Bandurski, R. S. (1964). *Biochem. biophys. Res. Commun.* **14**, 537.

Torii, K. & Bandurski, R. S. (1967). *Biochim. biophys. Acta* **136**, 286.

Traeger, J. & Linde, O. (1901). *Archs. Pharm.* **239**, 121.

Trautwein, K. (1921). *Zentbl. Bakt. ParasitKde* Abt II, **53**, 513.

Trautwein, K. (1924). *Zentbl. Bakt. ParasitKde* Abt II, **61**, 1.

BIBLIOGRAPHY

Trudinger, P. A. (1955). *Biochim. biophys. Acta* **18**, 581.
Trudinger, P. A. (1956). *Biochem. J.* **64**, 274.
Trudinger, P. A. (1958). *Biochim. biophys. Acta* **30**, 211.
Trudinger, P. A. (1959). *Biochim. biophys. Acta* **31**, 270.
Trudinger, P. A. (1961*a*). *Biochem. J.* **78**, 673.
Trudinger, P. A. (1961*b*). *Biochem. J.* **78**, 680.
Trudinger, P. A. (1964*a*). *Biochem. J.* **90**, 640.
Trudinger, P. A. (1964*b*). *Aust. J. biol. Sci.* **17**, 446.
Trudinger, P. A. (1964*c*). *Aust. J. biol. Sci.* **17**, 459.
Trudinger, P. A. (1964*d*). *Aust. J. biol. Sci.* **17**, 577.
Trudinger, P. A. (1964*e*). *Aust. J. biol. Sci.* **17**, 738.
Trudinger, P. A. (1965). *J. Bact.* **89**, 617.
Trudinger, P. A. (1967*a*). *Rev. pure appl. Chem.* **17**, 1.
Trudinger, P. A. (1967*b*). *J. Bact.* **93**, 550.
Trudinger, P. A. & Kelly, D. P. (1968). *J. Bact.* **95**, 1962.
Trüper, H. G. & Hathaway, J. C. (1967). *Nature, Lond.* **215**, 435.
Trüper, H. G. & Pfennig, N. (1966). *Antonie van Leeuwenhoek* **32**, 261.
Trüper, H. G. & Schlegel, H. G. (1964). *Antonie van Leeuwenhoek* **30**, 225.
Tsuruo, I. & Hata, T. (1967). *Agric. biol. Chem.* **31**, 27.
Tsuruo, I., Yoshida, M. & Hata, T. (1967). *Agric. biol. Chem.* **31**, 18.
Tsuyuki, H. & Idler, D. R. (1957). *J. Am. chem. Soc.* **79**, 1771.
Tudball, N. (1962*a*). *Nature, Lond.* **196**, 580.
Tudball, N. (1962*b*). *Biochem. J.* **85**, 456.
Tudball, N. (1965). *Biochim. biophys. Acta* **97**, 345.
Tudball, N., Noda, Y. & Dodgson, K. S. (1964). *Biochem. J.* **90**, 439.
Tudball, N., Noda, Y. & Dodgson, K. S. (1965). *Biochem. J.* **95**, 678.
Tudge, A. P. & Thode, H. G. (1950). *Can. J. Res.* **28**B, 567.
Turvey, J. R. (1965). *Adv. Carbohyd. Chem.* **20**, 183.
Tweedy, B. G. (1964). *Phytopathology* **54**, 910.
Tweedy, B. G. & Turner, N. (1966). *Contr. Boyce Thompson Inst. Pl. Res.* **23**, 255.
Tyulpanova-Mosevich, M. V. (1930). *Arch. Sci. biol. St. Petersb.* (*Arkh. biol. Nauk.*) **30**, 203.
Ugazio, G. (1960*a*). *Ital. J. Biochem.* **9**, 98.
Ugazio, G. (1960*b*). *Ital. J. Biochem.* **9**, 109.
Ugazio, G., Artizzu, M., Pani, P. & Dianzani, M. U. (1964). *Biochem. J.* **90**, 109.
Ugazio, G. & Pani, P. (1963). *Expl Cell. Res.* **31**, 424.
Ulrich, F. & Körmendi, N. (1967). *J. biol. Chem.* **242**, 3713.
Umbreit, W. W. (1951). In *Bacterial Physiology*, p. 569. Ed. by Werkman, C. H. & Wilson, P. W. New York: Academic Press.
Umbreit, W. W. (1954). *J. Bact.* **67**, 387.
Umbreit, W. W. & Anderson, T. F. (1942). *J. Bact.* **44**, 317.
Umbreit, W. W., Vogel, H. R. & Vogler, K. G. (1942). *J. Bact.* **43**, 141.
Underwood, E. J. (1956). *Trace Elements in Human and Animal Nutrition.* New York: Academic Press.
Unz, R. F. & Lundgren, D. G. (1961). *Soil Sci.* **92**, 302.
Urban, P. J. (1961). *Z. analyt. Chem.* **180**, 110.
Utermann, D., Lorenzen, F. & Hilz, H. (1964). *Klin. Wschr.* **42**, 352.
Valentine, R. C. (1964). *Bact. Rev.* **28**, 497.
van Delden, A. (1903). *Zentbl. Bakt. ParasitKde* Abt II, **11**, 81, 113.

van den Hamer, C. J. A., Morell, A. G. & Scheinberg, I. H. (1967). *J. biol. Chem.* 242, 2514.
van der Heijde, H. B. & Aten, A. H. W. (1952). *J. Am. chem. Soc.* 74, 3706.
van der Walt, J. P. & de Kruyff, C. D. (1955). *Nature, Lond.* 176, 310.
van Niel, C. B. (1931). *Arch. Mikrobiol.* 3, 1.
van Niel, C. B. (1935). *Cold Spring Harb. Symp. quant. Biol.* 3, 138.
van Niel, C. B. (1936). *Arch. Mikrobiol.* 7, 323.
van Niel, C. B. (1941). *Adv. Enzymol.* 1, 263.
van Niel, C. B. (1944). *Bact. Rev.* 8, 1.
van Niel, C. B. (1953). *J. cell. comp. Physiol.* 41, Suppl. 1, 11.
van Poucke, M. (1962). *Antonie van Leeuwenhoek* 28, 235.
Varandani, P. T., Wolf, G. & Johnson, B. C. (1960). *Biochem. biophys. Res. Commun.* 3, 97.
Vasudeva Murthy, A. R. (1953). *Curr. Sci.* 22, 371.
Verner, A. R. & Orlovskii, N. V. (1948). *Pochvovedenie* 9, 553.
Vestermark, A. & Boström, H. (1959a). *Acta chem. scand.* 13, 827.
Vestermark, A. & Boström, H. (1959b). *Expl Cell. Res.* 18, 174.
Viala, R. & Gianetto, R. (1955). *Can. J. Biochem. Physiol.* 33, 839.
Villarejo, M. & Westley, J. (1963a). *J. biol. Chem.* 238, PC1185.
Villarejo, M. & Westley, J. (1963b). *J. biol. Chem.* 238, 4016.
Villarejo, M. & Westley, J. (1966). *Biochim. biophys. Acta* 117, 209.
Vishniac, W. (1952). *J. Bact.* 64, 363.
Vishniac, W. & Santer, M. (1957). *Bact. Rev.* 21, 195.
Vishniac, W. & Trudinger, P. A. (1962). *Bact. Rev.* 26, 168.
Vlitos, A. J. (1953). *Contr. Boyce Thomson Inst. Pl. Res.* 17, 127.
Voegtlin, C., Johnson, J. M. & Dyer, H. A. (1926). *J. Pharmac. exp. Ther.* 27, 467.
Vogler, K. G. (1942). *J. gen. Physiol.* 26, 103.
Vogler, K. G., Le Page, G. A. & Umbreit, W. W. (1942). *J. gen. Physiol.* 26, 89.
Vogler, K. G. & Umbreit, W. W. (1941). *Soil Sci.* 51, 331.
Vogler, K. G. & Umbreit, W. W. (1942). *J. gen. Physiol.* 26, 157.
Volini, M., De Toma, F. & Westley, J. (1966). *Fedn. Proc.* 25, 412.
Volini, M., De Toma, F. & Westley, J. (1967). *J. biol. Chem.* 242, 5220.
Volini, M. & Westley, J. (1965). *Fedn. Proc.* 24, 530.
Volini, M. & Westley, J. (1966). *J. biol. Chem.* 241, 5168.
von Deines, O. (1933). *Naturwissenschaften* 21, 873.
von Wolzogen Kühr, C. A. H. & van der Vlugt, L. S. (1934). *Water* 18, 147.
Wainer, A. (1964). *Biochem. biophys. Res. Commun.* 16, 141.
Wainer, A. (1965). *Biochim. biophys. Acta* 104, 405.
Wainer, A. (1967). *Biochim. biophys. Acta* 141, 466.
Wainright, T. (1961). *Biochem. J.* 80, 27P.
Wainright, T. (1962). *Biochem. J.* 83, 39P.
Wainright, T. (1967). *Biochem. J.* 103, 56P.
Waksman, S. A. (1932). *Principles of Soil Microbiology*, 2nd ed. Baltimore: Wilkins and Wilkins.
Waksman, S. A. (1952). *Soil Microbiology.* New York: J. Wiley & Sons.
Waksman, S. A. & Joffe, J. S. (1922) *J. Bact.* 7, 239.
Waley, S. G. (1959). *Biochem. J.* 71, 132.
Walek-Czerneka, A. (1965). *Acta Soc. Bot. Pol.* 34, 573.
Walkenstein, S. S. & Knebel, C. M. (1957). *Analyt. Chem.* 29, 1516.

Wallace, E. & Silberman, N. (1964). *J. biol. Chem.* **239**, 2809.
Wang, D. Y. & Bulbrook, R. D. (1967). *J. Endocr.* **39**, 405.
Wang, S.-F. & Volini, M. (1967). *Fedn. Proc.* **26**, 390.
Wang, S.-F. & Volini, M. (1968). *J. biol. Chem.* **243**, 5465.
Warren, B. E. & Burwell, J. T. (1935). *J. Chem. Phys.* **3**, 6.
Warren, J. C. & French, A. P. (1965). *J. clin. Endocr. Metab.* **25**, 278.
Wassink, E. C. (1941). *Enzymologia* **10**, 257.
Webb, E. C. & Morrow, P. F. W. (1959). *Biochem. J.* **73**, 7.
Webb, E. C. & Morrow, P. F. W. (1960). *Biochim. biophys. Acta* **39**, 542.
Wedding, R. T. & Black, M. K. (1960). *Pl. Physiol.* **35**, 72.
Weidman, S. W., Mayers, D. F., Zaborsky, O. R. & Kaiser, E. T. (1967). *J. Am. chem. Soc.* **89**, 4555.
Weigl, J. & Yaphe, W. (1966). *Can. J. Microbiol.* **12**, 874.
Weitz, E., Becker, F., Gieles, K. & Alt, B. (1956). *Chem. Ber.* **89**, 2353.
Weitz, E., Gieles, K., Singer, J. & Alt, B. (1956). *Chem. Ber.* **89**, 2365.
Wellers, G. (1960). *Produits pharm.* **15**, 307.
Wellers, G. & Boelle, G. (1960). *Archs. int. Physiol. Biochim.* **68**, 299.
Wellers, G. & Chevan, J. (1959). *J. Physiol., Paris* **51**, 723.
Wellers, G., Boelle, G. & Chevan, J. (1960). *J. Physiol., Paris* **52**, 903.
Wengle, B. (1963). *Acta Soc. Med. upsal.* **68**, 154.
Wengle, B. (1964*a*). *Acta chem. scand.* **18**, 65.
Wengle, B. (1964*b*). *Acta Soc. Med. upsal.* **69**, 105.
Wengle, B. (1966). *Acta endocr., Copenh.* **52**, 607.
Wengle, B. & Boström, H. (1963). *Acta chem. scand.* **17**, 1203.
Werkman, C. H. (1929). *Bull. Io. St. Coll. agric. Exp. Stn.* No. 117, 163.
Westlake, H. E. & Dougherty, G. (1941). *J. Am. chem. Soc.* **63**, 658.
Westley, J. (1959). *J. biol. Chem.* **234**, 1857.
Westley, J. & Green, J. R. (1959). *J. biol. Chem.* **234**, 2325.
Westley, J. & Nakamoto, T. (1962). *J. biol. Chem.* **237**, 547.
Wheldrake, J. F. (1967). *Biochem. J.* **105**, 697.
Wheldrake, J. F. & Pasternak, C. A. (1965). *Biochem. J.* **96**, 276.
Wheldrake, J. F. & Pasternak, C. A. (1967). *Biochem. J.* **102**, 45 P.
Whistler, R. L., King, A. H., Ruffini, G. & Lucas, F. A. (1967). *Archs. Biochem. Biophys.* **121**, 358.
Whistler, R. L., Spence, W. W. & BeMiller, J. N. (1963). *Meth. carb. Chem.* **2**, 298.
Whitehead, J. E. M., Morrison, A. R. & Young, L. (1952). *Biochem. J.* **51**, 585.
Whittig, L. D. & Janitzky, P. (1963). *J. Soil Sci.* **14**, 322.
Wieland, H. & Vocke, F. (1930). *Justus Liebigs Annln. Chem.* **481**, 215.
Wight, K. M. & Starkey, R. L. (1945). *J. Bact.* **50**, 238.
Wilson, J. (1965). *Clin. Sci.* **29**, 505.
Wilson, L. G. (1962). *A Rev. Pl. Physiol.* **13**, 201.
Wilson, L. G., Asahi, T. & Bandurski, R. S. (1961). *J. biol. Chem.* **236**, 1822.
Wilson, L. G. & Bandurski, R. S. (1956). *Archs. Biochem. Biophys.* **62**, 503.
Wilson, L. G. & Bandurski, R. S. (1958*a*). *J. Am. chem. Soc.* **80**, 5576.
Wilson, L. G. & Bandurski, R. S. (1958*b*). *J. biol. Chem.* **233**, 975.
Winters, R. W., Delluva, H. M., Deyrup, I. J. & Davies, R. E. (1962). *J. gen. Physiol.* **45**, 757.
Winogradsky, S. (1949). *Microbiologie du Sol*, pp. 15–61. Ed. by Masson et Cie. Paris: Libraires de l'Académie de Médecine.

Witmer, F. J. & Austin, J. H. (1960). *Mikrochim. Acta*, p. 502.

Wolf, G. & Varandani, P. T. (1960). *Biochim. biophys. Acta* **43**, 501.

Wolf, G., Varandani, P. T. & Johnson, B. C. (1961). *Biochim. biophys. Acta* **46**, 59.

Wolf, P. L., Horwitz, J. P., Vazquez, J., Chua, J., Pak, M. S. Y. & Von der Muehll, E. (1967). *Proc. Soc. exp. Biol. Med.* **124**, 1207.

Wood, H. W. (1954). *J. photogr. Sci.* **2**, 154.

Wood, J. L. & Cooley, S. L. (1956). *J. biol. Chem.* **218**, 449.

Wood, J. L. & Fiedler, H. (1953). *J. biol. Chem.* **205**, 231.

Woodcock, J. T. (1967). *Proc. Aust. Inst. Min. Met.* p. 47.

Woodin, T. S. & Segel, I. H. (1968). *Biochim. biophys. Acta* **167**, 78.

Woohsmann, H. & Brosowski, K. H. (1964). *Acta histochem.* **18**, 179.

Woohsmann, H. & Hartrodt, W. (1964*a*). *Histochemie* **4**, 336

Woohsmann, H. & Hartrotd, W. (1964*b*). *Naturwissenschaften* **51**, 437.

Woolfolk, C. A. (1962). *J. Bact.* **84**, 659.

Woolley, D. (1962). Quoted by Woolley, Jones & Happold.

Woolley, D., Jones, G. L. & Happold, F. C. (1962). *J. gen. Microbiol.* **29**, 311.

Wortman, B. (1960). *Am. J. Physiol.* **198**, 779.

Wortman, B. (1961). *J. biol. Chem.* **236**, 974.

Wortman, B. (1962). *Arch. Biochem. Biophys.* **97**, 70.

Wortman, B. (1963). *Biochim. biophys. Acta* **77**, 65.

Wortman, B. & Schneider, A. (1960). *Anat. Rec.* **137**, 403.

Wusteman, F. S., Dodgson, K. S., Lloyd, A. G., Rose, F. A. & Tudball, N. (1964). *J. Chromat.* **16**, 334.

Wyler, H., Rosler, H., Mercier, M. & Dreiding, A. S. (1967). *Helv. chim. Acta* **50**, 545.

Yagi, T. (1964). *Biochim. biophys. Acta* **82**, 170.

Yagi, T. (1966). *J. Biochem., Tokyo* **59**, 495.

Yamagata, T., Kawamura, Y. & Suzuki, S. (1966). *Biochim. biophys. Acta* **115**, 250.

Yamagata, T., Saito, H., Habuchi, O. & Suzuki, S. (1968). *J. biol. Chem.* **243**, 1523.

Yamamoto, L. A. & Segel, I. H. (1966). *Archs. Biochem. Biophys.* **114**, 523.

Yamashina, I. (1951). *J. chem. Soc. Japan* **72**, 124.

Yamashina, I. & Egami, F. (1953). *J. biochem. Soc. Japan* **25**, 281.

Yoshida, H. & Egami, F. (1965). *J. Biochem., Tokyo* **57**, 215.

Yoshimoto, A., Nakamura, T. & Sato, R. (1961). *J. Biochem., Tokyo* **50**, 553.

Yoshimoto, A., Nakamura, T. & Sato, R. (1967). *J. Biochem., Tokyo* **62**, 756.

Yoshimoto, A. & Sato, R. (1965). *Koso Kagaku Shimpoziumu* **17**, 65.

Yoshimoto, A. & Sato, R. (1968*a*). *Biochim. biophys. Acta* **153**, 555.

Yoshimoto, A. & Sato, R. (1968*b*). *Biochim. biophys. Acta* **153**, 576.

Youatt, J. B. (1954). *J. gen. Microbiol.* **11**, 139.

Young, L. & Maw, G. A. (1958). *The Metabolism of Sulphur Compounds*, London: Methuen Co. Ltd.

Young, R. L. (1958). *Proc. Soc. exp. Biol. Med.* **99**, 530.

Yount, R. G., Simchuk, S., Yu, I. & Kottke, M. (1966). *Archs. Biochem. Biophys.* **113**, 288, 296.

Zaiser, E. M. & La Mer, V. K. (1948). *J. Colloid. Sci.* **3**, 571.

Zelnik, R., Desfosses, B. & Emiliozzi, R. (1960). *C. r. hebd. Séanc. Acad. Sci., Paris* **250**, 1671.

**BIBLIOGRAPHY**

Zgliczyński, I. M. & Stelmaszyńska, T. (1961). *Acta biochim. pol.* **8**, 123.
ZoBell, C. E. (1947*a*). *Wld. Oil* **126**, No. 13, 36.
ZoBell, C. E. (1947*b*). *Wld. Oil* **127**, No. 1, 35.
ZoBell, C. E. (1950). *Wld. Oil* **130**, No. 7, 128.
ZoBell, C. E. & Rittenberg, S. C. (1948). *J. mar. Res.* **7**, 602.
Zörkendörfer, W. (1935). *Biochem. Z.* **278**, 191.

# SUBJECT INDEX

# AUTHOR INDEX

386

# AUTHOR INDEX

52716